AF775531

Sustainable Development and Environmental Stewardship

Satinder Dhiman
Editor

Sustainable Development and Environmental Stewardship

Global Initiatives Towards Engaged Sustainability

Springer

Editor
Satinder Dhiman
School of Business
Woodbury University
Burbank, CA, USA

ISBN 978-3-031-28884-5 ISBN 978-3-031-28885-2 (eBook)
https://doi.org/10.1007/978-3-031-28885-2

© The Editor(s) (if applicable) and The Author(s), under exclusive license to Springer Nature Switzerland AG 2023
This work is subject to copyright. All rights are solely and exclusively licensed by the Publisher, whether the whole or part of the material is concerned, specifically the rights of translation, reprinting, reuse of illustrations, recitation, broadcasting, reproduction on microfilms or in any other physical way, and transmission or information storage and retrieval, electronic adaptation, computer software, or by similar or dissimilar methodology now known or hereafter developed.
The use of general descriptive names, registered names, trademarks, service marks, etc. in this publication does not imply, even in the absence of a specific statement, that such names are exempt from the relevant protective laws and regulations and therefore free for general use.
The publisher, the authors, and the editors are safe to assume that the advice and information in this book are believed to be true and accurate at the date of publication. Neither the publisher nor the authors or the editors give a warranty, expressed or implied, with respect to the material contained herein or for any errors or omissions that may have been made. The publisher remains neutral with regard to jurisdictional claims in published maps and institutional affiliations.

This Springer imprint is published by the registered company Springer Nature Switzerland AG
The registered company address is: Gewerbestrasse 11, 6330 Cham, Switzerland

This volume is dedicated to all seekers and practitioner of Environmental Stewardship, who are committed to sustainable living, for the common good!

Introduction

> Sustainable development is the pathway to the future we want for all. It offers a framework to generate economic growth, achieve social justice, exercise environmental stewardship, and strengthen governance. – Ban Ki-moon

Environmental stewardship deals with practical issues of ecosystem management and governance to address changes in socio-ecological systems to sustain the supply and availability of ecosystem services by society. It strives to answer the vital question: How can I be a good steward of nature? It means taking responsibility for our choices and entails protecting the environment through recycling, conservation, regeneration, and restoration. It means reducing our total footprint on the planet, not just carbon footprint.

This book builds on recent advances in the theory of sustainable development and links it with environmental stewardship in a dialectical manner, as envisioned by global scholars in this emerging field. This book will be of benefit to policy makers, managers of NGOs, and sustainability researchers in developing engaging strategies for a more sustainable planet.

Chapin III et al. (2010) define environmental stewardship as "the responsible use and protection of the natural environment through conservation and sustainable practices to enhance ecosystem resilience and human well-being." They conclude their fine study on sustainable development and environmental stewardship as follows, "The science of ecosystem stewardship is sufficiently mature to make important contributions to all social– ecological systems.... Sustaining ecosystem services and livelihoods will, however, require reconnecting people's perceptions, values, institutions, actions and governance systems to the dynamics of the biosphere through active ecosystem stewardship."[1]

[1]Chapin, F. S., 3rd, Carpenter, S. R., Kofinas, G. P., Folke, C., Abel, N., Clark, W. C., Olsson, P., Smith, D. M., Walker, B., Young, O. R., Berkes, F., Biggs, R., Grove, J. M., Naylor, R. L., Pinkerton, E., Steffen, W., & Swanson, F. J. (2010). Ecosystem stewardship: sustainability strategies for a rapidly changing planet. *Trends in ecology & evolution, 25*(4), 241–249. https://doi.org/10.1016/j.tree.2009.10.008

Truly global in scope, this volume is a humble offering of more than 12 scholars and practitioners from around the world to the exciting adventure of sustainable development and eco-stewardship.

The following section provides a brief overview of the various chapters in this book.

- In Chap. 1, the authors identify and analyze some of the best practices in relation to the impact of global warming on business planning in Mexico. The phenomenon of global warming as changes in the business environment has impacted on the strategic planning processes of Mexican companies in the creation of objectives and methodologies in various fields and sectors, highlighting the importance of adapting to the needs and the limitations that may arise with climate change in the various markets we are focused.
- In Chap. 2, Dr. Amanda L. Holst states that despite ethical leadership concepts being well studied, there is little research on ethical leadership training regarding sustainability. The purpose of this chapter is to explore literature from various current articles regarding the topic of ethical leadership development and sustainability, reveal the gaps, and show how transformational leadership, a theory of leadership in the field of global leadership, can be used as a framework for ethical decision-making. A qualitative comparison analysis was used to identify research gaps in ethical leadership training for ESD. It also revealed that making ethical decisions showed the most significant factor in creating a moral compass. The findings led to the development of an ethical decision-making conceptual framework using transformational leadership dimensions and the three main ethical approaches to normative ethics. This framework can be used for ethical leadership development training, and for educators and scholars to use as a guide in developing their framework when determining how to cultivate and lead with a reliable moral compass.
- In Chap. 3, the authors aim to analyze the implications between the declining and shrinking cities, urban vacant lands, urban land uses, green infrastructures, urban green areas, and their impact on climate hazards change. The analysis departs from the basic assumption that urban vacant land sites and spaces have a negative connotation but supported by the appropriate policies and programs of incentives can turn around and develop the essential green infrastructure to enable the mitigation of climate change hazards, economic growth, and socio-ecological development. The method used is the analytical-descriptive base on the theoretical and empirical literature review. It is concluded that the land uses of vacant land sites more vacationed towards urban green innovation infrastructure and forest areas contribute to mitigate the climate change hazards.
- In Chap. 4, Dr. José G. Vargas-Hernández analyzes the various features and elements related to strategic organizational sustainability. The analysis departs from the assumption that traditional organizations have to face a lot of challenges posed by the new global economy context confronting contradictory patterns of globalization and de-globalization processes. In order to create the new conditions for the organizations not only to survive but to succeed, it is proposed a new

model based on designing and implementing strategic organizational sustainability, abandoning the narrow focus on economic growth and profits to embrace the social inclusion and equity as well as the environmental sustainability issues.

- In Chap. 5, Dr. Nidhi Kaushal garners the view that environmental stewardship is not a contemporary practice, but it has explained through significant poetic works, mythological epics, folklore, and the framework of ecological-spiritual feelings as the duty of humans toward nature for centuries. Ancient sages have exquisitely classified the empirical knowledge of making harmony with life, by caring for flora, fauna, and Prakriti (mother nature) in their treatises, and ethnic methodologies have preserved the wisdom of moral practices toward maintaining the earth's sustainability by the cultural rituals, folktales, songs, poems, festivals, and devotional observances. The purpose of this work is to identify the classical facts of environmental stewardship and sustainable living and asserts the feasibility of indigenous techniques as well as their applicability for saving nature in the diversely changing modern era. It also gives an implication of noble approaches of the seers or intellectuals as ecologists through the ideology of reviewing the old concepts. Moreover, it advances an attitude to reverence for nature-oriented cultural reforms, epic endeavors, and narrative endowments.
- In Chap. 6, the authors analyze the individual, organizational, environmental, and contextual dimensions of organizational citizenship behaviors, supported under the assumption that organizational, individual, environmental, and contextual factors lead to the innovative behavior and organizational citizenship behavior as extra-role behaviors. The methods employed are the analytical-descriptive leading to the reflective analysis based on the theoretical and empirical review of the literature on these issues. The authors conclude that the individual, environmental, contextual, and organizational factors facilitate more innovative organizational citizenship behaviors that contribute to individual and organizational efficiency.
- In Chap. 7, the authors aim to analyze the elements of urban green innovation based on the guarantee that public interests decentralize the infrastructure to democratize the territory and innovating the institutional design to address the complexity of the challenges in the city. The method employed is the critical analysis supported by a review of the literature and consult to experts in the field. It is concluded that the urban green innovation capacity planning has a critical role in urban innovation development in specific areas of economic growth, social inclusion and equality, environmental sustainability, health, education, business, etc. To achieve these aims, urban green innovation requires to guarantee the public interest, the democratization of the territory, and the new institutional design.
- In Chap. 8, the authors analyze a review of the empirical literature on some important features of urban green spaces such as the components, functions, services, community involvement, initiatives, and actions from an ecosystem perspective. The analysis begins from the assumption that urban green spaces are ecosystems of vital importance in enhancing the quality of life in an urban environment and supplying ecosystem services, such as biodiversity and climate

regulation. Thus, the urban green space ecosystem is an important component of an ecosystem in any community development. Meeting the needs of users is related to the functions and services that urban green spaces provide to communities. Community involvement, engagement, and development require a methodology to ensure that the needs and aspirations of local users in the community are met. The methods employed in this analysis are a review of the empirical literature and documents, analysis of existing data on uses and users, interviews with authorities, and a more detailed examination of case-specific data. Also, as concluding remarks, some of the wider environmental, economic, and social initiatives for local authorities and communities are suggested that might justify any funding for the all the stakeholders that are represented and involved. Finally, the chapter proposes some of the opportunities, challenges, and further research.

- In Chap. 9, Drs. Younsung Kim and Daniel Ruedy observe that the challenge of global sustainability presents enormous opportunities for product innovation and creative response. Concerns about the rising volume of packaging wastes particularly request the redefinition and redesign of conventional products in the packaging industry. Ecovative Design, being founded by a young mechanical engineering student in 2007, develops an array of environmentally friendly materials that perform like plastics but are made from mushrooms. The mushroom packaging is renewable and biodegradable and can be made with crop waste brought from local farms. Drawing on the case of Ecovative Design, this chapter aims to describe a creative destruction approach to redress sustainability challenges and to explore pivotal factors that fostered the company's ability to decarbonize and dematerialize in support of sustainability. We find that a company's technical competency with systems-thinking skills and stakeholder engagement capacity can lead to the realization of ecological modernization principle.
- In Chap. 10, the authors maintain that the leaders need to consider multiple forces (climate change, increased demand, etc.) that impact (raw material scarcity, public pressure, security concerns etc.) businesses. The green business transformation process is comprehensive, which requires the active participation of stakeholders, planning, and implementation. The chapter focuses on the drivers and barriers for business organizations while following the green business practices. On the whole, the chapter will help in developing an understanding about the following: (a) understand leadership and sustainability, (b) explore the changing landscape of green business practices followed by leaders in various business organizations in the Indian context, (b) understand the organizational green transformation strategies adopted by leaders thereby leading to sustainable solutions, and (c) exploring the drivers and barriers toward implementation of green business practices, by the leaders in various industries.
- In Chap. 11, the authors aim to analyze a successful soft approach to new forms of protest as applied to the recovery and takeover of the Water Pedagogical Forest as an urban green innovation public space for knowledge and learning. In this case, it is analyzed a successful soft approach to new forms of protest with the implications in the recovery and takeover of the Water Pedagogical Forest. The method used is the empirical analytical based on the facts of the specific case

under study framed by the theoretical and empirical analysis of the literature reviews and the reflection applied to the situation. The analysis of this case concludes that the new forms of movements and protest actions are shifting toward using more soft power, non-violent, long-term processes, multifaceted, regulated through the institutional and legal framework being capable to process the conflicts and manage economic efficiency, social justice and inclusion, political participation, and sustainable environmental interests of all the stakeholders involved.

Satinder Dhiman
School of Business, Woodbury University
Burbank, CA, USA

Contents

About the Editor

Satinder Dhiman professor of Management at Woodbury University, Burbank, California, serves as the associate dean, chair and director of the MBA Program. With an interdisciplinary research agenda encompassing organization behavior, workplace spirituality, workplace well-being and fulfillment, sustainability, servant leadership, mindfulness, social entrepreneurship, education, organization development, and Eastern and Western philosophy in leadership, Professor Dhiman holds a Ph.D. in Social Sciences from Tilburg University, the Netherlands, an Ed.D. in Organizational Leadership from Pepperdine University, Los Angeles, an MBA from West Coast University, Los Angeles, and a master's degree in Commerce from Panjab University, Chandigarh, India, having earned the Gold Medal. He has also completed advanced Executive Leadership Programs at Harvard, Stanford and Wharton.

His academic leadership experiences include having served as the chair for a special MBA Program for the Mercedes-Benz executives, China; as a distinguished visiting professor at the Tecnológico de Monterrey, Guadalajara campus, Mexico; as E-Commerce curriculum lead advisor, Universidad Francisco Gavidia, El Salvador; coordinator for the MBA Student Los Angeles Fieldtrip Program for Berlin University for Professional Studies (DUW); accreditation mentor to Sustainability Management School (SUMAS), Gland, Switzerland. During December 12–15, 2019, he was invited by Monash University, Australia, to lead a track in Spirituality in Management in the 16th International Conference in Business Management. He has served as the president (2016–2018; 2022-present) and as a distinguished *Patron* (2019–2021) for the International Chamber for Service Industry (ICSI).

Recognized as a strategic thinker for his pioneering contributions to the field of transformational leadership, workplace spirituality, workplace well-being, sustainability and fulfillment in personal and professional arena, Professor Dhiman is a sought after keynote speaker at regional, national and international conferences such as the prestigious TEDx Conference @ College of the Canyons in Santa Clarita, California. Since then, he has led several major national and international conferences as co-organizer and/or as track chair.

Recipient of several national and international academic and professional honors and awards in teaching, scholarship and service, Professor Dhiman was awarded the Woodbury University Ambassador of the Year Award in 2015 and 2017 and MBA Professor of the Year Award in 2015; Scholarly and Creative Writing Award, 2019; Most Valuable MBA Professor Award, 2018; Most Inspirational and Most Charismatic MBA Teacher Award 2012, 2013/2014/2018; the Steve Allen Excellence in Education Award in 2006; and the prestigious *ACBSP International Teacher of the Year Award in 2004.* Most recently, he chaired a symposium at the Academy of Management that received the "2019 Best Symposium Proposal and Showcase Symposium" Award by the MSR Division.

Professor Dhiman's scholarly accomplishments include over 65 professional conference presentations, over 150 invitations to be a keynote speaker, over 100 online webinars, participation in plenary sessions, conference track chair sessions, leading symposiums and webinars, and scores of distinguished guest lectures and creative workshops—*nationally and internationally.* He has published over 70 peer-reviewed journal articles and book chapters. As author, translator, editor, co-author, co-editor of over 40 management, leadership, spirituality, sustainability and accounting-related books and research monographs, his most recent books include: *Leadership After COVID-19: Working Together Toward a Sustainable Future* (2022—Springer, with Marques), *New Horizons in Management, Leadership and Sustainability* (2021—Springer, with Samaratunge),

Bhagavad Gītā and Leadership: A Catalyst for Organizational Transformation (2019—Palgrave Macmillan), *Managing by the Bhagavad Gītā: Timeless Lessons for Today's Managers* (2018—Springer; with Amar), *Holistic Leadership* (Palgrave 2017), *Gandhi and Leadership* (Palgrave 2015) and *Seven Habits of Highly Fulfilled People* (2012); and co-editing and co-authoring, with Marques, *Spirituality and Sustainability* (Springer 2016), *Leadership Today* (Springer, 2016), *Engaged Leadership* (Springer, 2018), *New Horizons in Positive Leadership and Change* (Springer, 2020) and *Social Entrepreneurship and Corporate Social Responsibility* (Springer, 2020).

He has also translated several Indian spiritual classics into English, including the *Sahaja Gītā*. He is the editor-in-chief of seven multi-author Major Reference Works: *Springer Handbook of Engaged Sustainability* (2018—Springer International, Switzerland) and *Palgrave Handbook of Workplace Spirituality and Fulfillment* (2018—Palgrave Macmillan, USA); *Routledge Companion to Mindfulness at Work* (2020); *Palgrave Handbook of Workplace Wellbeing* (2021—Palgrave Macmillan); *Routledge Companion to Leadership and Change* (2023—Routledge, UK); *The Palgrave Handbook of Servant Leadership* (2023—Palgrave Macmillan, USA; with Roberts); *The Springer Handbook of Global Leadership and Followership* (2023—Springer International, Switzerland; with Marques, Schmieder-Ramirez, and Malakyan).

Additionally, he serves as the editor-in-chief of Palgrave Studies in Workplace Spirituality and Fulfillment and Routledge Frontiers in Sustainable Business; the general editor of a series entitled Routledge Frontiers in Sustainable Business Practice (the Series); and lead editor of Springer Series in Management, Change, Strategy and Positive Leadership.

Some of his forthcoming titles include *Leading without Power: A Model of Highly Fulfilled Leaders* (2022—Palgrave Macmillan); *Conscious Consumption: Healthy, Humane and Sustainable Living* (2022—Routledge, UK); *Wise Leadership for Turbulent Times* (2022—Routledge, UK); and *Creative Leadershift: Discover. Innovate. Enact.* (2022—Routledge, with Chandra Handa). He has published research with his colleagues in *Journal of Values-Based Leadership*, *Organization Development Journal*, *Journal of Management Development*, *Journal of Social Change*, *Journal of Applied Business and Economics* and *Performance Improvement*.

Professor Dhiman has served as accreditation consultant, evaluator and site visit team leader for the Accreditation Council for Business Schools and Programs (ACBSP) for more than 25 universities in America, Canada, Europe and Asia. He is the founder-director of Forever Fulfilled, a Los Angeles-based Well-being Consultancy that focuses on workplace wellness, workplace spirituality and self-leadership.

Chapter 1
Global Warming Affecting Mexican Firms

José G. Vargas-Hernández, M. C. Omar C. Vargas-González, and Pedro Antonio López de Haro

Abstract The aim of this chapter is to identify and analyze some of the best practices in relation to the impact of global warming on business planning in Mexico. The phenomenon of global warming as changes in the business environment has impacted on the strategic planning processes of Mexican companies in the creation of objectives and methodologies in various fields and sectors, highlighting the importance of adapting to the needs and the limitations that may arise with climate change in the various markets we are focused.

Keywords Global warming · Strategic planning · Mexican companies

JEL M100 · M110 · M160 · Q54

Small portions of the chapter were previous published in *The Basic Research Journal of Soil and Environmental Science*. ISSN 2345-4090 Vol. 3(4) pp. xx–xx September 2015. Available online http://www.basicresearchjournals.org

---

J. G. Vargas-Hernández (✉)
Postgraduate and Research Division, Instituto Tecnológico José Mario Molina Pasquel y Henríquez, Unidad Zapopan, Zapopan, Jalisco, Mexico
e-mail: jose.vargas@zapopan.tecmm.edu.mx

M. C. O. C. Vargas-González
Systems Department, Instituto Tecnológico Nacional de México, Cd. Guzmán, Jalisco, Mexico
e-mail: omar.vg@cdguzman.tecnm.mx

P. A. López de Haro
Universidad Autónoma Indígena de México, Unidad Los Mochis, Los Mochis, Sinaloa, Mexico
e-mail: pedrolopezh@uaim.edu.mx

© The Author(s), under exclusive license to Springer Nature Switzerland AG 2023

S. Dhiman (ed.), *Sustainable Development and Environmental Stewardship*,
https://doi.org/10.1007/978-3-031-28885-2_1

1.1 Introduction

The following chapter is exploratory research on the opinions of climate change on firms in Mexico. A total of 145 firms were surveyed using a website called surveymonkey.com, and the question referred to their stance regarding climate change, quality of life, and how to introduce it into the business practices of the firm. We found that only 33% of the firms surveyed said they took the threat of climate change seriously enough to factor it in their strategic planning. Some of the problems related to climate change that Mexican companies must deal with are pollution and deterioration in the environment, shortage of raw materials, an increase in health problems for workers and customers, water pollution, brand image due to environmental policies, among others.

Environmental management in business tends to evaluate the performance of business units with indicators of profitability and return on investment, in addition to their contributions to reduce carbon dioxide (CO_2) emissions, which has been identified as an especially important greenhouse gas. Global warming has been a major issue in business administration and public policy in recent years. Internationally, climate change has received attention in various economic and social forums.

In Al Gore's documentary (Pérez Salazar, 2007), *An Inconvenient Truth*, the US former vice president describes the relationship between global warming and CO_2 emissions and predicts that the phenomenon of global warming will get increasingly worse if greenhouse emissions are not reduced significantly. However, Gore also warns how certain interests' agents have spread doubt about global warming, and its relationship to human activity. Guhl Corpas (2008) talks about the role of mass media for the public's misunderstanding of the problem.

Global warming also operates within the context of the recent global change, associated with increased concentrations of certain gases, such as greenhouse gases (GHG), in the atmosphere. The last IPCC report (2007) warns that the biggest increases in GHG emissions between 1970 and 2004 come from power generation, transportation, and industry, followed by construction, forestry, and agriculture. There is incontrovertible evidence on how the dependence on fossil fuels is the most crucial factor contributing to global warming (Guhl Corpas, 2008; Sussman & Freed, 2008).

The aim of this research is to understand the impacts of climate change in the planning processes of companies and businesses in Mexico.

1.2 State of the Art

Although science provides clear and conclusive evidence of the impacts of climate change on the quality of life, both the private and public sectors are reacting slowly, privileging their own economic interests. In sum, collective efforts are insufficient to design and implement policies and strategies aimed at mitigating the effects of climate change on quality of life. Until now, the same market forces and the economic rationality of business have maximized profits with lower costs, which drive investment in the use of resources and technologies that produce greenhouse gas emissions, accelerating global warming with all its consequences for the earth.

Mexico is one of the most vulnerable nations to the effects of climate change, given that 15% of its territory, 68.2% of its population, and 71% of the gross domestic product (GDP) are prone to suffer negative consequences of this phenomenon (Sosa-Rodríguez, 2015).

Morales Lara (2021) mentions that several causes for climate change have been considered by scientists and experts over the last hundred years, such as natural causes, which affect temperature in a mostly cyclical manner, like volcanic eruptions, solar irradiance, the orbital position of the earth, etc.; the use of fossil fuels, which have been burned by humans constantly for over 80 years, they are limited and their exploitation generates energy, but also some derivatives that have been shown to cause global warming. Examples of this are natural gas, carbon, gasoline, etc.

Multiple productive activities, such as industry, farming, and agricultural activities, given their high concentration of nitrous oxygen, human consumption, and solid residue, given that a lot of organic waste generates methane, which is another especially important greenhouse gas, deforestation destroying trees that maintain natural balance and reduce the concentration of oxygen in the atmosphere.

Globalization and neoliberalism create a system that must keep producing and growing, not even allowing stagnation in the creation of increasingly polluting companies and organizations; this, we must realize, is not a sustainable model for the earth, or at least, for human beings. Given all of this or given that many of these factors are related to humanity, for some people and governments, it is simply easier to evade the problem, sending the consequences to the future generations; nobody wants to take responsibility, and, more importantly, take action to curve what is happening. There is no commitment with the environment, with biodiversity, and with climate change reduction, given that there is nobody compelling big companies to reduce their emissions.

The Paris Climate Accord was a non-binding agreement approved and signed by leaders of 195 countries in 2015, committing themselves to reduce their emissions and recognize the problem as the existential threat for humanity that it is (Morales Lara, 2021). Unfortunately, this agreement is considered by many just a show, given that it does not have any way to force countries to enforce it and they even put up their own goals, which they can simply fail to meet when the time comes. So, of course, the Paris Agreement is better than nothing, but it does not even come close to have the effect that it needs.

Climate change has and will continue to have significant cost for the entire Mexican economy in the future. Furthermore, the costs are exponential and not linear, which means that even if the cost may be manageable for now, it does not mean there is nothing to worry about. It is important to recognize that there are some sectors of the economy that could benefit, but the overall cost is a lot greater than the potential benefit for those specific sectors. There are also non-monetary costs that need to be considered, the uncertainty for the future, corrosion, food and water security, the effect of natural disasters, etc. (Galindo, 2009). Overall, the costs of creating and following international joint, and binding, agreements are a lot lower than the impacts that we are already seeing, not only for Mexico, but all over the world.

In Mexico, the manifestations on climate change are added to the structural conditions of poverty. According to official data, 46.2% of the population live in poverty and 10.4% live in extreme poverty; in the context of economic crisis, population that lack access to proper food has increased and the real income of households has reduced; 80% of Mexicans that live in rural areas are classified as poor. Such levels of social inequity, along with the geographical situation of extreme hydrometeorological events, which have been presenting with a greater frequency and intensity over the last years, have exacerbated a very delicate situation.

While drought and erosion are covering the north and center of the country, the south is being impacted by tropical storms, hurricanes, and flooding that have never been seen before, and the infrastructure is not prepared for these kinds of events. It is calculated that around 20 million people live in areas that are susceptible to the impact of hurricanes, and infrastructure may not be enough to resist these phenomena (Landa et al., 2008). The most meaningful impacts of climate change are observed in the agricultural sector, with serious effects on job and food security, given that this is a sector that is particularly susceptible to gradual changes in temperature and precipitation.

Climate change is one of the greatest challenges that Mexico is facing, but it also represents an opportunity to promote a more sustainable development if the capabilities for mitigation and adaptation are developed and strengthened. Currently, there is evidence that the atypical changes in temperature and precipitation have increased the frequency and intensity of floods, droughts, and heatwaves, among other extreme hydrometeorological events.

Since these impacts will be distributed in an uneven manner among society, meaning that they affect the economically and socially disadvantaged at a much higher severity, a successful climate change policy from the government must consider the differences in capabilities, levels of exposition, and vulnerability of diverse groups and sectors. Given the limitation in human, financial, and technological resources, the integration of the strategies and programs that will contribute to face this phenomenon could, if applied correctly, reduce costs, and increase the efficacy of the responses from the government (Sosa-Rodríguez, 2015).

So far, the advancements in mitigation and adaptation efforts in Mexico have not been enough to reduce greenhouse gas emissions, which have increased significantly from 1990 to 2010, and have been accumulating in the atmosphere, creating an exponential problem. Also, the development of adaptation capabilities has been limited by an increase in poverty and a decrease in opportunities for people to have lasting and sustainable jobs.

The long economic recession and lack of opportunities for the population has forced many of the most marginalized groups to locate in risky areas, where they often lack access to basic services, such as electricity, running water or sewage, making them even more vulnerable to climate risk. Therefore, an essential strategy to face this phenomenon is to reduce poverty through the creation of job opportunities, as well as the construction of safer housing, more resistant to natural disasters and with all the basic services.

Unfortunately, variation in climate and its impact will affect vulnerable groups the most. Despite this, there are some important breakthroughs in mitigation and adaptation; among them, we have the reduction in electricity consumption, construction of cleaner means of transportation, the development of alternative sources of energy, improvements on systems of alert, reforestation, and improvements on systems of waste management. Just like it was possible to stop using chlorofluorocarbons to protect the ozone layer in the atmosphere, it is possible to reduce our carbon footprint to mitigate the effects of climate change, even though it is not an easy task.

In Mexico, despite an increase in GDP of 75.03% between 1998 and 2010, the consumption of fossil fuels was reduced in 24.29%, this situation underscores the importance of the promotion of the development and use of renewable and clean energy. We, as a species, are very energy intensive; we need it, but we also need to take care of the planet we live on. The process to make systems more efficient is terribly slow and complicated, and that is why there needs to be a joined effort, between academia, scientists, engineers, politicians, and the public, to keep developing this technology and deploying solutions to help.

Also, one of the requirements for the success of a policy on climate change, for the entire country, is the effective regulation on the use of soil to avoid an accelerated demographic and urban growth, which would overcome the capacity of the environment to provide basic needs like food and water. If urban development continues growth of irregular settlements in protected natural areas of high environmental value, no measures would be enough to reduce the vulnerability of the affected groups; these protected regions will be destroyed, and this will continue to cause irreparable damage and affect everyone in the short and long terms (Sosa-Rodríguez, 2015).

Active and informed participation from the general public must be promoted to create and strengthen capabilities of mitigation and adaptation that the country requires; for this, it is of vital importance that the general public is aware of the severity of the problem, cautioning them not to panic, but to start doing their part and demanding their politicians to do right by them. So far, climate change is still a problem that is only being considered when something like a natural disaster, heatwaves, or mega droughts happen, and instead of trying to do something about it, many people simply resort to complain. This is a problem that is unfortunately still restricted to political, intellectual, and economic elites in Mexico, which are, paradoxically, the ones that are causing most of the damage but, at least in the short term, the least affected by the problem.

As a result of the evaluation of the impact of climate change, several reports have been produced. This problem has been addressed in the past by the oil companies (Van den Hove et al., 2002; Levy & Kolk, 2002), and three strategies were developed: (1) prioritizing trade implications while weakening perception that anthropogenic greenhouse gas emissions are causing climate change, (2) avoiding responsibility, and (3) prioritizing the need for a change in the business process while limiting the negative effect in terms of commercial consequences. Boiral (2006) mentions that managers, especially in large industries, should try to develop an inventory of alternatives and sources of GHG, to determine which options are more efficient for replacement.

The implementation of a proactive strategy can help improve employee involvement and the image of a corporation. Many factors such as the sector, technological innovation, the market price of CO_2 emissions, social pressures, and public policy, can significantly change the assessment of the opportunities and threats arising from the Kyoto Protocol. This depends on the vision, values, and skills of managers. Jorqueta-Fontena and Orrego Verdugo (2010) mention that addressing the impact of climate change will allow the industry to define some strategies that could be useful for future planning, including sustainability strategies.

According to Sussman and Freed (2008), the projected effects of climate change are as follows:

A. Increase in temperature, with fewer chilly days and warmer nights, and the duration, intensity, and duration of heat.
B. Rising sea level, between 18 and 59 centimeters.
C. Increased precipitation and humidity, including at high latitudes, and winter precipitation.
D. Extreme winds, tropical storms, and other events, including the activity of hurricanes, floods, and fires in arid and semi-arid areas; and.
E. Other related effects, including reduction in the duration of snow seasons, loss of glaciers, and permafrost. All of them are labeled as elements of climate risk.

Meanwhile, Proverbs (2010) states that global warming increases the risk of flooding, provided that the construction industry must consider as an element in its planning risk and consider their impacts to the sustainability of locations in the real estate developments, such as: reduction in price of the land, increased costs of formalities for the construction, increasing construction costs, reduced salability, and difficulty in securing long-term viability of skills construction.

According to García Fernández (2006), when applying the methodology of cost-benefit analysis (CBA) to the phenomenon of global warming, there are several complications:

A. Proper data collection. Because there is still uncertainty about what will be the environmental impact that causes greenhouse gases and about the social, economic, and probable environmental impacts, along with a lack of agreement on the economic costs of those impacts.
B. The use of gross national product (GNP). The greenhouse effect damages existing production flows and welfare, measuring the cost and loss of total annual production, and is detrimental to the results of past production and existing stocks of natural resources.

C. The use of the discount rate. The use of CBA does not lead to a policy of sustainability. The cost-benefit analysis is related to total quantities (net costs or benefits), while sustainability is related to equity in the distribution over time.
D. The uncertainty. Scientists still do not understand some of the key mechanisms that determine climate change.

De la Vega-Salazar (2011) mentions that raising the average temperature of the planet's surface does not occur uniformly, and given its geographical position, Mexico can be very vulnerable. The global average temperature has increased by more than 1 °C over the last 30 years, with some regions having increased 3 °C. And while this can be explained by natural processes, such as atmospheric circulation of the planet and the relationship of ocean currents, the phenomenon is aggravated by overexploitation of natural resources and damage to ecosystems by deforestation. According to De la Vega-Salazar (2011), Mexico is having significant impacts due to global warming: higher maximum temperatures, duration, frequency, and intensity of extreme events such as warmer heat waves causing decrease in the availability of water.

Currently, for many organizations, climate change is a priority issue on their agenda. The threat of climate change is, for Ernst & Young Business Risk Report, the fourth place in business risks (Ernst & Young, 2010). This danger was exceeded only by the risks of the credit crisis, regulation, compliance, and recession. This issue concerns global leaders and high-growth companies. In a recent study of global organizations with more than $25 billion of market capitalization, performed by Ernst & Young, 73% of the companies had been committed to reducing its greenhouse gas (Ernst & Young, 2010).

So far, most of the efforts to combat climate change are disjointed and uncoordinated; actors meet frequently, but they don't seem to agree on much and, because they are negatively affected by the same policies they are deciding on, they fail to propose meaningful solutions, in fact, actors have formed strong coalitions against carbon tax and the General Law on Climate Change (Ortega Díaz & Casamadrid Gutiérrez, 2018).

1.3 Research Methodology

The methodological design of the study is a non-experimental, quantitative, field research, transversal, and exploratory-descriptive. Quantitative and exploratory research was selected because of the scope of the investigation: to determine the impact on business planning processes of global warming.

Business planning is a process usually written in a document that determines and describes, in detail, the goals, objectives, strategies, policies, and actions of the whole firm, including human resources, financial, production, operations, and marketing activities and how they are going to be achieved. Business planning is a fundamental tool that provides a road map to promote the business development and growth, focusing on dealing with the business risks to make profits and ensuring business prosperity and success.

The research is considered transversal because no historical comparisons are to be made over a period. In addition, the survey technique was chosen because the source of information is the executive and managerial personnel of firms.

1.4 Survey Design

For the survey design, we developed a model that relates the variables to be measured (Fig. 1.1).

The applied model (Fig. 1.1) was developed from the contributions of Ernst and Young (2010), Boiral (2006), and Sussman and Freed (2008), organizing the factors according to the components of the external environment of a company according to Thompson et al. (2008). The variables identified as impact of global warming are grouped into two dimensions:

A. Identification of risks.
B. Identification of opportunities, both elements in the environment analysis as input for strategic planning and business.

In this case, the variables in the models were measured using Likert scales.

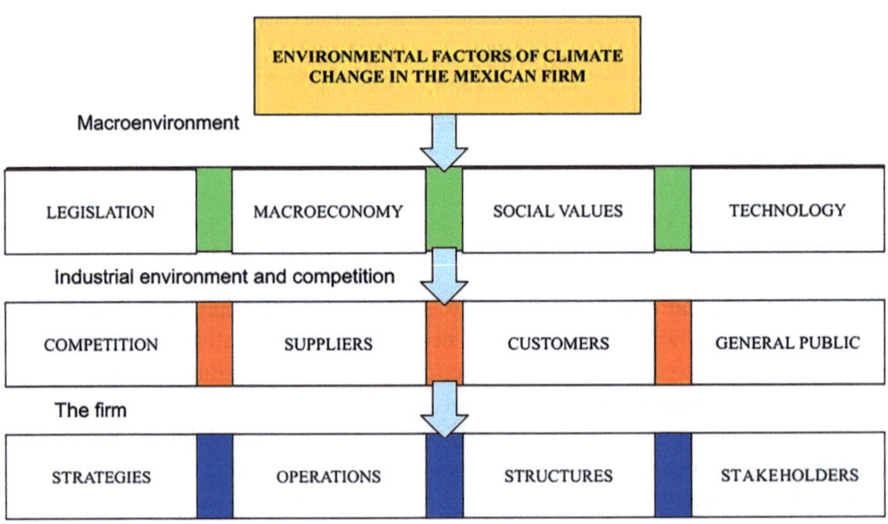

Fig. 1.1 Environmental factors of climate change in the Mexican firm. (Source: Own construction)

1.5 Data Collection

The target population of this research are the companies located in Mexico, of any size, business orientation, and location, this geographic coverage being chosen by the exploratory nature of the study. The universe of companies was determined as infinite population, given that they are more than 500,000 units of study, according to the Mexican Business System (Secretaría de Economía, 2012).

Survey technique using online information systems (www.surveymonkey.com) was used for data collection. The strategy for contacting these companies was a combination of personal visits, telephone contact, and email. The general intention was that managers or business owners were the ones who provided the information.

In total, 145 surveys were conducted, gathering information from 15 states. Applying the criteria of the Ministry of Economy (Secretaría de Economía, 2009), of these 145, 25% were micro enterprises, 13% small enterprises, 30% were medium, and 33% were large. On the other hand, 31% mentioned having an international geographical coverage, 36% national, 26% regional, and only 7% local. Most of those surveyed belong to the service sector (70%), while 20% to industry and 10% to trade. Of those surveyed, 78% were private and 22% public.

1.6 Data Analysis

Once the information was collected, we proceeded to process and analyze the data, using SurveyMonkey's tools themselves. The first activity was to build a description based on the responses, identifying profiles according to different respondents.

The second activity involved the construction of a relevance tree (Chevalier & Buckles, 2009; Godet, 2000).

1.7 Results

Practice of Strategic Planning

Sixty-one percent of the respondents regularly are performing strategic planning, while 17% said they engage in strategic planning but in periodicity. Nine percent said they did once, and 13% said no strategic planning is done.

Of the respondents who engage in strategic planning, 80% answered were participating directors, 56% the board of directors, 48% managers, and 2.5% financial advisors' entities.

About Climate Change

Of the respondents who perform strategic planning, 93% answered that they know about the issue of climate change. Moreover, when asked if they considered its effects on the strategic planning of the organization, 33% said yes, they have done analysis of risks and opportunities, 30% mentioned that have been touched on, and 38% mentioned that have not been considered.

On the risks that are considered in strategic analysis, cost increases are the elements that are considered: the cost of energy (85%), supplies (50%), damage to facilities and infrastructure (35%), and the cost of public health factors (30%). In the second place, respondents cited increased uncertainty in planning elements, such as water supply (55%), energy (45%), supplies (40%), and food (20%). Third, respondents expressed increased compliance requirements of regulations (50%) and pressures on the public image of the organization (45%).

Among the less considered risks, investors said discounts (5%), tax compliance from climate change (10%), and changes in the behavior of stakeholders (10%), product and service competition substitutes (10%), and changes in expectations of provider (15%), dissatisfaction with climate change (15%), brand awareness (20%), economic sanctions (20%), and changes in the segment of customers (25%).

On identified opportunities, 75% of the respondents mentioned the development of new products and services; 70%, implementation of environmental education programs; 60%, training; 55%, improvement of corporate image as well as the application of technologies, green and ecological values of employees; 45%, implementation of green strategies; 40%, improvements in energy and operational efficiency for power generation, as well as greater participation and responsibility of employees; 35% new customer segments.

Of those surveyed, 33% said that, derived from the analysis of climate risks and opportunities, they have invested in several strategic projects; 5% said that they had invested in a single project, 33% replied that they had not invested but have analyzed various initiatives; and 29% replied that they had not invested.

1.8 Discussion

The problem tree (Fig. 1.2) identifies the main causes and effects found in this research. As the main problem, we considered the effects of climate change of the firm.

In this way, the scheme was developed grouping the effects of the global warming such as external public, social, and political pressures; compliance requirements of global, regional, local regulations, and other sectors; uncertainty factors in the effect on public health; and uncertainty in the supply of raw materials. And the causes that have led directly to the effects mentioned above are: There has been a shortage of resources (raw materials) caused by climate change, social groups can

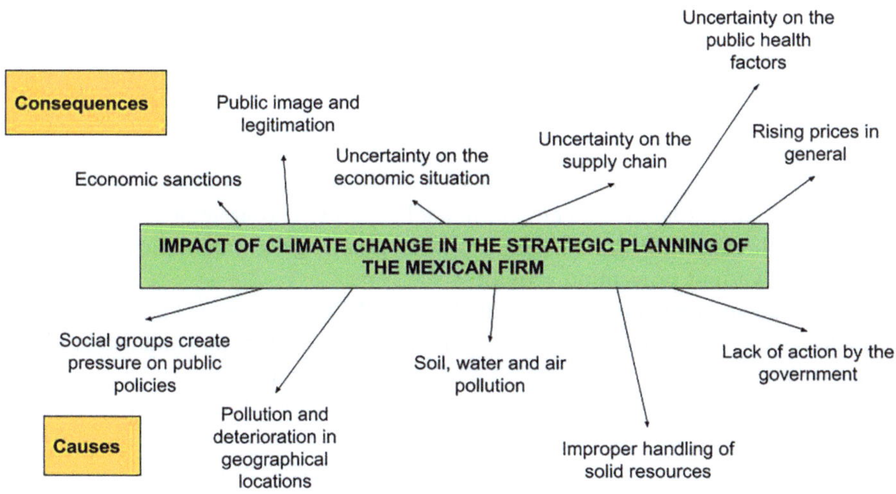

Fig. 1.2 Problem tree for climate change on the Mexican firm. (Source: Own creation)

begin to put pressure on companies that pollute, and it has had increased health effects from climate change and pollution in vulnerable areas where the companies are established.

As we can see, there is a clear relationship between the effects and causes which are based on conditions of social, health, and scarcity of resources (raw materials) for organizations. These indicators have permeated the effects on business planning which has led to strategies in various sectors, such as the development of new products and services, green brand strategy, enhancement of corporate image, improvements in energy efficiency, generation of clean energy, developing ecological values on employees, and conducting environmental education programs.

These strategies are part of the effects that global warming has generated directly in planning organizations, as indicated by the surveys. Another essential element in organizations is the creation of brands focused on specific segments of green consumers, as well as diversification in green bonds and green technologies management.

1.9 Best Practices on Quality of Life Regarding the Impact of Climate Change on Business Planning in Mexico

All business units in Mexico should adopt some of the best practices already tried and tested and learn advanced technology innovations that can add greater efficiency to resources and thereby increase the quality of life. The efficiency of resource use to reduce global warming in the business units is crucial to improving the quality of life and economic development.

In Mexico, for climate protection the priority is to reduce emissions from building operations and economic units. Rational use of other energy sources such as geothermal waters, which do not contribute to global warming, are some of the many options to increase the environmental performance of business units that also brings benefits to improving environmental performance, quality life and public health. The prevention of climate change by establishing protective measures has an impact on local economic and environmental benefits. The strategic and direct confrontation threat of climate change, by governments and communities make them involved in various prevention activities.

Other recommended measures considered as the best practices in action plan for counteract the global warming phenomenon, are the development and use of biofuels, promoting the use of alternative and renewable energy, the development and promotion of clean development mechanisms and projects, design and implementation of a watershed development strategies, development policy for reservoirs, waters and irrigation capacities, promotion of sustainable forests and green urban areas, etc.

Among the best practices to mitigate global warming of economic units in Mexico is the creation of green spaces and recreational green parks and gardens around the most polluting manufacturing plants. These areas and green gardens will help promote the quality of life by providing opportunities for further relaxation, sports, and health. Rational use of other energy sources such as geothermal waters, which do not contribute to global warming, are some of the many options to increase the environmental performance of business units that also brings benefits to improving environmental performance, quality life and public health.

To reduce the contribution of economic units in Mexico to climate change, it is evident that they require reducing energy consumption based on fossil and increase the use of green energy through processes and innovative measures and cost effectiveness, contributing to improve air quality and a better quality of life. There is not an inclusive list of best practices that local governments and communities have in terms of protection activities that demonstrate their potential for prevention activities of global climate through local action. Each case is different (O'Neill & Rudden, 2013).

Some of these best practices to improve the standard of quality of life are: The establishment of standards of sustainable energy through mechanisms of energy codes for use in buildings and processes in businesses as mechanisms supervised by local governments. A procurement program of environmentally friendly products is important to be implemented such as purchasing green energy for operations. Reducing greenhouse gas emissions in operations and buildings of economic units is also critical by increasing standards of energy efficiency of the processes using machinery and heavy equipment.

Other practices require the development and implementation of programs of energy conservation, the establishment of audits, quotas, and incentives for improvements in energy efficiency and the use of other green energy such as solar, wind, etc. Cooperative programs can start purchasing green power for more efficient use of machinery and equipment. Programs for planting trees and green areas around the buildings of the plant to improve climate and that has other benefits. Firms must reduce the use of vehicles transporting goods and people by promoting improvements in pedestrian areas and transit bicycles and other vehicles, as well as to establish incentive programs for green transit-oriented development. Land development programs in planning city through urban policy areas and discouragement of activities and permit fees that invade the occupation of floor space.

Climate changes severely affect companies due to the potential of environmental conditions associated with the risks of adverse conditions, such as natural disasters, which are not always identified as vulnerable or considered in the planning of business operations situations. Small and medium businesses that have limited resources and are located only in a few places are very vulnerable to the ravages of natural disasters caused by climate change. With limited access to resources and capital, small businesses may suffer from severe economic damage. The impacts of climate change on small and medium businesses have a long reach to break the chains of supply and consumer markets.

Among other recommended measures to improve the environment and mitigate the effects of climate change on small and medium-sized business practices, we recommend working in partnership with communities and local governments to design and implement a joint plan and programs to share data and information, promote actions that build resilience of the value chain so that vulnerabilities are reduced, to sustain with less economic damage and recover quickly from disasters and calamities. These plans and programs should include strategies and policies for cost effectiveness of measures to reduce emissions and bring great benefits to the quality of life of people and the economy in general.

Thinking about economic growth using the scope of sustainability, as well as the importance of considering the conditions related to nature, may lead to better ways of social organization, and it is especially important for society and the private sector. It may, in a way, decrease human and economic loss at several levels. Which is why it is especially important to design public policies that are more appropriate to promote sustainable activity in the private sector, looking for the need to plan and act on mitigation schemes related to climate change in this region, which is highly vulnerable to this phenomenon (Delgado et al., 2013).

Despite all the scientific information circulating on the media on climate change, the controversy on its origin and severity has introduced deliberate confusion and doubt on the population, because the presence of the phenomenon questions the belief systems that push the current dominant political and economic model. Several studies show the complex relationship between educational level and the development of climate-conscious attitudes. In this way, the people with egalitarian ideologies (fight for social justice, impartiality, and non-discrimination, among others) tend to accept the science on the topic of climate change.

Conversely, those with hierarchical and individualist personalities (people who oppose policies toward minorities and against poverty, who back private enterprise and with strong convictions toward the idea that everyone gets what they deserve) reject the scientific consensus. In these people with strong individual convictions, concern for climate risks is inversely proportional to their scientific knowledge.

Even though it may seem paradoxical, in these cases, a greater education on the topic does not generate greater environmental concern. So, there is no indication that more scientific literacy will automatically lead to change in behavior toward a proper environmental action (González Gaudiano & Meira Cartea, 2020). This is particularly important for us because many of the higher ups at the firms subscribe to these ideas, and they tend to see economic progress as more important than environmental protection even if all of what the experts are saying is true; other countries are polluting even more because they are more industrialized.

1.10 Conclusion

Climate change is a complex phenomenon, which has been very polemic, and has presented various effects on societies and businesses.

The implementation of complex environmental innovations as a distinct organizational strategy encourages entrepreneurship to mitigate the effects of global warming and increases the quality-of-life standards of all workers and other stakeholders, while preserving environmental conditions in favor of long-term life.

The literature review showed how various organizations have considered the issue at international and national level. This research shows that the issue has been considered by Mexican firms in their planning processes and strategic projects are being generated from this consideration.

Climate change has had an impact on the strategic planning process of Mexican companies as a factor that is changing the economic, political, and legal environment in which it operates, and has included three key elements: the increase in costs, negative externalities and the increase in the uncertainty of key factors such as the development of new products and markets.

This exploratory study opens the door for further descriptive and explanatory analysis studies highlighting success stories of strategies in Mexican and foreign companies.

References

Boiral, O. (2006). Global warming: Should companies adopt a proactive strategy? *Long Rabge Planning, International Journal of Strategic Management, 39*(2), 315–330.

Chevalier, J. M., & Buckles, D. J. (2009). *SAS2 Guía para la Investigación Colaborativa y la Movilización Social*. Plaza y Valdés.

De la Vega-Salazar, M. Y. (2011). Posibles factores de riesgo de los peces epicontinentales de México ante el calentamiento global. *Avances en Investigación Agropecuaria, 15*(2), 65–78.

Delgado, G., Gay C., Imaz, M., & Martínez, M. (2013) *México frente al cambio climático*. Colección El Mundo Actual. Universidad Nacional Autónoma de México.

Ernst & Young. (2010). *La respuesta de las empresas al cambio climático*. Elegir el camino correcto. Recovered from: http://www.ey.com/MX/es/Services/Specialty-Services/Climate-Change-and-Sustainability-Services

Galindo, L. M. (2009). La economía del cambio climático en México. *Síntesis, 21*.
García Fernández, C. (2006). El análisis coste-beneficio y la dificultad de su aplicación al cambio climático. *Estudios de Economía Aplicada, 24*(2), 751–762.
Godet, M. (2000). *La caja de herramientas de la prospectiva estratégica* (4a. ed.). Laboratoired'Investigation Prospectiva et Stratégique y Prospektiker – Instituto Europeo de Prospectiva y Estrategia.
González Gaudiano, E., & Meira Cartea, P. (2020). Educación para el cambio climático: ¿Educar sobre el clima o para el cambio? *Perfiles educativos, 42*(168), 157–174.
Guhl Corpas, A. (2008). Aspectos éticos del calentamiento climático global. *Revista Latinoamericana de Bioética, 8*(2), 20–29.
Jorqueta-Fontena, E., & Orrego Verdugo, R. (2010). Impacto del calentamiento global en la fonología de una variedad de vid cultivada en el sur de Chile. *Agrociencia, 44*(4), 427–436.
Landa, R., Magaña, V., & Neri, C. (2008). *Agua y clima: elementos para la adaptación al cambio climático*. SEMARNAT.
Levy, D. L., & Kolk, A. (2002). Strategic responses to global climate change: Conflucting pressures on multinatrionals in the oil industry. *Business and Politics, 4*(3), 275–300.
Morales Lara, B. (2021). Calentamiento global y desarrollo sostenible, avances y retos actuales en la política pública de México. *Ciencia Económica, 9*(15), 25–41.
O'Neill, K., & Rudden, P. J. (2013). *Environmental Benchmarking & Best Practice Report RPS Group, Ireland*. European Green Capital Award 2012 & 2013.
Ortega Díaz, A., & Casamadrid Gutiérrez, E. (2018). Competing actors in the climate change arena in Mexico: A network analysis. *Journal of Environmental Management, 215*, 239–247.
Pérez Salazar, B. (2007). Reseña de "Una verdad incómoda. La crisis planetaria del calentamiento global y cómo afrontarla" de Al Gore. *Revista de Economía Institucional, Colombia, 9*(17), 385–395.
Proverbs, D. (2010). The impact of climate change on local scale flood risk for Indivual developments. School of Engineering and the Built Enviroment, University of Wolverhampton, United Kingdom. *Systematic Review* 89.
Secretaría de Economía. (2009). Acuerdo por el que se establece la estratificación de las micro, pequeñas y medianas empresas. *Diario Oficial de la Federación*, June 30th, 2009.
Secretaría de Economía. (2012). *Sistema de Información Empresarial Mexicano*. Recovered from http://www.siem.gob.mx/siem/
Sosa-Rodríguez, F. (2015). Política del cambio climático en México: avances, obstáculos y retos. *Realidad, Datos y Espacio. Revista Internacional de Estadística y Geografía, 6*(2), 4–23.
Sussman, F. G., & Freed, J. R. (2008). Adapting to climate change: A business approach. *Pew Center on Global Climate Change*. Recovered from http://www.c2es.org/business-adaptation
Thompson, A. A., Jr., Strickland, A. J., III, & Gamble, J. E. (2008). *Administración Estratégica. Teoría y casos* (15th ed.). McGraw Hill.
Van den Hove, S., Le Menestrel, M., & de Bettignies, H.-C. (2002). The oil industry and climate change: Strategies and ethical dilemmas. *Climate Policy, 2*, 3–18.

José G. Vargas-Hernández is research professor at Instituto Tecnológico Mario Molina Unidad Zapopan, and earlier at University Center for Economic and Managerial Sciences, University of Guadalajara. Member of the National System of Researchers of Mexico. Professor Vargas-Hernández has a Ph.D. in Public Administration and a Ph.D. in Organizational Economics. He has been visiting scholar at Carleton University Canada, University of California Berkeley and Laurentian University, Canada. He holds a Ph.D. in Economics, Keele University; Ph.D. in Public Administration, Columbia University; studies in Organizational Behavior at Lancaster University and has a Master of Business Administration. He has Published 9 books and more than 300 papers in international journals and reviews (some translated to English, French, German, Portuguese, Farsi, Chinese, etc.) and more than 300 essays in national journals and reviews. He has obtained several international awards and recognitions. He has also experience in consultancy. His main research is in organizational economics and strategic management. He teaches for several doctoral programs.

M. C. Omar C. Vargas-González serves as professor and head of the Department of Systems and Computing at the Technological National of Mexico Campus Ciudad Guzmán and professor of Telematics Engineering at the University Center of the South of the University of Guadalajara, with a master's degree in Computer Systems. He has received training in Entrepreneurship and Multidisciplinary Innovation at Arizona State University (2018) as well as a diploma in Ecosystems of Innovation and Entrepreneurship from the Harvard University School of Business. He currently conducts research in diverse areas such as Entrepreneurship, Economics, Statistics, Mathematics and Information and Computer Sciences, has collaborated in the publication of more than 15 articles in magazines and directs innovation and technological development projects.

Pedro Antonio López de Haro is research professor at the Universidad Autónoma Indígena de México. He is a member of the National Research System (SNI) in Mexico. He got his Ph.D. in Economics and International Business at UAIM and his master's degree in Industrial Engineering at the Instituto Politécnico Nacional. He has been an engineering, mathematics and English professor for 13 years. He was the editor of the *Ra Ximhai* scientific journal for 4 years. He has published two books and several scientific articles in intercultural education, economics and research methodology.

Chapter 2
Finding a Moral Compass: Grounded Theory Research on Ethical Leadership Training for Education for Sustainable Development (ESD)

Amanda L. Holst

Abstract Leaders today are tasked to be future-focused; work in ambiguity, and be environmentally conscious, while at the same time expected to promote ethical behavior and standards Despite ethical leadership concepts being well studied, there is little research on ethical leadership training regarding sustainability. Furthermore, the relationship between ethical leadership and sustainability, and the role of leaders creating a moral climate has been underrepresented in the education for sustainable development (ESD) leadership training field according to many scholars. The purpose of this chapter is to explore literature from various current articles regarding the topic of ethical leadership development and sustainability, reveal the gaps, and show how transformational leadership, a theory of leadership in the field of global leadership, can be used as a framework for ethical decision-making. A qualitative comparison analysis was used to identify research gaps in ethical leadership training for ESD. It also revealed that making ethical decisions showed the most significant factor in creating a moral compass. The findings led to the development of an ethical decision-making conceptual framework using transformational leadership dimensions and the three main ethical approaches to normative ethics. This framework can be used for ethical leadership development training, and for educators and scholars to use as a guide in developing their framework when determining how to cultivate and lead with a reliable moral compass.

Keywords Ethical leadership development · Education for sustainable development (ESD) · Transformational leadership · Ethical decision-making · Sustainability · Moral compass · Sustainability curriculum · Moral decision-making · Ethical leadership · Virtue-based approach

A. L. Holst (✉)
Global Leadership Program, Indiana Institute of Technology, Fort Wayne, IN, USA
e-mail: alholst01@indianatech.net

© The Author(s), under exclusive license to Springer Nature Switzerland AG 2023

S. Dhiman (ed.), *Sustainable Development and Environmental Stewardship*,
https://doi.org/10.1007/978-3-031-28885-2_2

2.1 Introduction

What Does Ethics Have to Do with Sustainability?

Has society lost its ethical, moral compass? Some researchers argue that a push for technology has left a gap in the moral aspect of ethical learning, and thus leading to educators' lack of exposure to the reflection of ethics (England, 2014). Current theories have elaborated on the sustainable development (i.e., social, economic, and environmental). However, there is little research on ethical leadership training regarding sustainability (Omer, 2014). Education for sustainable development (ESD) is a United Nations program first notably presenting itself in the 1980s World Conservation Strategy in the Brundtland Report. This report brought together environmental and socioeconomic questions with the aim of coming up with the goals of educating and promoting the "development of the knowledge, skills, understanding, values, and actions" required to create a sustainable world as well as ensure environmental protection and conservation, promotion of social equity, and encouragement of economic sustainability into teaching and learning (UNESCO, 2018). These goals became Sustainable Development Goals (SDGs) created by an international community, which included 169 aims to be met by 2030 (UNESCO, 2018). While there is a great deal of literature on ethics and ESD (Jickling et al., 2006), it is marginalized in the discussion on views of attitudes and ethics in the environment (Kopnina, 2013). To add to this, the SDGs have been under scrutiny for being "inconsistent, difficult to quantify, implement and monitor" (Swain, 2018, p. 341), leading to no clear path of how leadership styles and choices can influence decision-making in the process of understanding the complexity within these challenges. More specifically, there are gaps in connecting the SDG goals to educational outcomes, and a lack of connection or responsibility to the environment and consensus of terms related to sustainability, further exacerbating confounding challenges that must be addressed and causing additional barriers to leadership training (Jackson et al., 2017). While the SDG collectively is regarded as a fresh way of thinking and working on issues such as education, climate, and poverty, it is widely accepted that sustainability overall is a complex problem and today remains a challenging problem. Some researchers go one step further to explain the behavior and relationships of organizations, leaders, and decisions about practical sustainability. Metclaf and Benn (2013) wrote: "Effective whole Earth sustainability may be less about human moral decision making and more about complex problem-solving, with most leaders either unable or discouraged to explore the full complexity of the organisation's role or impact within its wider systems environment" (p. 370).

Furthermore, there is little known about sustainability being infused throughout the education curriculum, making it difficult for ethical leadership training to find a place in the SDG's achievement. To add to this, there is a limited professional development and help with curriculum to educate instructors how to teach Education for Sustainability (EfS) in higher education as well as little pedagogical tools for sustainability material, further complicating the dissemination of ethical principles in attaining sustainable development course study (Michel, 2019). There is also limited research on leadership values, ethics, and style in corporate social responsibility, showing the lack of moral teachings. Metclaf and Benn (2013) pointed out, "Leadership for sustainability requires leaders of extraordinary abilities. These are leaders who can read and predict through complexity, think through complex problems, engage groups in dynamic adaptive organisational change and have the emotional intelligence to adaptively engage with their own emotions associated with complex problem solving" (p. 381). While many training efforts enact training to enhance making ethical decisions to protect organizations' reputation and doing what is right, there is no specific knowledge for ethical leadership training in ESD. A large amount of literature exists in educational philosophy, and moral, educational leadership regarding universal principles, especially justice, care, and virtue. However, there is a lack of research on the challenges that universities face when creating a sustainability curriculum and implementation (Wilson, 2000). Ultimately, leaders face decisions and ethical questions, and these decisions determine whether they are moral leaders. Since ethical leadership requires a clear and reasonable ethical framework that leaders need to use to put into play, the ethical structure is necessary to make ethical decisions. Research shows that there are many connected leadership types to ethical leadership (See Table 2.1), and thus can be used with ethics and implications for ESD.

Aligning the leadership type to ESD concepts can help develop a moral compass framework and align ethics with leadership. This ethical leadership will be responsible for influencing the process and implementation of ethical practices in the 2030 Agenda for Sustainable Development and should align with the values of the leader to serve in decision-making to advance the common good of the organization and society.

Understanding the Role of Ethical Decision-Making in ESD

The growing urgency to fulfill the 2030 Global Goals aimed at ending poverty, hunger, and acting on climate change, and the environment (https://www.globalgoals.org/) ensues. It is time to motivate action with global responsibility and integrity efforts through ethical leadership. It is needed to take on the vastly complex challenges lying before the next generation. Eyes are looking to higher education institutions (HEIs) and the global leadership fields to equip college students and leaders with the capability and competency skills related to social and environmental responsibility. Scholars and practitioners across disciplines agree that the challenges for achieving Sustainable Development Goals are intricate and cumbersome. In many cases, views on ethics are not clearly defined or difficult to understand. Therefore, it is essential to keep this in mind and work to adopt an interdisciplinary mindset. Schwartz (1987) points out the discord that while philosophical ethics were once thought used to make decisions and taught in higher education just over a century ago, by contrast, today, the idea has shifted to moral individualism, where

Table 2.1 Leadership styles, relation to ethics, and implications for ESD

Type of leadership	Definition	Relation to ethics	Used in practice	Implications for leadership development training for ESD
Relational leadership	Involves everyone and their ideas	Being unbiased to yourself as a leader, other leaders, and staying fair in the process of meeting goals	Participants leverage both positional and relational influence and understand one's unique strengths and weaknesses	Agency important to integrate leadership and share power; Driven by values and standards which is "good" or "moral" in nature (Komives et al., 1998)
Spiritual leadership	Focuses on attitudes and behaviors in leadership	Shows care, compassion, and responsibility	Contributes to leader and follower efficacy	Research shows that there is a clear consistency between spiritual values and practices, and leadership effectiveness and values such as integrity, honesty, and humility have positive influences on health and psychological well-being and can help to create vision and value congruence across the organization (Jody & Egel, 2018)
Transformational leadership theory	Focuses on passions and desires of follower to match desire of leader	Values and ethics are the main approach, same as moral-based leadership, idealized influence, inspirational motivation, intellectual stimulation, and individualized consideration. Organic worldview and moral altruistic motives grounded in deontological perspective (Kanungo, 2001)	Being a role model, future vision, individual support, promotion of the goals of the group, intellectual stimulation, and high-performance expectation	Leads through example, inspires certain behavior, empowers people and uses their strengths, acts like a coach and aware of personal challenges and endeavors, has moral reasoning and moral identity, positive related strong moral identity is more likely to accentuate moral values in communication and decision-making

Table 2.1 (continued)

Type of leadership	Definition	Relation to ethics	Used in practice	Implications for leadership development training for ESD
Transactional leadership theory	Exchange of rewards and targets between employees and management focuses on performance of organization and supervision and leaders; obtain compliance through rewards and punishments, usually through extrinsic motivation	Altruistic motives grounded in teleological perspective (Kanungo, 2001)	Keep projects moving and work in the same fashion, focus on specific tasks and processes, and look for faults and deviations	Effective during emergency and crisis situations and specific projects that need to work by a specific process
Authentic leadership	Connected to ethical leadership: “Confident, hopeful, optimistic, resilient, transparent, moral/ethical, future-orientated, and gives priority to developing employees to be leaders”	Having to do with ethical development and impact on employees, and fulfilling their full potential of their plan I upbringing and spirituality, two factors that lead to ethical decisions in the workplace (April et al., 2010, as cited by Salter et al., 2014)	Focuses on performing in alignment with their moral principles, on the common good, and with respect and transparency to all stakeholders involved	Ethical development and impact on leading in challenging environments
Sustainability leadership	Leadership that embraces the triple-bottom-line concept: Mindful actions and behaviors that embraces a global worldview to recognize the connection between planet and humanity	Leaders are expected to serve and protect their communities and engage in sustainability and environmental responsibility	Taking mindful actions that embraces a global worldview of the connection between the planet and humanity producing positive environmental and social change	Leading by example in your actions Leader works within holistic interconnections that are present between people and natural world and encourages others to come together to generate answers (Ferdig, 2007, as cited by Diamond, 2012)

serving private interests supersedes the coordination of collective culture. Morality, according to Schumaker (2001), is based on individualized perception. This ethical relativism of making up one's rules for one's good within the norm of one's own society (Smith et al., 2012) results in what Velasquez et al. (1992) assert as a dilemma that culminates into issues with reaching agreements and having no collective thinking of whether an action is right or wrong. Despite ethical efforts in schools and organizations to train in the implementation of sustainable work combined with the increased pressure to meet the increased expectations of the SDG goals, ethical decision-making can frequently be caught between a rock and a hard place (Wood, 2001). Therefore, exploring what it means to have a moral compass and how it can guide a person regarding ethical behavior and decision-making can further assist leaders in handling and managing complex problems that they are faced with today.

What Do Ethical Leaders Do, and Why Do They Do What They Do?

Lawton and Páez (2015) argue that integrity and authenticity are the closest tied to how a leader will use their beliefs in the decision-making process and their ethical leadership behavior and are about "doing, not just being." Since leaders must make decisions that need judgment outside of rules, they must lean on ethical underpinnings to get them there (Dukerich et al., as cited by Lawton & Páez, 2015). As a result, the outcome leads to a positive response in employees. Lawton and Páez's framework draw together the facets of ethical leadership and includes *practices, purposes, and virtues*. In this framework, leaders cannot separate virtues from practices and purposes, and the combination of all three will lead to different forms of ethical leadership. Also, this framework is holistic in that ethical decision-making is a part of a practice, while the virtues to accomplish the method may require judgment and integrity. A limitation to this framework and future research area is that there needs to be more explanation and justification for leaders' actions in ethical leadership training in education for sustainable development (Lawton & Páez, 2015). Knowing how leaders acted, why they did, and their reasons could help make decisions and create ethical standards.

Ethical Theories and Ethical Theories on Leadership

Levi and Rothstein (2018) define being ethical as a "set of principles about the right way to behave." According to Neubert et al. (2009), ethical theories provide the decision-making foundation and represent the viewpoints from which individuals seek navigation in making choices. Thus, ethical theories fall into two categories: one relating to the character of leaders and the other that links to the leaders' behaviors. Three approaches to ethical theory include teleological, deontological, and virtue-based approaches (Kapur, 2018). In the teleological approach, outcomes of a leader's actions related to their behavior or conduct are emphasized by determining whether effects were ethical. These effects include the categories of ethical egoism, utilitarianism, and altruism. Kanungo (2001) posited that the actions of leaders have no intrinsic moral status, and the basic standard of what an effort brings is an "end" outcome orientation (Kanungo, 2001). By contrast, deontological ethical theories are duty-based, and state that people should follow their obligations and duties when occupied with making ethical decisions. The leaders focus on their actions and their responsibility in doing what is right and moral (Kapur, 2018). Finally, the character of the leader and the virtues that are learned and practiced through involvement within the community and understanding are considered the virtue-based approach. Virtue-based approaches are approaches taken from the character of a leader and are explained by deontological theories. Kapur (2018) posited ethical leaders' virtues: generosity, kindness, courage, temperance, self-control, discipline, honesty, fairness, equality, modest manner, decency, and justice. Kant's universal ethics principle theory asserted that all people are responsible for all other people under their belief to be a part of humanity.

Additionally, in the principle of morality, all people should respect the humanity in others and only act in the moral law, acting rationally and in the truth of reason, which should be upheld by everyone (Immanuel Kant, n.d.). Finding a moral compass requires consideration in seeing the different relationships and interrelationship of circumstances when making an ethical decision. Therefore, in ESD training and development, educators must consider different ethical theories and theories of leadership.

What Is Ethical Leadership?

Ethical leadership is the use of personal conduct that is ethically appropriate in decision-making and is carried out through two-way communication such that others are heartened to comply (Lawton & Páez, 2015). Ethical leadership is defined as affecting employees' behavior with the leaders' value system, attitudes, decision-making, and influence of processes (Yi et al., as cited by Rao, 2019). Ethical leadership is broken down in literature and studied by the following categories: characteristics of an ethical leader, the relationship between being a good leader ethically and being an effective leader, and how leaders keep self-interests, the interests of the organization, and the interests of the community balanced. An ethical leader's characteristics include the ability to empower colleagues, promote values and motivations, and encourage growth and self-realization (Enderle, 1987, as cited by Rao, 2019).

There is a connection between individual responsibility and structural transformation (Kioupi & Voulvoulis, 2019). Research shows that behaviors are driven from an ethical point of acting with a civic sense to prioritize the common good. Collective values such as moral judgment, value clarification, and critical reflection on personal and collective values are needed for ESD, especially competencies and value orientations between individual responsibility and structural transformation (Schank & Rieckmann, 2019). Also, Rao (2019) argues that the ethics of leaders are the key to promoting sustainability in higher education institutions (HEIs) (Rao, 2019). In any case, both organizational culture and trust are needed to encourage the development of ethical leadership (Brown et al., as cited by Rao, 2019).

Ethical leadership focuses on features of moral reasoning, locus of control, conscientiousness, and traits that include being kind, sympathetic, cooperative, warm, and considerate (Wolfgramm et al., 2015). Four central positions of ethical leadership are humane, justice, responsibility and sustainability, and moderation orientation. Therefore, ethical leaders must serve others and make beneficial decisions in the best interest of equality, fairness, and justice (Kapur, 2018). Since ethical leaders are honest, truthful, and contribute to building a productive community, they are more likely to consider values, standards, and ethics when performing job duties and take on the responsibility of communicating the influence of ethical values through their behavior (Kapur, 2018). There is a lack of ethical leadership research on antecedents, purpose, and moral theory in the areas of ethical leadership research (Lawton & Páez, 2015). The Global Leadership and Organization Effectiveness (GLOBE) is a study of cross-cultural leadership that studies organizational norms, values, and beliefs of leaders in various societies; over 60 countries recognize the same deficit and identify and define what it means to be an ethical leader. The four conceptual dimensions of ethical leadership, according to the GLOBE study, are *character/integrity, altruism, collective motivation, and encouragement* (Resick et al., 2006). Using the GLOBE's recommended conceptual dimensions and combining its critical characteristics of ethical leadership, Resick et al. (2006) tested these qualities in a cross-cultural setting from 17,000 middle managers in 951 organizations and 62 societies across the world and identified the following four themes: *character and integrity*; *community/people orientation*; *motivating, encouraging, and empowering ethical awareness*; *and managing ethical accountability*. In their results, Resick and colleagues found that attributes mentioned by the GLOBE study were universally supported across cultures, and can be viewed as contributors to effective, ethical leadership.

2.2 Theoretical Framework

Transformational Leadership Theory Linked to Ethics

Transformational leadership is a leadership style under the Full Range of Leadership Model (FRLM) that focuses on the passions and desires of the follower to match the leader (Bass & Bass, 2008). According to Bryman et al. (2011), transformational leadership is a process of leaders using emotional attachment and organizational goals with followers. A transformational leader acts as a role model and leads by example through idealized influence, which makes followers aspire to be like them. Through inspirational motivation, the leader inspires certain behaviors and actions in others through high and reasonable goals. As an outcome, followers become committed and passionate about the organization and share the organization's shared vision (Odumeru & Ogbonna, 2013). A transformational leader puts effort into laying the groundwork for the organization's strategies, policies, and procedures as well as implementing the vision (Kanungo, 2001). Concerning ethical underpinnings, transformational leadership has mutual altruistic motives grounded in a teleological perspective and provides moral influence over followers over the long term, according to Kanungo (2001). Therefore, this type of leadership is shown to be ethical due to the empowering strategies of satisfying the followers' "higher-order growth needs", the ability to transform the followers' self-interest into collective concerns, and the style's overall ability to "engage the full person of the follower" (Burns, 1978, p. 4, as cited by Kanungo, 2001). Transformational leadership can also lead to positive impacts on followers in ways that improve organizational environments and performance (Bryman et al., 2011). However, this is only if the leader can push past the reasonable expectations of a given group or organization by harnessing their collective performance into dedication, through motivation and shared values (Bryman et al., 2011). One meta-analytic study of 57,000 participants from 34 countries studying cultural norms within the GLOBE project showed that employee performance was more influential in cultures whose values were in alignment with the transformational leader (Crede et al., 2019). The study further concluded that the effectiveness was not consistent cross-culturally. The impact was more significant on followership in developing countries such as sub-Saharan Africa and the Middle East, showing that more review for this leadership style across cultures is needed (Crede et al., 2019). The Full Range Leadership Model (2019) lists the six dimensions that make up transformational leadership as the following: creating a vision through inspiration and being a role model, future vision, individual support, promotion of goals of the group, intellectual stimulation, and high-performance expectation. If the follower can be transformed into thinking along the same line of the leader's vision, desire, and goals, the follower will be more motivated to pursue the same vision and goal (Anderson & Peter, 2017). Transformational leadership's four main concepts and attributes are *idealized influence, inspirational motivation, intellectual stimulation, and individualized consideration* (Choi et al., 2017). Inspirational motivation has to do with an individual who can inspire certain behaviors and actions in others through high and reasonable goals. The transformational leader recognizes that people have skills, talents, and knowledge to contribute to the organization's goals. This acknowledgment enables followers to become empowered and work together to think outside of the box to come up with a solution, even if it requires a risk (Bryman et al., 2011). In individualized consideration, the leader encourages the followers to think for themselves and use their strengths to put in to help in the overall success of the organization. The leader acts as a coach during their growth and mistakes, creating opportunities and steer them away from practices that do not work. Additionally, the leader is generally open to learning and aware of followers' challenges and endeavors. Transformational leaders can secure trust, admiration, loyalty, and respect among their followers. One of the most significant criticisms of transformational leadership is the potential for abuse of power, through tendencies such as narcissism, thrift for strength, and manipulation (Hay, 2006).

Differences between moral reasoning and moral identity affect leadership behavior (Olsen et al., 2006, as cited by Salter et al., 2014). Research has found that moral reasoning and honorable character are positively related to transformational leadership. Leaders who possess a strong moral identity would be more likely to accentuate moral values in their communication and decision-making than other leadership types. In transformational leadership, values and ethics are the primary approach (Rowold, 2008) and considered the same as moral-based leadership (Burns, 2003 as cited by Rowold, 2008). As a result, followers are lifted to perform without self-interest for the goals of the company.

Additionally, transformational leaders motivate others to follow the higher good's moral-based vision, which drives them to go above and beyond. A transformational leader creates an environment that considers individuals' viewpoints and considers openness to needs and values with the notion that everyone is a unique human being (Rowold, 2008). Rowold (2008) stated that since transformational leadership is based on morals, the leader must articulate shared values over their values when it comes to deciding. There is much research on the behaviors of transformational leadership contributing to the effectiveness of a transformational leader in political leaders, nonprofit agency leaders, educators, business managers, and health care directors, to name a few (Bass & Bass, 2008). Still, leadership characteristics outlined by Brown and Trevino (2006) showed that ethical leaders in moral management traits were not found for transformational leadership. Therefore, additional research is required to determine the use of transformational leadership as a theoretical framework for ethical implications, such as making moral decisions.

2.3 Research Questions

To address the need to understand better what it means to cultivate a solid moral compass for a transformational leader in ESD, a classic grounded theory (CGT) study using an inductive approach to generate a theory was applied. Grounded theory is a "set of integrated conceptual hypotheses systematically generated to produce an inductive theory about a substantive area" (Glaser & Holton, 2004), and is an analytical method used to generate a theory for a process or an action by using three steps of the coding process: open or initial coding; axial or focused coding; and theoretical coding (Creswell, 2003). The conceptual analysis study reviewed the literature on ethical leadership for education for sustainable development and then used patterns and themes to identify categories and concepts. Other frameworks developed in this area focus on the understanding of sustainable development and its complexities (Jabareen, 2008); however, it only focuses on distinctive meanings and aspects of "sustainable development" and not in ethical leadership for education for sustainable development. Transformational leadership theory, ethical theories, and ethical theories on leadership approaches served as the theoretical framework for the study and provided almost a decade of the flow of research. The overarching research question was: RQ 1: What is a theory that explains what it means to cultivate a solid moral compass for leaders in education for sustainable development? Additional questions were centered around identifying current ethical leadership training for education for sustainable development (ESD), and what ethical theories and leadership styles could serve as a framework for ethical leadership training. Further subquestions that guided the study were: RQ 2: How do leadership training facilitators participate in developing a moral compass? RQ 3: What do ethical leaders do? RQ 4: What ethical leadership types help to promote establishing a moral compass? RQ 5: What do ethical leaders do not do? RQ 6: What are the gaps in research concerning the development of ethical judgment and educators? RQ 7: What are the factors affecting ethical leadership training for ESD? and RQ 8: What are some theoretical frameworks that educators and students can use to help in cultivating a solid moral compass?

2.4 Stages of Analysis

A constant comparative method of data analysis was used, taking data from the gaps of research and comparing it to emerging categories (Creswell, 2003). The grounded theory approach for this study was the systematic procedure based on Corbin and Strauss (2007), which takes an analytic method to develop a theory that explains "process, action, or interaction" on a topic (Creswell, 2003). A grounded study was chosen because it was in line with the research purpose, and inductive methods to analyze and expand knowledge on the topic denote its use (Corbin & Strauss, 2007). This study aimed to synthesize a theoretical framework for finding a moral compass using multidisciplinary literature on ethical development for education for sustainable development by detailing the gaps in ethical leadership development, exploring leadership theories, current models and frameworks, and finding emergent themes.

The dataset for this research, conducted in 2020, included 15 articles on constructs of ethical dimensions of sustainable development from 66 publications. They were pulled from EBSCO host and ProQuest searches identifying articles using the keywords *sustainable development*, *education*, *ethical judgment*, *and global leadership*. Most of the reviewed articles were published within the last 10 years. The study reviewed journals in the field of sustainability, social and behavioral sciences, education, business, and management. Countries of origin included the USA, Canada, Romania, Australia, Sweden, Slovenia, and Russia.

There were four stages to the qualitative data analysis and three distinct processes of analysis involved in grounded theory. These are initial/open coding, axial coding, and selective coding. The first phase included the collection of journal articles that utilized sustainable development, education, and ethical judgment and global leadership as keywords and where the authors wrote about leadership development, ethical leadership, and education for sustainable development. The original search yielded 1573 searches. After going through and finding the relevant keywords again, the list was reduced to 208. A quality check disqualified 77 that were eliminated for nonapplicability. For each article, the review analyzes the construct, origin of country, education level, conceptual framework, method of study, tools used, the sample of the study, key issues studied, main findings, gaps in research, and implications of the study.

Phase two identified the set of patterns within the results of the first phase by looking for similarities and then coding the results according to the categories of gaps in research. This was achieved by identifying *initial codes* from previous research theories, research questions, and gaps and implications from the study to determine connections among categories.

The third phase of coding was the *focused/axial coding* of the data, in which initial codes were further analyzed into categories. Creswell's (2003) comparative approach was taken for this, and the categories were reviewed, as well as the identification of causal conditions related to the central phenomenon. Patterns were identified based on frequency, magnitude, structures, processes, causes, and consequences. The comparative process involved three types of comparisons:

1. Incident to incident for the emergence of concepts.
2. Concepts to more incidents for further concepts and themes.
3. Concepts to concepts for theoretical coding and integration (Evans & John, 2013).

Phase four involved the focused/axial coding of the data in which initial codes are further analyzed into categories. Corbin and Strauss's (2007) paradigm model for the development of relational statements and theoretical coding was used, pulling in causal conditions, context, intervening circumstances, action, and consequences about and in response to the phenomenon based on the results of the third phase.

In the fifth stage, a hypothesis was created through *selective coding* to develop a moral compass theoretical framework. This was done by describing the connections among the resulting concepts, resulting in a conceptual framework that identified research gaps in ethical leadership training for ESD. The relational statements used for the theory were developed from the axial coding and selective charts. The conceptual codes were derived from the axial codes. They included the main topics of evaluating a situation, choosing among alternative decisions in a way that is consistent with ethical principles, and choosing the best decision in the situation. *Making ethical decisions* emerged as the selective code, as it was the central category that correlated to all the other categories. A conditional relationship guide was used for transformational leadership and a relational matrix to develop the storyline (Table 2.2). The conditional relationship guide when making ethical decisions based on a transformational leadership style depended on the choice and process of evaluating and choosing the best ethical decision: when, where, why, and how the decision was made. When to make decisions, how decisions can be applied and under which expectations, circumstances, and agency of ESD were some of the big codes that emerged in axial coding. Therefore, this led to ethical decision-making

Table 2.2 Conditional relationship guide using transformational leadership

Category	What	When	Where	Why	How	Consequence
Making ethical decisions	The process of evaluating and choosing among alternative decisions in a way that is consistent with ethical principles Choosing the best ethical alternative	During times of crisis During everyday practice During the present and the future Outside of rules During ambiguity	At school At work At the local level At state level At the global level	Consequentialist ethical decision-making framework Duty ethical decision-making framework Virtue ethical decision-making framework	Lead by example and set expectations Encourages learning and growth Inspires to go to new heights Coaches and empowers success	Choice

emerging as the selective code for the theoretical framework. Understanding the choice of a source of ethical standards and helping a leader determine what standard of behavior can be considered ethical in a certain circumstance are synthesized using the conditional relationship guide in Table 2.2.

2.5 Conceptual Framework Toward a Moral Compass for Transformational Leadership

In response to a call for providing a guiding moral compass for students and leaders, a model of a moral compass that embeds ethical decision-making into the transformational leadership dimensions is offered (see Fig. 2.1). In doing so, the model draws from the emerging field of global leadership and ethical leadership frameworks and proposes that the dimensions of transformational leadership provide the foundation for leaders to hone in their skills and competencies to develop a moral compass for education for sustainable development. The model does this by drawing on the ethical framework of consequentialism, duty, and virtue—which is essential for ethical decision-making. A consequentialist ethical approach is concerned with the moral consequences of a particular action. It can be drawn from the utilitarian (the ethical action that provides the most good and reduces harm), egoistic (high concern for self and low concern for others), or the common right approach (guided by the general will of the people). The duty-based approach is doing due to obligation to perform the action and following rules, expectations, and responsibilities that meet what should be done or not done due to acting morally (Immanuel Kant, n.d.).

The core phenomenon was used to create an illustration to describe the theory that explains what it means to cultivate a solid moral compass for transformational leaders in education for sustainable development and how their ethical

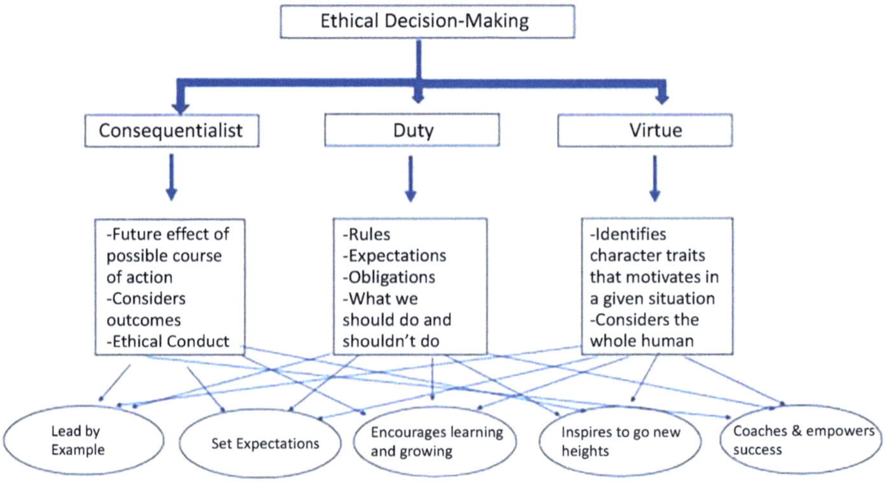

Fig. 2.1 Conceptual framework toward a moral compass to guide transformational leadership development for ESD

decision-making plays a role in a consequentialist, duty, and virtue framework. The model includes relationships among concepts, consequences of the phenomenon, and causal conditions and intervening conditions. The ethical theoretical framework consists of three approaches to take, weighing the consideration that impacts the choice of a course of action and then provides guidance for implementing the decision based on the dimensions of transformational leadership. In this framework, an individual can recognize an ethical issue, choose a source of the ethical standards of approach (*consequentialist*, *duty*, *or virtue*), and then implement the decision by using each of the dimensions of transformational leadership (*leads by example*, *sets expectations*, *encourages learning and growing*, *inspires to go new heights*, *and coaches and empowers success*). Making ethical decisions requires a practiced method, but when practiced regularly, the technique will become automatic.

2.6 Discussion and Findings

The presented conceptual framework toward a moral compass to guide transformational leadership development for ESD will help students and educators to develop their foundation for making ethical decisions as future leaders confronting the next generation of global challenges. The lack of ethical leadership training or lack of leadership style studied along with ESD poses a challenge and remains scarce in research. In the 15 articles explored, the presence, role, duties, and qualities of ethical leadership of sustainable development were discussed as constructs. However, there was no specific discussion on ethical leadership development or training. This supports Jickling et al. (2006)'s claim of many articles on the topic of ethics and

ESD but reiterates Kopnina (2013)'s claim that the information about ethics in ESD is marginalized. The other articles for this paper supported this marginalized concept seen by topics explored including efficacy of the Dynamic Interaction of Agency in Leadership (DIAL) framework for analyzing agency in sustainability leadership (Wolfgramm et al., 2015); literacy of, the value in, and attitudes toward the environment; and the evaluation of ethical leadership in sustainability. This reiterates that a framework or construct for ethical leadership on leadership styles is needed. These findings also raise the fact that there is still confusion and lack of consensus regarding sustainability, agency, and value-based practices regarding the role of ethical leadership in HEIs (Rao, 2019). There is also a lack of consensus on the efficacy (Wolfgramm et al., 2015), literacy within the frameworks of ESD (Franck & Osbeck, 2018), general ethic concern, and integrating sustainability with a focus on ethics in education and at a leadership development level. Findings show that there is an overall concern for fulfilling the SDGs due to vagueness and difficulty in connecting the SDGs to educational outcomes (Kioupi & Voulvoulis, 2019). The findings indicate that the gaps and interpretation of SDGs should be dealt with at the local level first and to focus on how to connect the SDGs to educational outcomes and then how they play a part in the grander SDG realm should be explored after. The consensus of sustainability definitions and biases and gaps that prevent development in moral judgment are areas that others can help join in the effort, and HEIs, organizations, and other groups should work together to coordinate efforts.

Making Ethical Decisions

This study showed that making ethical decisions was dependent on the dimensions of leaders. Aspects included the leadership style used, how sustainability and higher education play a role in judgment and moral foundation, and using ethical standards, norms, and rules in typical situations, dilemmas, risk, and ambiguity. Axial coding revealed that in making ethical decisions, leaders and individuals must meet the standards, judgments, and actions for ethics, and believe that their mindset, values, actions, and responsibility will guide outcomes. Also, understanding the expectations, connectedness, and agency in the ESD curriculum will help promote the development of education in achieving sustainability. Other factors in making decisions include the use of methodologies, practices, personal reflections in which complex problems arise, and different educational and cultural environments make ethical decisions. An individual choosing the best ethical alternative uses the approach that best suits them while considering stakeholders. All factors can use dimensions of a leadership style to apply the implementation of their decision. Since transformational leadership is proactive, the transformational leader will search for universal values or principles to guide them in their organization (Kanungo, 2001). Additionally, since an ethic of duty guides the influence strategy of empowerment, the transformational leader will motivate followers by encouraging them to put groups first. The transformational leader achieves objectives through higher ideals and moral values, making it easy to apply the new theoretical framework toward a moral compass and implement leadership dimensions.

Additional findings showed the following factors in leadership development training for ESD: management challenges, scaling practicing, commitment and individual responsibility to the SDGs and agency, ethical literacy and ethical concern for the planet, leadership styles, and collective visions for sustainability. These factors should be considered when developing training programs. On-duty performance and communication of ethical standards can also affect ethical leadership training for ESD. Therefore, scaling ethical aspects as a learning process and identifying what is ethically acceptable and what can be changed to scale are valuable for setting moral foundations.

There are many connected leadership types to ethical leadership; however, transformational leadership showed the most accentuated in decision-making and communication than any other kind of leadership (Rowold, 2008). Transformational leadership uses the most altruistic motives grounded from a teleological perspective. This aspect provides moral influence over followers for the long term and is most ethical due to its empowering strategies of the higher-order need growth (Kanungo, 2001). Also, transformational leadership can mutually uplift followers in morality (Yasir & Mohamad, 2016). Additionally, the components of being able to act as a role model, having ethical conduct, and behaving in a way outside of self-interest with a collective mission show how transformational leaders can apply ethics to practice (Avolio & Bass, 2004, as cited by Yasir & Mohamad, 2016). This aspect allows others to follow ethical conduct and follow the same mission. Furthermore, transformational leaders must focus on the ethical policies and procedures and a culture that reinforces ethical behavior and practices to flourish in ESD. From the relational matrix (Table 2.3) causes that influence making ethical decisions can include choosing the best ethical alternative; the central position and orientation of ethical leadership; the relationship between a good leader ethically and being an effective leader; decision-making process; the ethical leadership type; the approach, character, personality, characteristics, dimension, and behaviors of leaders; and the value-based strategies taken.

Intervening conditions cause leaders to reevaluate their beliefs and values at any time and make ethical decisions, including stakeholders, employees, and society. Other factors include values, interdisciplinary curriculum, complex problems, decision-making process, voluntary compliance, vested interest, and dilemmas. Therefore, having training geared toward understanding what to do in certain circumstances and with whom and across different disciplines can support the leadership development efforts.

Other factors while in a practice setting included peer-assisted models, feedback from supervisor/teacher, readiness to teach, decision-making model, social influence pressure, dual roles, and role-modeling (Passi & Neil, 2016). Universities and organizations should be aware of these factors and work together to facilitate an environment that aids in development growth.

This study showed that those leaders acting with ethical decision-making have different viewpoints. Decision-making factors include mindset, attitudes

Table 2.3 Relational matrix

Phenomenon	Making ethical decisions
Causal conditions	Choosing the best ethical alternative; factors affecting ethical leadership training for ESD; scaling ethical aspects as a learning process and identifying what is ethically acceptable and what can be changed in order to scale; central positions and orientation of ethical leaders; relationship between a good leader ethically and being an effective leader; what ethical leaders do and do not do; decision-making process for ethical leaders (when new challenges came up, when it dealt with what values to use, and how judgment can be made and applied during a crisis); intention-based morality; ethical leadership type that promotes establishing a moral compass; approach of ethical leaders; character, personality, characteristics, dimensions, and behaviors of leaders; and the value-based approaches for ethical decision-making
Context	Knowing when and how to make decisions and knowing if it is the right decision; study of the importance of ethics in schools, ethical foundation in university curricula; culture as sustainable development; types of leadership styles; relationship between a good leader ethically and being an effective leader; responsibilities of ethical leaders (self and others); on-duty performance and communication of ethical standards; and responsibilities of ethical leaders (self and others)
Intervening conditions	Society, stakeholders, employees, values, interdisciplinary curriculum, complex problems, decision-making process, when challenges and crisis come up, voluntary compliance, vested interest, knowing what is ethically acceptable, scaling ethical aspects, role of sustainability, moral theory with risk, and dilemmas that cause leaders to reevaluate their beliefs and values; consensus of sustainability definitions; biases in moral judgment; the role of values, beliefs, and behavioral norms in new culture of leadership; factors affecting ethical leadership training for ESD; early grade reading; gaps in research that prevents development of ethical judgment; readiness to teach, decision-making model, social influence pressure, dual roles, role-modeling (Passi & Neil, 2016)/peer-assisted models, and feedback from supervisor/teacher; readiness to teach, decision-making model, social influence pressure, dual roles, role-modeling (Passi & Neil, 2016)/peer-assisted models, and feedback from supervisor/teacher
Action/interaction	Tools to carry out judgment, understanding motives, understanding the power of social movements in developing countries, and personal reflection; effective teaching methods; types of training; concepts, methodologies, and practice that can enable transformational learning; learning approaches; early grade reading; case studies on the successful implementation of programs; decision-making factors; mindset, attitudes (ecocentric versus anthropocentric), culture, views about interconnectedness with natural world and sustainability; finding ways to meet ethical standards of participation and adherence to ethical norms, rules, and high level of conduct; ethical leadership testing hypothesis and having longitudinal studies; theoretical frameworks for educators and students to develop moral compass; efforts at a societal level that will improve social values; and participation of leadership training facilitators
Consequences	Justification of the leaders for their actions in ethical leadership training in ESD; effectiveness of ethical factors relating to ethical leadership training; concepts, methodologies, and practice that can enable transformational learning; intention-focus

(ecocentric versus anthropocentric), culture, views about interconnectedness with the natural world, and sustainability. The leader should find outlets to consider their perspective and how it relates to their decision-making and find ways to incorporate their viewpoints while leading positively, but also within the scope of their leadership and understanding how it plays into the vision. Also, finding ways to meet ethical standards of participation and adherence to ethical norms, rules, and a high level of conduct can be an area of focus when creating materials for leadership development and is vital for facilitators and leaders. Therefore, understanding motives, power of social movements in developing countries, and having a personal reflection on when to act or interaction during ethical decision-making could benefit teachings. Having access to effective teaching methods, tools to carry out judgment, and explicit concepts, methodologies, and practice that can enable transformational learning can assist educators in ethical leadership development and leaders looking to reinforce their moral compass. Ethical leadership testing and having longitudinal studies can assist in the understanding of participation and how that participation affects efforts and outcomes.

Additionally, considering different learning approaches, concepts, methodologies, and practice on enabling transformational learning can be beneficial to carry out judgments, understanding of motives, and implementation of programs. Early grade reading, readiness to learn and teach, role-making, and the influence of social factors can help the leadership development field in this area. Other areas include the decision-making model, social influence pressure, dual roles, and role-modeling (Passi & Neil, 2016). Case studies on the successful implementation of programs and understanding of biases and dilemmas cause leaders to reevaluate their ideas and values. Additionally, the role of values, beliefs, and behavioral norms in transformational leadership and factors affecting ethical leadership training for ESD should be studied more. Theoretical frameworks for educators and students to developing moral compass such as the one provided to include different leadership styles would be valuable to the field.

This research shows that central positions and orientation of ethical leaders and their relationship to the development of ethical judgment can be explored in more detail, particularly from voluntary compliance versus vested interest manner. Intention-based morality, and transformational leaders basing moral decisions on intentions or outcomes, is another area to investigate. Kaptein et al. (2005) argue that ethical leaders can influence followers if they can measure their actions. Yukl (2010) offers ethical criteria; however, this criterion may not take into all complex situations, especially in the complexities of sustainable development. Therefore, understanding the conditions making ethical decisions can be useful for students, transformational leaders, and practitioners (Table 2.2). Knowing what ethical decision to make, and when, where, why, and how it will best identify and quantify each choice that is created can help in any situation that is faced. Table 2.3 outlines the phenomenon and how to approach or what to consider when making ethical decisions. These findings answered the research questions that were posed, which will be discussed in the following section.

2.7 Research Question Findings

Central Question *RQ1: What is a theory that explains what it means to cultivate a solid moral compass for leaders in education for sustainable development?*

Open coding revealed that although leadership styles in cultivating a moral compass were not immediately apparent, other components such as the character, personality dimensions, behaviors, and characteristics of leadership applied helped to determine whether a good outcome was achieved. Causal conditions such as choosing the best ethical alternative, scaling ethical aspects as a learning process, identifying what is ethically acceptable, and what can be changed to scale can help leaders make ethical decisions, in addition to knowing what central position and orientation of ethical leaders should help to identify how a leader will act, what part they will play, and how effective they will become. Decision-making came up in coding three times, which resulted in the most apparent area of need for leaders when it comes to ethical leadership and training. The decision-making process for ethical leaders appeared when new challenges arose, when it dealt with what values to use, and how judgment can be made and applied during a crisis. Transformational leadership revealed the most promising leadership style that promoted establishing a moral compass and became a theme in the last half of the study. Furthermore, making ethical decisions was directly linked to values. Knowing when to manage challenges effectively while maintaining a strategic vision is key to cultivating a solid moral compass.

Subquestion *RQ2: How do leadership training facilitators participate in the process of developing a moral compass?*

Sustainability

Mochizuki and Bryan (2015) argue that education is an ethical means of addressing climate change and should be central to climate change's role in ESD. Current research shows that courses and programs on sustainability only focus on sustainability content or issues and do not foster the practices, ways of being, and skills needed to be an effective sustainability leader (Burns & Wolf, 2014). Sustainability leadership identity and skills can be developed in sequence with experiential learning, reflective practice, and community building (Burns & Wolf, 2014). Having more education means that people are more equipped to make better decisions. Furthermore, considering other education modalities such as environmental education, climate change, as well as global and development education can be valuable in knowing when and how to implement ethical leadership development. Education is a way for new generations to understand the causes and consequences of climate change and the actions to address it (Mochizuki & Bryan, 2015). Training programs development to further enhance leader's levels of moral-based leadership would be valuable in strengthening followers' performance (Rowold, 2008). There is a deeper understanding of different worldviews on sustainable development. Article 6 of the United Nations Framework Convention on Climate Change (UNFCCC) urges countries to consider education, training, and public awareness on climate change (Mochizuki & Bryan, 2015).

Ethical Development

Research shows that there is a lack of empirical research on ethics in educational leadership (Glanz, 2010, as cited by Salter et al., 2014). Key learning capabilities for ethical development can include the following skills: (1) critical thinking; (2) problem-solving; (3) systems thinking; (4) reasoning and questioning patterning; (5) dealing with uncertainty; (6) evaluating and analyzing information; (7) planning and management skills; (8) learning how to adapt knowledge to new contexts and include self-directed learning; and (9) critical thinking skills in technology, information, and media skills (Cohen, 2007). These skills are ethical that learners will analyze the relationships among their practices and make informed decisions while considering the implications of an unknown future. Also, identifying which skills and learning will promote action, critical thinking, and thoughtful choices, including discussion applying values and analysis conducting, is useful to ethical leadership development. An example is the Climate Change Education for Sustainable Development (CCESD), which uses climate change as an "entry point for promoting the principles and practice of sustainable development through education" (en.unesco.org). CCESD prepares students to have ethical dimensions of climate change. Fundamental principles include the following: (1) ensuring an integrated approach to learning about climate change; (2) addressing local and global perspectives on climate change; and (3) taking a climate justice perspective. Climate change education (CCE) can be promoted through training, just like ethical leadership. The educational paradigm should include the following aspects expressing the integrity of the common and the specific goals: connection of education with cultural tradition, human striving to be capable of cross-cultural communication, and absorbing values of the new globalized civilization (Nasibulina, 2015).

Ethical Development Pedagogies

Regarding ethical development pedagogies, sustainability projects on campus are an excellent example of promoting the advancement of sustainability, but they are not being taken advantage of. Challenges to developing sustainability in the curriculum of other disciplines due to difficulties of education within disciplines and education that is cocurricular involve assessments and evaluation for competencies in coursework and general education credit (Wilson, 2000). Another future pedagogy for business school is that it should increase confidence ability for ethical decision-making through business forums (Rao, 2019). Reflection is necessary to understand one's self-belief by allowing a space to understand attitudes and actions and help individuals develop into leaders (Diamond, 2012). Therefore, it can be used to create a clearer vision and gain confidence in leading. Agency or the "ability to have a secure sense of oneself and one's values" can be used with reflection as a learner moves through self-authorship and listening to their internal voices to set personal values in the developmental process of becoming a leader (Diamond, 2012). One area that leadership training facilitators can help participation is in the problem of the interdisciplinary and holistic curriculum with multiple approaches and taking complex issues and integrating them with other environments.

Subquestion *RQ3: What do ethical leaders do?*

In open coding, it was found that there is confusion in the literature about the responsibilities of ethical leaders about self and others and a discrepancy between being a good ethical leader and being an effective leader. This stemmed from not knowing what type of leadership could be used to establish and function as using a moral compass. A collection of traits of what ethical leaders do is shown in Table 2.4 and it encompasses the character, custom, and conduct under the Greek word "ethos" (Northouse, 2018). Ethical leaders' approach could be dampened by the issues of taking ESD to interdisciplinary and holistic curriculums, further complicating how to take complex problems and integrate with them with other educational environments. While the main objective of ESD is the development of nurturing high moral values and acquiring new skills for sustainability (Nasibulina, 2015), Burford et al. (2016) argue there is a "value-discourse" gap. In this gap, the current model does not lend itself to critical thinking and needs a more value-based approach. This gap can be filled by taking a virtue-based approach, taken from the character of leaders. In this approach, the leader knows which ethical principle to use and is guided by their conduct. This guidance comes from either a teleological view of ethics (doing the greatest good for the greatest number and maximizing the most benefit with the least cost) or a deontological approach (where the leader focuses on their actions and responsibility in doing what is right and ethical) (Kanungo, 2001). Leaders who are ethical under the virtue theory are looked on by their behavior or conduct (Northouse, 2018) and are known to have qualities of self-control, courage, temperance, honesty, sociability, modesty, fairness, and justice (Choi et al., 2017).

Furthermore, Lawton and Páez (2015) argue that virtues cannot be separated from practices and purposes. Integrity and authenticity are centrally crucial to ethical leadership because they will be used in their decision-making and judgment calls. Therefore, ethical leaders using virtue ethics operate in a way that is good for themselves and others in a proper way with the future in mind. Open coding revealed that the relationship between a good leader ethically and being an effective leader was needed to develop a moral compass. The research gaps and implications showed having a decision-making process for ethical leaders, especially when new challenges came up. A need to know what values to use and how to apply judgment

Table 2.4 Checklist for ethical leaders

What ethical leaders do
Have qualities of self-control, courage, temperance, honesty, social ability, modesty, fairness, and justice (Choi et al., 2017)
Characteristics include the ability to empower colleagues, promote values and motivations, and encourage growth and self-realization (Enderle, 1987, as cited by Rao, 2019)
Virtues include generosity, kindness, courage, temperance, self-control, discipline, honesty, fairness, equality, modest manner, decency, and being just (Kapur, 2018)
Ethical leadership in Big 5 factors of personality includes conscientiousness and agreeableness to the most relevant traits (Kalshoven et al., as cited by Wolfgranum et al., 2013)
Being reflexive
Kant's universal ethical principle—All people should respect humanity in others and only act in moral law, acting rationally and in the truth of reason and upheld by everyone
Ethical behaviors of environment orientation—fairness, power sharing, ethical guidance, role clarification, people orientation, and integrity (Kalshoven et al., as cited by Wolfgranum et al., 2013)
Altruistic (take actions to consider the goodwill of others despite self-interests and in opposition to others' interest (Kapur, 2018))
Virtue-based approach—Involvement within the community and understanding
Transformational leadership uses mutual altruistic motives grounded in a teleological perspective that provides moral influence over followers for long term and is ethical due to empowering strategies of higher-order needs growth (Kanungo, 2001)
Acting in a way that is good for oneself and others in a sensible and judicious way with future in mind (virtue ethics)
Being responsible for humanity and other's being toward the guiding actions of doing good (the agency in sustainability leadership) (Wolfgramm et al., 2015).
A positive promotion of organizational culture and trust (Brown et al., as cited by Rao, 2019)
Features of moral reasoning, locus of control, conscientiousness, kind, sympathetic, cooperative, warm, and considerate (Wolfgramm et al., 2015)
Act within the four positions of ethical leadership—Humane, justice, responsibility, sustainability, and modern orientation
Duty to serve and make beneficial decisions in the best interest of equality, fairness, and justice (Kapur, 2018)
Honest, truthful, and contributes to building a useful community (Kapur, 2018)
Uses communication in the influence of ethical values through behavior (Kapur, 2018)
Four aspects of ethical leadership as conceptual dimensions: (1) character/integrity, (2) altruism, (3) collective motivation, and (4) encouragement (GLOBE Project) (Resick et al., 2006)

during a crisis would create a moral compass. Intention-based morality emerged as an emergent theme, and more research on ethical decision-making on moral obligation, intent, and judgment would help create a moral compass. Creating a culture of adaptive "complex intelligence" (Boyer & Levine, 1981) around ecological literacy and sustainability can guide character. Using the conceptual framework in Fig. 2.1, the transformational leader can lead by example, set expectations, encourage learning and growing, inspire followers to go new heights, and coach and empower their success when making ethical decisions. Using the virtue-based tier of the new conceptual framework of ethical decision-making, a transformational leader helps identify character traits that motivate in a given situation while considering the "whole human" and then inspire the follower by leading by example and setting expectations for that virtue. Additionally, encouraging learning and growing in righteousness, inspiring to go to new heights, and coaching and empowering followers' success by transformational leaders can help the follower have a foundation for ethical decision-making.

Subquestion *RQ4: What ethical leadership types help to promote establishing a moral compass?*

Since ethics concerns itself with morals and a set of rules to determine what's good or bad in a situation (Northouse, 2018), leadership styles that emphasize leaders' moral character or encourage a high moral standard would be beneficial when establishing a moral compass. Other similarities to ethical leadership include a concern for others (altruism, integrity, and role-modeling). Authentic leaders and spiritual leaders stress a leader's selflessness and integrity in practice. Ethical decision-making is a similarity that authentic leadership, spiritual leadership, and transformational leadership share (Brown & Trevino, 2006). The difference in ethical leadership compared to leadership types has to do with the implication and follow-through of ethical standards and moral management. For example, in authentic leadership, leaders focus on authenticity and self-awareness, while ethical leaders emphasize host management through transactional and managing other awareness.

Similarly, spiritual leaders emphasize vision, hope, and faith, and transformational leaders emphasize vision, values, and intellectual stimulation. Relational leadership or leadership driven by values and standards that are "good" or "moral" in nature (Komives et al., 1998) involves everyone and their ideas with the leader being unbiased and staying fair in the process of meeting goals through the framework of an exemplary leader (Wolfgramm et al., 2015). Ethical leadership has a focus on management. Therefore, all four leaderships can contribute to helping to promote establishing a moral compass. Transformational leadership did show up as the more promising style of leadership due to the positive underpinnings of moral reasoning and moral identity, which are more likely to be accentuated in decision-making and communication than any other type of leadership (Rowold, 2008). Transformational leaders motivate others to follow the moral-based vision and create an environment that considers individuals' viewpoints and appeals to higher-order needs. As an outcome of the inspirational motivation of individuals' behaviors and actions, followers are more committed and passionate to the organization in sharing the shared vision (Odumeru & Ogbonna, 2013). Arranging the axial codes from this research with the gaps in the literature for leadership training in ESD appears that transformational leadership best aligns with finding a moral compass (See Table 2.5). An ethical framework should underpin competencies, and ethical leadership should align with skills for sustainability.

Additionally, the collective vision for sustainability needs to be set by educators, practitioners, stakeholders, and learners. Therefore, transformational leadership

Table 2.5 Axial codes and transformational leadership

Axial codes	Transformational leadership
Meeting the standards, judgments, and actions for ethics	Leadership is proactive I Always searches for universal values or principles to guide in the vision for the organization (Kanungo, 2001). I Theory Y (Douglas McGregor): Managers work to encourage their workers and assume the best of their employees. Managers believe employees to be trusting, respectful, and self-motivated and leaders help followers with the tools that they need to excel I Motivate followers by encouraging them to put group interests first
Believing that mindset, values, actions, and responsibility will guide outcomes	Works to change the organizational culture by implementing new ideas. I Influence strategy of empowerment is guided by ethic of duty I Always searches for universal values or principles to guide in the vision for the organization (Kanungo, 2001) I Theory Y (Douglas McGregor): Managers work to encourage their workers and assume the best of their employees. Managers believe employees to be trusting, respectful, and self-motivated and leaders help followers with the tools that they need to excel I Employees achieve objectives through higher ideals and moral values
Understanding expectations, connectedness, and agency in ESD curriculum	Motivates followers by encouraging them to put group interests first I works to change the organizational culture by implementing new ideas I influence strategy of empowerment is guided by ethic of duty I individualized consideration: Each behavior is directed to everyone to express consideration and support. Intellectual stimulation: Promote creative and innovative ideas to solve problems
Factoring in concepts, methodologies, practices, personal reflections, complex problems across and integrated with other educational environments and cultures	Always searches for universal values or principles to guide in the vision for the organization (Kanungo, 2001) I Influence strategy of empowerment is guided by ethic of duty I Works to change the organizational culture by implementing new ideas I Leadership is proactive I Influence strategy of empowerment guided by ethic of duty I Motivates followers by encouraging them to put group interests first

must take consideration of meeting these standards and requirements in the ESD field. In the educational leadership field, leaders should take moral positions that guide actions to sustain life and create a culture of integrating a green literate and sustainable framework.

Subquestion *RQ5: What do ethical leaders do not do?*

A leader's character and conduct are studied in leadership ethics and presented as what leaders are and what they do (Northouse, 2018). However, there is a lack of ethical leadership research on antecedents, purpose, and moral theory (Lawton & Páez, 2015), limiting effectiveness and efficiency in leadership or performance of employees. For example, in one study, researchers found that employees' ethical leadership standards were perceived as overly demanding or blocking work (Quade et al., 2019). The potential limitations of ethical leadership should be explored in ESD. Other topics to explore include conditions in boundaries (Quade et al., 2019), times when ethical leaders become less effective (Babalola et al., 2019), the outcomes of ethical leadership, how elements of ethical leadership are viewed across the world, and how moral management is addressed in other cultures. Also, since social learning theory can be used to explain the antecedents and outcomes of ethical leaders and why and how ethical leaders influence their followers, more research in the area should be explored (Brown & Trevino, 2006).

Additionally, central positions and orientation of ethical leaders and their relationship to the development of ethical judgment can be explored in more detail, particularly from voluntary compliance versus vested interest manner. Intention-based morality and transformational leaders basing moral decisions on intentions or outcomes is another area to investigate. However, this criterion may not take into all complex situations, especially in the complexities of sustainable development. Therefore, understanding the conditions for making ethical decisions can be useful for students, transformational leaders, and practitioners (Table 2.2). Knowing what ethical decision to make, when, where, why, and how will best identify and quantify each choice that is created and can help in any situation that is faced. Table 2.3 outlines the phenomenon and how to approach or what to consider when making ethical decisions.

The character of a leader and the virtues that are learned and practiced through the involvement within the community is an area that should be studied. Other areas of further study include an intrinsic and extrinsic moral status and motivation. The ethical leadership behaviors of environment orientation are another area to be explored. Knowing how leaders perceive their responsibility to other people and the environment can be valuable to the ESD field.

Subquestion *RQ6: What are the gaps in research about the development of ethical judgment and educators?*

There are gaps in connecting the SDG goals to educational outcomes, a lack of connection or responsibility to the environment and consensus of terms related to sustainability (Jackson et al., 2017), exhibiting confounding challenges that must be addressed, and causing barriers to leadership training. Therefore, leaders' core values must align with and is what promotes sustainability in the role of ethical leadership in HEIs. Rao (2019) suggests having a cultural awareness also helps implement sustainability initiatives and the results. Omer (2014) argues that sustainability and ethical principles are entwined because they cannot be applied without strong moral principles and ethical behavior connected to society's welfare. Additionally, creating a culture of adaptive "complex intelligence" (Boyer & Levine, 1981) can help with development and learning. There is a gap in research in the relationship to ethical judgment and educators' development, especially in the on-duty performance and communication of ethical standards (Avolio & Bass, 2004, as cited by Yasir & Mohamad, 2016). Therefore, studying the lack of leadership study in ESD as a general conceptualization of ethical leadership, the scaling aspects as a learning process, and identifying what is ethically acceptable and can be changed to scale are valuable for setting moral foundations. Connection of education with cultural tradition, human striving to be capable of cross-cultural communication, and absorbing values of the new globalized civilization in a modern world can be an area of study (Nasibulina, 2015). Ethical competence to ESD is another area that should be studied, including ethical influence and job satisfaction. Additional areas in ethics such as practitioners and ethical conflict, applying ethical standards on practice, promoting ethics, moral reasoning, and moral identity in leadership training, and communication in decision-making can be helpful to leadership development. It is shown that value and virtue of ethics is missing from the literature and needs to be explored in the educational leadership field. More research on learning and rational reflection that causes leaders to reevaluate their beliefs and values would be helpful in ethical leadership development. Spiritual leadership came up as an emergent theme. Since religious leaders stress a leaders' altruism and integrity in practice (Brown & Trevino, 2006) and have a component of ethical decision-making, spiritual leadership as an approach of ethical leadership and research on ethical decision-making in this leadership style should be explored. Finding ways to meet ethical standards of participation and adherence to ethical norms, rules, and high level of conduct and encouragement of leadership training facilitators should be explored since there is little research in this. Since reflection came up in axial coding in using ethical standards, norms, and rules, more research on when reflection is used, what a leader can accomplish in reflection, and the right way to be reflexive to reap the benefits as a leader would be helpful. Also, knowing what kind of thoughts are more beneficial and in what circumstances, dilemmas, and risks would help create methodologies and practices and assist in ethical leadership development. Since ethical situations and problem contexts can always emerge (Kioupi & Voulvoulis, 2019), there is a need for ethical leadership training focused on ethical situations and problems. There should be more study on the situational judgment tests in decision-making (Bender & Haller, 2017), moral theory with risk (Erman & Möller, 2018), and how tools can be used to improve judgment practices (Ell & Haigh, 2015). Understanding the power of social movements in developing countries and their motives and impediments to manifest on the green movement and what it means to society (Ziniel & Bradley, 2018) are all areas of further research and should be considered when training individuals in ethical decision-making. Therefore, finding ways to achieve shared vision and goals for sustainability to include ethical decisions should be set at a local orientation of education and implementation of interactive teaching first before trying to solve or move forward (Nasibulina, 2015). Regarding learners of transformational leadership development, rational reflection, social processes, and dilemmas that cause leaders to reevaluate their beliefs and values are gaps that appeared. Therefore, understanding how tools can be used to improve judgment practices (Ell & Haigh, 2015), understanding the links between ethical decision-making and leadership in ethical leadership training for ESD, and the effectiveness of ethical leadership training through ethical factors (and not organizational factors) are beneficial to study (Lawton & Páez, 2015).

Subquestion *RQ7: What are the factors affecting ethical leadership training for ESD?*

Scaling the ESD practice over time to meet the emerging challenges and increasing ownership and accountability should be implemented in leadership development. Examining the tools and efficacy of ethics and agency can be an area where sustainability actions are used (Wolfgramm et al., 2015). Also, understanding the links between ethical decision-making and leadership in ethical leadership training for ESD (Lawton & Páez, 2015) can help develop leadership development programs. Another strong component desperately missing from research includes use cases. Use cases can show definite success and failure, especially in analyzing agency in sustainability leadership (Wolfgramm et al., 2015). Therefore, the same study of what is successful and not successful in ethical leadership and especially in each of the leadership styles would also aid in ethical leadership development. Participatory research should be explored to identify the skills required and develop appropriate learning strategies using ethical circumstances and problem context.

Additionally, scaling ESD practices and learning over time (to include ethical decision-making) can help meet the emerging challenges. This includes increased ownership and accountability since there is a lack of ethical leadership research on antecedents, purpose, and ethical theory. There are more explanation and justification of the leaders for their actions in ethical leadership training in ESD (Lawton & Páez, 2015).

Subquestion *RQ8: What are some theoretical frameworks that educators and students can use to help in cultivating a solid moral compass?*

Additional theoretical frameworks, in addition to the one shown in this chapter, can include other leadership styles.

2.8 Implications

This study's main aim was to develop new perspectives on a theoretical model and theory for how students and educators can build their foundation for future leaders' ethical decisions. A conceptual framework toward a moral compass to guide transformational leadership development for ESD was developed, showing how this can be achieved while observing the core phenomenon of making a choice and understanding the gaps in ethical leadership development as it relates to the field of ESD. The practical use for this study is that it provides much-needed empirical data on understanding the gaps in ethical leadership development. This research can offer students, leaders, schools, HR, and the ESD field with further information on the development of ethical perspective and ethical decision-making skills when there is no immediate framework. This research also contributes to the body of literature by filling in the knowledge gap about cultivating a solid moral compass for leaders in education for sustainable development. Additionally, factors revealed can help direct ethical decision-making efforts to relevant areas resulting in better planning. Courses in sustainability leadership and spiritual leadership can be helpful to the development of a meaningful and relevant leader. Facilitating training and courses on leadership based on experiential education and transformational leadership (Diamond, 2012) can create underpinnings for ethical leadership development. Additionally, grounding training in theories about self-authorship, identity development, participatory learning, and teaching a rights-based approach can help in the event of ethical leaders. There are already organizations committed to establishing sustainable development principles in their education and line with the Sustainable Development Goals. The conversation about how specific actions will affect future generations and how we can work with others to achieve broad and specific goals have started and are laggardly rolling out, with no real way to measure its success. While competencies and laying out of values are sprinkled through ESD approaches and strategies, a more concrete ethical leadership development program must ensue. Ambiguity about sustainability terms and end goals pose a problem (Swain, 2018); therefore, having a mutual set of definitions can help in the policymaking process (Gober et al., 2015). Case studies showed to be the most critical resources (after academic papers) for teaching courses (graduate and postgraduate) to understand biases and dilemmas that cause leaders to reevaluate beliefs and values (Passi & Neil, 2016). There could be different outcomes for each case, so seeing the decisions, thinking, and results can be valuable. A combination of moral imperative with the practicality of decision-making in business is the key component in creating resources for teaching sustainability in management. Students should be encouraged to participate in hands-on participation of situations with the ability to analyze problems. Solutions understand the consequences of their actions as leaders to social, environmental, and the economic development.

In transformational leadership, values and ethics are the main approaches. Leaders that motivate others to follow the moral-based vision for the higher good with a strong moral identity are more likely to accentuate moral values in communication and decision-making and create an environment that considers viewpoints. The leader articulates shared values over their importance when making a decision (Bryman et al., 2011). As long as the leadership is ethical, transformational leaders can lead through example and inspire followers to have a specific behavior. Like a coach, they can empower people, use their strengths to act like a coach, and be aware of personal challenges and endeavors. Elements of transformational leadership include building a commitment to the vision, developmental to a strategic vision, modeling the concept, and encouraging experimentation. Since change is necessary, transformational leaders must be able to handle change. Transformational leaders model vision and can implement a plan and blueprint and can put it into practice. Transformational leaders deal with experimentation and can step out of their comfort zone. Transformational leaders can also build commitment, get people on the same page, and work toward an ongoing process. A transformational leader can link a strategic vision to personal values, fulfill stakeholder needs, meet challenging objectives, and help to join principles.

Courses in sustainability leadership and spiritual leadership can be helpful to the development of meaningful and relevant leaders. Facilitating training and courses on leadership based on experiential education and transformational leadership (Diamond, 2012) can create underpinnings for ethical leadership development.

Additionally, grounding training in theories about self-authorship, identity development, participatory learning, and teaching a rights-based approach can help in the development of ethical leaders.

2.9 Future Studies

Ethical leadership development can be utilized by active leaders in ESD in communities, regions, and countries worldwide in workshops, foundational concepts, guidelines, and structures for action. Integrating ethical leadership development into such training can provide a framework for working with other curriculums such as environmental ethics, conflict resolution, corporate responsibility, ecological values, values for ESD, being an ethical citizen, and more. Additionally, colleges and universities have an unprecedented opportunity to cultivate students' moral compass and ethical perspective and behavior in ESD. This study can also lay down the foundation for future studies on how HEIs can implement ethical leadership development through their curriculum and across interdisciplinary studies. This leadership development model can be used to connect educators and students with a useful methodology and increase the quality of grounded research published in the ethical leadership training for the ESD field. We are more interconnected than before, and global leadership is integral to the sustainability of ethical leadership development across borders. There is evidence that a manager's ethics and environmental values may play a role in environmental leadership, uniquely when they are aligned with society's expectations and within the organization's belief system (Bansal et al., as cited by Boiral et al., 2009). Therefore, knowing what other contributions and ethical traits are valuable in leadership styles can contribute to the successful implementation of the development of a moral compass. Moreover, understanding the types of leadership types that best match ethical leadership can help students, educators, and active ESD leaders in their training and framework development. This research would add to the body of knowledge on what is needed for ethical leadership development and how transformational leadership type can be useful in ethical leadership training.

More studies are required on how transactional and authentic leadership can be useful in ethical leadership training. More research is needed on other types of research on ethical leadership development for ESD. Current leaderships affect decision-making and how change management plays into it. Further research should aim to discover the motives, values, norms, and self-concepts of transformational leadership. Additional research should build on the moral groundwork of leadership behavior and the guiding of leadership training and education in ethics, in areas such as changes in motives, values, worldviews, and the preparation of ethical leaderships in different leadership styles. Finally, research on the impact of leadership training on the moral development of leaders and followers is essential to this field (Kanungo, 2001). Areas of future research include additional study on determining how tools can be used to improve judgment practices (Ell & Haigh, 2015), more explanation and justification of the leaders for their actions in ethical leadership training in ESD; links between ethical decision-making and leadership in ethical leadership training for ESD; the effectiveness of ethical leadership training (Lawton & Páez, 2015); moral theory dealing with risk (Erman & Möller, 2018); decision-making factors (Jackson et al., 2017); assessment/situational judgment tests (SJT); culture as sustainable development (Bender & Haller, 2017); mindset, attitudes (ecocentric versus anthropocentric), and culture; views about interconnectedness with natural world and sustainability; case studies on successful implementation of programs; survey of the importance of ethics in schools; ethical foundation in university curricula; scaling ethical aspects as a learning process and identifying what is ethically acceptable and what can be changed in order to scale; a teacher's readiness to teach, shared decision-making model, social influence pressure, dual roles, and role-modeling/peer-assisted models (Passi & Neil, 2016); and feedback from supervisor/teacher. Future research can be conducted on how the current ethical leadership framework in this study could guide developing a moral compass for other leadership styles.

Additionally, further research on the links between ethical decision-making and leadership in ethical leadership training for ESD would help know what decisions individuals within organizations make versus at what different levels (O'Fallon and Butterfield; Tenbrunsel and Smith-Crowe; Tevino et al., as cited by Lawton & Páez, 2015). Another area is to look at the effectiveness of ethical leadership training. Currently, the effectiveness of leaders is determined by organizational, not ethical factors. Ethical leadership development in ESD is an important topic today, and is interdisciplinary; therefore, it will have implications for generations to come.

Another area to investigate is seeing what current places are training and how it aligns with the curriculum. Also, encouraging youth leaders to promote leadership development while modeling can be another strategy for training. Offering a certificate program can help leaders to benefit from having one in their careers while promoting ethical leadership development.

Future follow-up studies may focus on the study of ethical competence to ESD, including moral influence and job satisfaction, practitioner and ethical conflict, and value and virtue of ethics in the educational leadership field. Furthermore, applying ethical standards on practice, promoting ethics, moral reasoning, and moral identity in leadership training and communication in decision-making can be helpful to leadership development. It would be interesting to explore spiritual leadership as an approach to ethical leadership. Future studies can include finding ways to meet ethical standards of participation and adherence to moral norms, rules, and a high level of conduct and encouragement of leadership training facilitators. There should be more studies on what a leader can accomplish in reflection and the right way to be reflexive to reap the benefits as a leader. Also, knowing what kind of thoughts are more beneficial and in what circumstances, dilemmas, and risks would help create methodologies and practices, and help in ethical leadership development.

2.10 Conclusion

ESD is complex and requires effective leaders skilled at leading in complex situations and adaptable to change. Prior studies show a lack of understanding of the concept of sustainability and the lack of research on effective teaching practices in newer interdisciplinary fields (Michel, 2019). There is a lack of exposure to the reflection of ethics (England, 2014). Moral teachings relate to limited research on leader values, ethics, and style in corporate social responsibility (Metclaf & Benn, 2013). Furthermore, there is a lack of ethical leadership research on antecedents, purpose, and ethical theory (Kalshoven et al., Eisenbeiß and Gissner; Kempster et al., Rost, as cited by Lawton & Páez, 2015) as well as empirical research on ethics in educational leadership (Glanz, 2010, as cited by Salter et al., 2014). The consensus between what is right and wrong and deciding on a set of moral rules is a persistent problem. The lack of leadership training and understanding what leadership style works best when focusing on ethical situations and issues is a much-needed area of concern. Leadership for achieving sustainable development will require a joint effort among international partners, governments, and individuals who must be ready to face uncertainty and chaos at any given time. These leaders must be able to accept and deliver upon shared responsibility and maintain and utilize a working knowledge of their own set of skills and be prepared to exhibit and actively encourage ethical standards and behavior (Iftakhar & Bahauddin, 2018). Therefore, ethical leadership training and development need to be integrated as a core competency within this framework to develop a moral compass when conflicts arise, and governments cannot help. There is no doubt that there must be ethical training in education for sustainable development beneficial to companies and leaders that are good for the interest of all—beyond the human.

2.11 Cross-References

- Business Youth for Engaged Sustainability to Achieve the United Nations 17 Sustainable Development Goals (SDGs)
- Education in Human Values: Planting the Seed of Sustainability in Young Minds
- Ethical Decision-Making Under Social Uncertainty: An Introduction of Überethicality
- Moving Forward with Social Responsibility: Shifting Gears from Why to How
- Sustainable Decision-Making: Moving Beyond People, Planets, and Profits
- Sustainable Higher Education Teaching Approaches
- Transformative Solutions for Sustainable Well-Being: Designing Effective Strategies for Addressing Our Planetary Challenges

References

Anderson, M., & Peter, S. (2017). Reviewing leadership styles: Overlaps and the need for a new 'full-range' theory. *International Journal of Management Reviews, 19*, 76–96. https://doi.org/10.1111/ijmr.12082

Babalola, M. T., Stouten, J., Camps, J., et al. (2019). When do ethical leaders become less effective? The moderating role of perceived leader ethical conviction on employee discretionary reactions to ethical leadership. *Journal of Business Ethics, 154*, 85–102. https://doi.org/10.1007/s10551-017-3472-z

Bass, B. M., & Bass, R. (2008). *The Bass handbook of leadership: Theory, research, and managerial applications*. Free Press.

Bender, O., & Haller, A. (2017). The cultural embeddedness of population mobility in the Alps: Consequences for sustainable development. *Norwegian Journal of Geography, 71*(3), 132–145.

Boiral, O., Cayer, M., & Baron, C. M. (2009). The action logics of environmental leadership: A developmental perspective. *Journal of Business Ethics, 85*(4), 479–499. https://doi.org/10.1007/s10551-008-9784-2

Boyer, E., & Levine, A. (1981). *A quest for common learning: The aims of general learning*. Carnegie Foundation for the Advancement of Teaching. https://doi.org/10.1080/00091383.1981.10569804

Brown, M. E., & Trevino, L. K. (2006). Ethical leadership: A review and future directions. *The Leadership Quarterly, 17*(6), 595–616. https://doi.org/10.1016/j.leaqua.2006.10.004

Bryman, A., Colinson, D., Grint, K., Jackson, B., & Uhl-Bien, M. (Eds.). (2011). *The Sage handbook of leadership*. SAGE.

Burford, G., Tamás, P., & Harder, M. K. (2016). Can we improve indicator design for complex sustainable development goals? A comparison of a values-based and conventional approach. *Sustainability, 8*, 861. https://doi.org/10.3390/su8090861

Burns, H., & Wolf, J. R. (2014). Leadership for sustainability: Pedagogical practices that empower learners to become leaders. *Educational Leadership and Policy Faculty Publications and Presentations, 46*. Retrievable at https://pdxscholar.library.pdx.edu/cgi/viewcontent.cgi?article=1045 & context=elp_fac

Choi, S. B., Kim, K., & Kang, S.-W. (2017). Effects of transformational and shared leadership styles on employees' perception of team effectiveness. *Social Behavior and Personality, 45*(3), 377–386. https://doi.org/10.2224/sbp.5805

Cohen, B. (2007). Developing educational indicators that will guide students and institutions toward a sustainable future. *New Directions for Institutional Research, 2007*(134), 83–94. https://doi.org/10.1002/ir.215

Corbin, J., & Strauss, A. (2007). *Basics of qualitative research: Techniques and procedures for developing grounded theory*. Sage Publications.

Crede, M., Jong, J., & Harms, P. (2019). The generalizability of transformational leadership across cultures: A meta-analysis. *Journal of Managerial Psychology, 34*(3), 139–155. https://doi.org/10.1108/JMP-11-2018-0506

Creswell, J. W. (2003). *Research design: Qualitative, quantitative, and mixed methods approaches*. Sage.

Diamond, H. (2012). Empowering sustainability leaders: Developing an authentic leadership identity. *Leadership for Sustainability Education Comprehensive Papers, 4*.

Ell, F., & Haigh, M. (2015). Getting beyond "gut feeling": Understanding how mentors judge readiness to teach. *Asia-Pacific Journal of Teacher Education, 43*(2), 143–155. Retrieved from https://search.ebscohost.com/login.aspx?direct=true & db=eric & AN=EJ1052492 & site=ehost-live & scope=site

England, D. L. (2014). Leader perceptions of ethical/moral leadership in community college business school curriculum (doctoral dissertation). Retrieved from ProQuest Dissertations & Theses Global. (order no. 3583238).

Erman, E., & Möller, N. (2018). The interdependence of risk and moral theory. *Ethical Theory & Moral Practice, 21*(2), 207–216. https://doi.org/10.1007/s10677-018-9881-y

Evans, G., & John, L. (2013). A novice researcher's first walk through the maze of grounded theory: Rationalization for classical grounded theory. *Grounded Theory Review, 12*(1).

Franck, O., & Osbeck, C. (2018). Challenging the concept of ethical literacy within education for sustainable development (ESD): Storytelling as a method within sustainability didactics. *Education 3–13, 46*(2), 133–142. https://doi.org/10.1080/03004279.2016.1201690

Full Range Leadership Model. (2019). Retrieved February 6, 2020, from https://en.wikipedia.org/wiki/Full_Range_Leadership_Model

Glaser, B. G., & Holton, J. (2004). Remodeling grounded theory. *Forum Qualitative Sozialforschung/Forum: Qualitative Social Research, 5*, 1–22.

Gober, P., Strickert, G., Clark, D., Chun, K., Payton, D., & Bruce, K. (2015). Divergent perspectives on water security: Bridging the policy debate. *The Professional Geographer, 67*(1), 62–71. https://doi.org/10.1080/00330124.2014.883960

Hay, I. (2006). Transformational leadership: Characteristics and criticisms. *EJournal of Organizational Learning and Leadership, 5*, 2. Retrievable at http://ww.weleadinlearning.org/ejournal.htm

Iftakhar, N., & Bahauddin, K. M. (2018, January 10). *Why leadership is essential for achieving sustainable development goals*. Retrieved February 18, 2020, from https://intpolicydigest.org/2018/01/10/why-leadership-is-essential-for-achieving-sustainable-development-goals/

Jabareen, Y. (2008). A new conceptual framework for sustainable development. *Environment, Development and Sustainability, 10*(2), 179–192. https://doi.org/10.1007/s10668-006-9058-z

Jackson, S. A., Kleitman, S., Stankov, L., & Howie, P. (2017). Individual differences in decision making depend on cognitive abilities, monitoring and control. *Journal of Behavioral Decision Making, 30*(2), 209–223. https://doi.org/10.1002/bdm.1939

Jickling, B., Lotz-Sisitka, H., ODonoghue, R., & Ogbuigwe, A. (2006). *Environmental education, ethics and action. A workbook to get started*. UNEP. http://openjournal.lakeheadu.ca/public/journals/22/Ethics_book_english.pdf

Jody, W. L., & Egel, E. (2018, February 13). *Spiritual leadership*. Retrieved February 20, 2020, from https://gbr.pepperdine.edu/2017/12/spiritual-leadership/

Kant, I. (n.d.). Retrieved February 28, 2020, from https:/www.iep.utm.edu/kantview/

Kanungo, N. (2001). Ethical values of transactional and transformational leaders. *Canadian Journal of Administrative Sciences, 18*(4), 257–265. https://doi.org/10.1111/j.1936-4490.2001.tb00261.x

Kaptein, M., Huberts, L., Avelino, S., & Lasthuizen, K. (2005). Demonstrating ethical leadership by measuring ethics: A survey of US public servants. *Public Integrity, 7*(4), 299–311.

Kapur, R. (2018). *Ethical leadership*. ResearchGate. https://www.researchgate.net/publication/323829602_Ethical_Leadership

Kioupi, V., & Voulvoulis, N. (2019). Education for sustainable development: A systemic framework for connecting the SDGs to educational outcomes. *Sustainability, MDPI, Open Access Journal, 11*(21), 1–18.

Komives, S., Lucas, N., & McMahon, T. (1998). *Exploring leadership for college students who want to make a difference* (pp. 68–72). Jossey-Bass.

Kopnina, H. (2013). Evaluating education for sustainable development (ESD): Using ecocentric and anthropocentric attitudes toward the sustainable development (EAATSD) scale. *Environment, Development & Sustainability, 15*(3), 607–623. https://doi.org/10.1007/s10668-012-9395-z

Lawton, A., & Páez, I. (2015). Developing a framework for ethical leadership. *Journal of Business Ethics, 130*(3), 639–649. https://doi.org/10.1007/s10551-014-2244-2

Levi, L., & Rothstein, B. (2018, November 9). *Universities must lead on sustainable development goals*. Retrieved from https://www.universityworldnews.com/post.php?story=20181106131352348

Metclaf, L., & Benn, S. (2013). Sustainability leadership: An evolution of leadership ability. *Journal of Business Ethics, 112*, 369–384. https://doi.org/10.1007/s10551-012-1278-6

Michel, J. O. (2019). An assessment of teaching and learning about sustainability across the higher education curriculum (order no. 13805642). Available from ProQuest Dissertations & Theses Global. (2195494539). Retrieved from https://search-proquest-com.proxy.indianatech.edu/docview/2195494539?accountid=42681

Mochizuki, Y., & Bryan, A. (2015). Climate change education in the context of education for sustainable development: Rationale and principles. *Journal of Education for Sustainable Development, 9*(1), 4–26. Retrieved from http://search.ebscohost.com/login.aspx?direct=true & db=eric & AN=EJ1064014 & site=ehost-live & scope=site

Nasibulina, A. (2015). Education for sustainable and environmental ethics. *Social and Behavioral Sciences, 214*, 1077–1082.

Neubert, M., Carlson, D., Kacmar, K., Roberts, J., & Chonko, L. (2009). The virtuous influence of ethical leadership behavior: Evidence from the field. *Journal of Business Ethics, 90*(2), 157–170. https://doi.org/10.1007/s10551-009-0037-9

Northouse, P. G. (2018). *Leadership: Theory and practice* (7th ed.). SAGE Publications.

Odumeru, J., & Ogbonna, I. (2013). Transformational vs. transactional leadership theories: Evidence in literature. *International Review of Management and Business Research, 2*(2), 355–361.

Omer, S. (2014, December 11). *Sustainable development and the ethical issue of human morality: An overview*. Retrieved from https://medinanet.org/2014/12/sustainable-development-and-the-ethical-issue-of-human-morality-an-overview/

Passi, V., & Neil, J. (2016). The hidden process of positive doctor role modelling. *Medical Teacher, 38*(7), 700–707. https://doi.org/10.3109/0142159X.2015.1087482

Quade, M. J., Perry, S. J., & Hunter, E. M. (2019). Boundary conditions of ethical leadership: Exploring supervisor-induced and job hindrance stress as potential inhibitors. *Journal of Business Ethics, 158*, 1165–1184. https://doi.org/10.1007/s10551-017-3771-4

Rao, I. (2019). A brief note on the role of ethical leadership in higher education institutions for sustainability. *IUP Journal of Management Research, 18*(4), 70–79. Retrieved from https://search-proquest-com.proxy.indianatech.edu/docview/2322070753?accountid=42681

Resick, C. J., Hanges, P. J., Dickson, M. W., & Mitchelson, J. K. (2006). A cross-cultural examination of the endorsement of ethical leadership. *Journal of Business Ethics, 63*, 345–359. https://doi.org/10.1007/s10551-005-3242-1

Rowold, J. (2008). Relationships among transformation, transactional, and moral-based leadership: Results from two empirical studies. Leadership review, Kravis Leadership Institute, Claremont McKenna College, Vol. 8, Winter 2008.

Salter, C. R., Harris, M., Woodhull, M., & McCormack, J. (2014). *Bass & Avolio's full range leadership model and moral development*. E-Leader Milan.

Schank, C., & Rieckmann, M. (2019). Socio-economically substantiated education for sustainable development: Development of competencies and value orientations between individual responsibility and structural transformation. *Journal of Education for Sustainable Development, 13*, 67–91. https://doi.org/10.1177/0973408219844849

Schumaker, J. F. (2001). *The age of insanity: Modernity and mental health*. Praeger.

Schwartz, B. (1987). *The battle for human nature: Science, morality and modern life Barry Schwartz 1987* (p. 1986). Norton.

Smith, C., Christoffersen, K., Davidson, H., & Herzog, P. (2012). Lost in transition: The dark side of emerging adulthood. *Sociology of Religion, 73*(3), 349–351. https://doi.org/10.1093/socrel/srs048

Swain, R. B. (2018). A critical analysis of the sustainable development goals. In W. Leal Filho (Ed.), *Handbook of sustainability science and research* (pp. 341–355). Springer. https://doi.org/10.1007/978-3-319-63007-6_20

UNESCO. (2018). *Education for sustainable development goals learning objectives*. https://unesdoc.unesco.org/ark:/48223/pf0000247444

Velasquez, M., Andre, C., Shanks, T. S., & Meyer, M. J. (1992, August 01). *Ethical relativism*. Retrieved July 05, 2020, from https://www.scu.edu/ethics/ethics-resources/ethical-decision-making/ethical-relativism/

Wilson, K. M. (2000). Communicating climate change through the media. In S. Allan, B. Adam, & C. Carter (Eds.), *Environmental risks and the media* (pp. 201–217). Routledge.

Wolfgramm, R., Flynn-Coleman, S., & Conroy, D. (2015). Dynamic interactions of agency in leadership (DIAL): An integrative framework for analysing agency in sustainability leadership. *Journal of Business Ethics, 126*(4), 649–662. https://doi.org/10.1007/s10551-013-1977-7

Wood, J. (2001). Ethical decision making. *Journal of Perianesthesia Nursing, 16*(1), 6–10. https://doi.org/10.1053/jpan.2001.18202

Yasir, M., & Mohamad, N. (2016). Ethics and morality: Comparing ethical leadership with servant, authentic and transformational leadership styles. *International Review of Management and Marketing, 6*(S4), 310–316.

Yukl, G. A. (2010). *Leadership in organizations*. Prentice Hall.

Ziniel, C., & Bradley, T. (2018). Greening British businesses: SMEs and the new wave of the environmental social movement. *International Journal of Social Quality, 8*(2), 1–26. Retrieved from http://search.ebscohost.com/login.aspx?direct=true & db=lkh & AN=137244120 & site=ehost-live & scope=site

Amanda L. Holst maintains the highest performance standards with a diverse range of leadership functions to identify complex problems and evaluate options to implement solutions.

With over 20 years in technology, marketing, and operations, Amanda has experience in streamlining operations, strategic implementation and cross-functional collaboration in leadership roles spanning tech, education, health, and real estate. Currently, Amanda manages operations, processes and programs at Cisco Systems, where she has received Webexcellence Operational Excellence and Connect Everything awards for her work, dedication and contribution to efficiently executing the common goals of the organization. Amanda is a Ph.D. in Global Leadership Candidate at the Indiana Institute of Technology and holds Six Sigma Master Black Belt and Lean Culture Certified certifications. In her free time, she's exploring trying new food with her boy/girl twins, ages 14, submitting patent applications with her Cisco WINN++ patent team, mentoring college students from her alma mater, San Jose State University (SJSU), and running.

Chapter 3
Analysis of the Implications of Declining Cities, Urban Vacant Land Uses, and Green Infrastructure, and Their Impact on Climate Change Hazards

José G. Vargas-Hernández and M. C. Omar C. Vargas-González

Abstract This chapter aims to analyze the implications between the declining and shrinking cities, urban vacant lands, urban land uses, green infrastructures, and urban green areas, and their impact on climate hazards change. The analysis departs from the basic assumption that urban vacant land sites and spaces have a negative connotation but if supported by the appropriate policies and programs of incentives, they can turn around and develop the essential green infrastructure to enable the mitigation of climate change hazards and bring about economic growth and socio-ecological development. The method used is analytical-descriptive based on the theoretical and empirical literature review. It is concluded that the land uses of vacant land sites more oriented toward urban green innovation infrastructure and forest areas contribute to mitigate the climate change hazards.

Keywords Climate change · Hazards · Declining cities · Forest areas · Green infrastructure · Land use · Shrinking cities · Urban vacant land

3.1 Introduction

Urban processes linked to shrinking cities feature characteristics including economic decline associated to urban land uses, and other socio-economic dimensions that may lead to the adaptive capacity building which in turn increases the urban

J. G. Vargas-Hernández (✉)
Postgraduate and Research Division, Instituto tecnológico José Mario Molina Pasquel y Henríquez, Unidad Académica Zapopan, El Bajío, Zapopan, Jalisco, Mexico
e-mail: jose.vargas@zapopan.tecmm.edu.mx

M. C. O. C. Vargas-González
Instituto Tecnológico de México, Cd. Guzmán, Jalisco, Mexico
e-mail: omar.vg@cdguzman.tecnm.mx

© The Author(s), under exclusive license to Springer Nature Switzerland AG 2023
S. Dhiman (ed.), *Sustainable Development and Environmental Stewardship*,
https://doi.org/10.1007/978-3-031-28885-2_3

green infrastructure and moderates the impact of climate hazards change. The vacant land is used as green infrastructure to improve the health of urban ecosystems.

City regions experiencing economic growth and increases in urban population are subject to the adversity of losing green infrastructure resources leading toward a disequilibrium in adaptive capacity to climate change hazards and the quality-of-life goals.

The objective of this analysis is to find the relationships of the implications between the declining and shrinking cities, urban vacant lands, urban land uses, green infrastructures, urban green areas, and their impact on climate hazards change. First, it begins analyzing the processes of declining and shrinking cities to establish the relevance of the major urban vacant land sites and the land uses derived. The land uses of vacant land sites more oriented toward urban green innovation infrastructure and forest areas are the ones that lead to mitigate the climate change hazards.

3.2 Declining and Shrinking Cities

City regions and urban areas develop, expand, or decline, remaining sensitive to environmental and climate change agendas guided by spatial planning regulations. Cities that are experiencing shrinkage and decline face different challenges. Around the globe, cities are either growing and expanding or shrinking and declining because of some dynamic processes such as deindustrialization, globalization, and demographic change (Martinez-Fernandez et al., 2012; McKinsey and Company, 2016; Sassen, 2001; Hall, 1993). Besides the complexity of these challenges, there are the global warming and climate change affecting urban settings (Ruth & Coelho, 2007).

The economic growth processes of cities that benefited from the development of some comparative and competitive advantages in the past have led to the corresponding deindustrialization processes while unplugged from the global value chain and resulting in urban population and economic decline (Hall, 1993; Pike et al., 2016; Martinez-Fernandez et al., 2012: 220). Patterns of economic decline and depopulating in communities caused by contextual conditions result in vacant land with implications in institutional, fiscal, and programmatic responses that may bring some potential benefits.

Urban areas in large cities of the developed world regions experience population decline (McKinsey and Company, 2016). Economic and urban decline is linked to limited regulation over the land use in the different locations and forms of new urban development supported by deficient urban spatial planning, small changes in green infrastructure cover, and lack of strategies aimed for the mitigation of socio-economic inequality and environmental degradation. All these factors have an impact on the related implications for adaptive capacity to cope with the risks from urban weather and climate change.

The concept of shrinking cities is widely accepted as a global phenomenon which dates to the urban sociology theories of cyclical growth and decline (Martinez-Fernandez et al., 2012). Shrinking cities are places characterized by economic and population decline that harbor vacant properties and land parcels (Dietersdorfer et al., 2012: 9; Burkholder, 2012; Schilling & Logan, 2008).

The urban expansion, decline, and shrinkage processes are projected to continue influencing the global landscape of city spaces, urban land uses, and green infrastructure. Declining urban land uses, derelict and disused land experiencing declines in surface area, may replace informal open green spaces and can be considered prospects for urban green infrastructure.

3.3 Urban Vacant Land

Many cities are dominated by abandoned and vacant land. Cities that have experience of land abandonment use vacant land for creating open green spaces and sites for productive purposes (Alexander, 2005; U.S. Department of Housing and Urban Development, 2009; Wachter et al., 2010).

The term vacant land has a negative connotation considered as a problem (Jakle & Wilson, 1992; Coleman, 1982). Vacant land in physical blight conveys negative images that contribute to disinvestment, reducing reinvestment and development, and increases urban vacancy rates (Neukrug, 2009). Vacant land deters and discourages investments in the community due to the costs of maintaining vacant properties. Urban vacant land, empty urban spaces, and abandoned buildings have become a wasteland undervalued as a normal part of the urban fabric despite the lack of public policy and economic investment for renewal.

Most cities do not have adequate programs of economic incentives to support maintenance and improvements of abandoned buildings and vacant lands. Vacant lands and abandoned buildings are not maintained and have a negative impact on the quality of life and property values of the community.

Urban vacant lands are structural assets with economic and socio-ecological value relative to other urban land uses. Urban vacant lands are ecological habitats for a wide range of biodiversity of the wildlife of plants and animals and provide ecological corridors and green network systems, wetland and storm-water management, regulation services to disperse pollutants (Kamvasinou, 2011).

The emergence of open vacant land spaces provides some benefits and opportunities associated to secure adaptive capacity with new land uses and green infrastructure. Revitalization of urban vacant land and reuse of urban wasted land are critical to urban land use in urban spaces. The capacity of urban vacant land to design, plan, and manage promotes the public socio-ecological value of urban landscape as natural resource and community asset to improve the quality of life in the surrounding neighborhood. Vacant urban land covers a significant spatial territory of the urban landscape that provides green infrastructure for ecosystem services used to enhance health and better quality of life of residents (Pagano & Bowman, 2000; Kim et al., 2015).

Urban vacant land is a valuable resource that creates a variety of urban open spaces providing opportunities available to city residents. Urban vacant land is a valuable ecological resource as green infrastructure used to enhance the urban socio-ecosystems and promote the health and quality of life of residents (Kim et al., 2015). Vacant land is a valuable urban landscape and source of ecosystem services that benefits the urban communities and provides transformation opportunities for community development (Burkholder, 2012; Little, 2008).

Urban vacant land has a public value in terms of the potential to enhance socio-ecological benefits if they are redeveloped and integrated with other urban green infrastructure (Kim, 2016). Planning for urban vacant land leads to better appreciation of public value and to design strategies to optimize the utilization of these spaces in landscape design, urban environmental regeneration, and renewal. Urban vacant land has potential as a valuable socio-ecological resource to provide socio-ecosystem services in the form of public open and urban green spaces etc. Urban vacant land has relevant functions as a strategic use and reuse in the urban landscape provided by the ecological opportunities and benefits in creating a more prosperous, livable, healthy, and sustainable cities.

From the economics point of view, the vacant space should be subject to development and contribute to socio-ecological benefits (Németh & Langhorst, 2014; Kim et al., 2015, 2016). Urban vacant land benefits provided as green infrastructure are not well recognized since it has both economic and political problems.

There are different types of urban vacant land sites which can be identified and explored by a wide range of typologies (Blakeman et al., 2008). These are different types of urban vacant land habitats, among other vacant lots and areas alongside parking lots and roads, abandoned buildings in industrial areas, wastelands, old town centers, railway sites, etc., where can be seen sustained diversity of urban plant communities and animal populations depending on the environmental conditions, natural habitats, and management practices (Robinson & Lundholm, 2012; Darlington, 1969; Jehlík, 1986; Klimeš, 1987; Sukopp et al., 1979; Brandes, 1995; Hope et al., 2003; Burkholder, 2012).

The different types of urban vacant land and lot have potential values and uses to contribute to urban socio-ecology and landscape as integrated green infrastructure, despite the minimal considerations for design, planning, and management. A typology of vacant land is a planning and designing tool to plan for better utilization of the spaces and open creative approaches to envisioning urban space and landscape design in enhancing the ecological and social benefits.

A typology of urban vacant land has implications for policy design, planning, and development to utilize for an advantage seeking to identify and catalog post-industrial land; derelict, unattended land with vegetation; natural land; and transportation-related land. This typology categorizes the vacant land in terms of its potential uses, as natural sites, unattended sites, post-industrial sites, derelict sites, and transportation-related sites. The derelict sites were previously developed but wasted, abandoned, or unused sites. The unattended vacant land with vegetation is inactive.

A typology of urban vacant land may support urban planning policies to design different policies, regulations, and programs in advancing redevelopment to support design, maintenance, and implementation of green infrastructure for the different types of vacant urban land and abandoned buildings. Urban vacant residential, industrial, and commercial sites with historical relevant background can be developed, preserved, and enhanced in their ecosystem values.

Vacant land is an alternative to public spaces because it has easy access to accommodate social groups, including marginalized and young people hanging around, provides playgrounds accommodating a wide range of activities for creative uses, and in the social dimension, families enjoying contact with nature, matching the loose and open spaces as a resting area and extend choice for other activities. Public maintenance of the different types of vacant land should be supported by local authorities by enforcing public and private maintenance requirements, registration fees, procuring green infrastructure credits for socio-ecological and environmental benefits.

Urban vacant land has the potential to contribute to socio-ecology and environmental system besides the best economically viable use by promoting green infrastructure and open urban green spaces when it is managed properly. Urban vacant land offers creative landscape design alternatives for alternative open spaces and opportunities in built-up city environment. Caring for the vacant land increases people's engagement and promotes neighborhood identity.

Properly managed urban vacant land provides beneficial socio-ecological services that have a positive impact to the surrounded communities by creating and developing infrastructure services, community and wildlife gardens, public plantings, and recreational areas (Bonham et al., 2002; Bonham Jr. & Spilka, 1995). Socio-ecological values and benefits provided by urban vacant land as a resource, improve urban policies through urban planning, development, and management aimed to enhance environmental quality, human health, and quality of life for residents in the community.

In highly vacant urban land areas, the real estate market is weak, although space planning and development for vacant urban land with an adequate socio-ecological systems knowledge can be used for green infrastructure and agroecology to provide ecosystem services, adaptive capacity to climate change, healthy fresh food, and water, among other benefits.

Clusters of urban vacant lots with overlapping urban areas of low concentration of green spaces and ecological value require socio-ecological functions associated to ecosystem services. Vacant land can be reused by its socio-ecological value and the financial value is an obstacle for redevelopment. Land vacancy rates in developed urban areas tend to increase due to de-industrialization processes, declining city population, urban sprawl, and preferences for new residential choices (Kremer et al., 2013).

The transformation of vacant land includes projects that may be components of urban community development comprehensive plans that may include greenways, waterfront parks, and pathways development (City of Pittsburgh. Greening Resources, 2012; The Greening of Detroit, 2012). The transformation of different types of vacant land and abandoned buildings into green infrastructure can have the goal to promote community stabilization, economic growth, socio-ecological well-being, and quality of life.

City authorities and private market must be able to establish a redevelopment program of vacant land, empty spaces, and abandoned buildings and to reuse them.

Investment policies, incentive programs, and regulatory strategies on rehabilitation on vacant land and abandoned buildings such as enforcement codes on maintenance requirements and increasing ownership rates may encourage owners to improve their vacant properties and neighborhood stability. Local authorities can maintain enforcement codes if the owners fail to improve their vacant land and abandoned buildings (Accordino & Johnson, 2000).

A positive incentive program and graded property tax system on the requirement of vacant land, empty spaces, and abandoned buildings as primary assets for productive speculation, restructuring is promoted encouraging redevelopment (Kim et al., 2015; Bowman & Pagano, 2004).

3.4 Urban Land Uses

Vacant land as green infrastructure is used to perform the function of providing ecosystem services and socio-ecological values and benefits for the welfare of communities. The socio-ecosystem values of urban green vacant sites can be developed and protected for a variety of types of land uses and revitalized those that are wasted, underused, and undervalued compared to other types. Transformation of expanding uncontrolled urban land uses leads to loss of natural forest and vegetation followed by open spaces, a decline in ecosystem services provision, and environmental and climate changes (Sackey, 2008; Weng, 2001).

Urban vacant land sites are a valuable ecological resource despite that it has fewer impervious ground cover than other urban land uses. To assess the value of the urban vacant land, it can be classified into categories based on their physical, biological, and socio-ecological characteristics to support urban planning and developing.

Uses of urban land spaces have contrasting implications for cities, which in turn is a driver committed to adapt green infrastructure to changing climate; this in turn can influence and enhance the capacity of urban land uses of green spaces and infrastructure to provide some functions to adapt to climate change, such as for example rainwater infiltration. Land use changes focusing on green infrastructure may contribute to develop the capacity building for urban spaces to adapt to global warming, climate change, and extreme weather hazards.

Transdisciplinary approaches can be applied to address vacant urban land uses and urban design and planning to implement strategies aimed to improve the urban green areas and urban green infrastructure which may mitigate the hazards of climate change. Urban land use planning must be based on a comprehensive benchmark for ecosystem services productivity considering the benefits of other land uses and management of vacant land. Urban land uses planning and development, urban forestry, and green spaces can help to mitigate the problems associated with climate change hazards. Vacant land policies and programs may alter the land use structure and nature to achieve higher redevelopment potential utility (Blakeman et al., 2008).

Uses of different types of vacant land should be supported by local authorities, communities, land trusts, and greening organizations to maintain green programs and crops initiative aimed to enhance the socio-ecosystem. Social organizations may manage special programs for the improvement and maintenance of different types of vacant land using project funding from tax levies, foundations, endowments from private donations, and personal funds.

3.5 Green Infrastructure and Forest Structure

Urban vacant land sites are a component of green infrastructure systems that improve the health of the socio-ecosystems with an enduring value for the community. Urban vacant land is a vital source of green infrastructure and creates opportunities for urban green open spaces and creative landscapes.

Vacant parcels contain unimproved green areas and forests. Cultivated and vacant land uses, and expansion of urban built-up spaces are the result of the decrease in vegetation, wetlands, and water. As the result of urban expansion, the environmental functions of vegetation change and the urban land uses change, altering windspeeds, evaporation of water, storage, and exchange of heat, influencing local climate changes in urban areas with their consequences in human comfort and quality of life.

Greening vacant parcels can be transformed into public green spaces to support economic development, social and environmental well-being of local neighborhoods by encouraging reinvestment and decreasing gun violence (Branas et al., 2011). Some cities have begun greening their urban vacant land and public open spaces supported by management tools incorporating the urban storm-water management strategy to reduce the rainfall drains.

A city can utilize the greening of vacant lots as integral urban planning strategy for an urban green infrastructure system with benefits for the socio-ecologically underserved regions and communities while utilizing green infrastructure. Greening of vacant land projects may be developed on a lot-by-lot basis and through site aggregation as, for example, next to major roadways, urban development of open public spaces, community gardens, etc.

Urban farming in vacant lots may not be the optimal application to improve urban agroecology and mitigate climate change. Traditional low-tech farming holds more advantages than high-tech farming in terms of climate change and resource use in urban vacant land sites. Conventional and traditional production methods of tomatoes have potentially lower environmental burdens than greenhouse production in relation to climate change in urban vacant lots. The use of solar panels has greater benefits than other urban benefits to mitigate climate change. Planning and implementation capacities to acquire the ownership of different types of vacant land until construction through multiple strategies can be completed by local authorities and local market.

Urban vacant land providing ecosystem services for urban forest structure and green infrastructure may differ from those provided among the different types of vacant land with different land uses. Urban vacant land can be used for urban forests including park systems, street trees, natural resources, and vegetation to create a more sustainable, healthy, and livable city.

Some of the functional benefits of urban vacant land used for forest structures are the structural value of trees, the value of air pollution removal, carbon sequestration and storage, energy savings, avoided runoff, etc. (Nowak et al., 2002). Forest infrastructures in urban vacant land sites help to counteract the climate change by removing carbon dioxide through photosynthesis and alter the energy consumption by reducing the emissions from the fossil-fuels burned by power plants (Abdollahi et al., 2000).

Vacant land can be a strategically valuable ecological resource that can be utilized as green urban forests infrastructure for storm water with an impact on the runoff for socio-ecological services and environmental benefits. Forest structure on current urban vacant land helps cities to manage urban storm water to prevent floods and filter the polluted water running off, leading to recharge ground water systems.

Urban vacant land infrastructure is a cost-effective way of reducing an expensive infrastructure for storm-water management system. The assessment of urban vacant land forest structure functions as urban green infrastructure to provide ecosystem services is required for an effective and efficient urban forest management.

Vacant land vegetation is a cost-effective way of reducing storm-water management infrastructure, able to mitigate urban runoff (Crauderueff et al., 2012). Urban vacant land cover vegetation and bare soil is a major driver to regulating services and climate regulation, provisioning socio-ecosystem services among others food production, air pollution removal, water supply and supporting services, such as nutrient cycling and soil building, etc.

The urban forest structures provide economic values through the ecosystem services from urban vacant land and other land uses. Well-maintained urban green vacant land vegetative structures affect both the land rising property values by increasing tax receipts and the improvement of the quality of life in the community.

Urban green and forest structures have a structural value in providing substantially greater environmental and socio-ecosystem services, including among others the climate change adaptation, decreased urban heat island effect, energy saving, air pollution removal, carbon sequestration and storage, rainfall interception, and other benefits (Rosenthal et al., 2008). Determining the characteristics of the different forest structures of urban vacant land is vital for a detailed evaluation and accurate estimate of the ecosystem services productivity and value of the urban green infrastructure.

Forest infrastructure on urban vacant land reduces the energy consumption by evaporative cooling in nearby buildings and contributes to regulate climate changes.

In forested areas, vacant lands may produce ecosystem benefits in similar or greater amount of ecosystem services to other forested vacant lands. The plantable space available on urban vacant land sites is a valuable ecological resource when it is high relative to other urban land uses and planting on this unused land increases, have considerable potential of being utilized strategically for reducing surface runoff. These urban forests become ecological resources utilized as urban green stormwater infrastructure, intercepting precipitation, and promoting infiltration and storage in the soil to reach the vegetation root systems.

Green infrastructure projects on urban vacant land have an impact on urban design properties focusing more on functions of the socio-ecosystem services considering their impact on sustainability, social responsibility, and health in socio-ecologically underserved communities. Many cities are greening urban vacant land sites as a strategy of storm-water management infrastructure and potentially becoming a useful component.

Urban infrastructure as an urban green innovation can contribute in the long term to capacity building aimed to adapt to global warming and climate change in urban areas given their severity, increased risks, and related impacts. Urban green landscapes have an influence either to intensify or to moderate climate change impacts.

Urban vacant land requires the design and implementation of policies, planning, and management of urban open green spaces with the inclusion of urban forests, street trees, and any other vegetation as valuable natural and ecological resources to provide ecosystem services. Urban forest structures as well as green roofs and other green innovation areas have a duality of providing benefits but also some catastrophic events, disease spreading, and other negative effects critical for urban planning, designing, and development. This can be prevented by implementing an integrated approach for urban forest structure management of vacant land sites as part of the urban green infrastructure, aimed to achieve an optimum balance and maximize benefits and advantages (Baró et al., 2014).

Spatial planning policies may enhance urban green infrastructure which can be interrupted and lost due to changes driven by pressures in economic development, population growth, and urban expansion (Lennon & Scott, 2014; Angel et al., 2005). The design and implementation of policy mechanisms aimed to greening urban vacant lots improve the environmental benefits to the community such as the promotion of urban agriculture to grow food crops for self-consumption and sale (Hodgson et al., 2011; Ackerman, 2012; Lawson, 2005).

Urban planning policies have an impact on green infrastructure of vacant lots based on their physical, biological, and sociological characteristics, which may be of help to identify and catalog on an inventory many different types of vacant land spaces. An inventory to identify vacant lots, open green spaces, green infrastructure, and underutilized green areas and parks with agricultural potential is a step to determine vacant land to be committed to urban agroecology and forest infrastructure. However, it is necessary to have a site assessment before a negotiation process of potential conflicting interests of the stakeholders.

Planning the greening of vacant lots is cost-effective and a potential for converting to a natural asset for the benefit of public health and supports the climate change adaptation of the urban community and promotes the quality of life. The use of vacant land forest must be encouraged to preserve it where it would be beneficial and healthier for the community, and the owners must be supported by a structure of tax credits and incentives.

3.6 Socio-ecosystem Services

Vacant land contributes to ecosystem services that result in benefits for local communities (Schilling & Mallach, 2012). Ecosystem services are the benefits obtained from ecosystems as provisioning services, regulating services such as climate regulation, supporting services, recreation, and education (Millennium Ecosystem Assessment, 2003). Urban vacant land has potential value in socio-ecosystem benefits based on actively greening vacant lots using urban planning and development of management tools (Lynch, 1995). Vacant land supports the ecosystem services to obtain benefits depending on the natural systems, environmental conditions, and management practices. Vacant land has species diversity in the ecosystem relative to other land uses providing natural sustainability and resilience to recover from natural hazards (Shah, 2015).

Urban vacant land provides ecosystem services more than other urban land uses, such as biocontrol in gardens and lawns, soil food web, rainwater retention, habitat provision services, carbon capture and climate regulation, among others (Yadav et al., 2012; Grewal et al., 2011; Shuster et al., 2014; Robinson & Lundholm, 2012).

To assess the ecosystem services productivity, the benefits from the forest structures of urban vacant land uses must be considered that may have different results and outcomes depending on the type and extent of residential, industrial, commercial, etc., characteristics of the surrounding urban areas.

3.7 Climate Change Hazards

Rapid urbanization processes have increased the socio-ecological and environmental threats resulting in climate change and its consequences in alterations of biochemical cycles, ozone depletion, chemical pollutants, flooding, modifications of the water balance, etc. (Fohrer et al., 2001; Rosa et al., 2004). The new context of urban areas requires to build and maintain an adaptive capacity as the ability to anticipate and influence the urbanization features of their landscapes aimed to moderate and manage climate related risks and climate change impacts (Sun et al., 2013). Climate hazards lead to health risks and the vulnerability of the urban area. In urban areas, climate change patterns of vulnerability and risk are uneven (Carter et al., 2015).

It is broadly accepted that climate change is occurring assuming that the different potential of land use has effects on the production of temperature and precipitation

(IPCC, 2001). The surface temperatures of land uses exchange moisture and energy with the environment with rapid variations depending on time and space and having an impact on the climate change. These land surface temperatures have an impact on heat and radiation balance in global climate changes (Wubet, 2003; Ifatimehin, 2007; Weng, 2001).

The long-term relationship between urban green infrastructure and urban land use change has implications on the adaptive capacity of urban spaces to enhance the spatial context of climate change impacts on related biophysical processes (Blanco et al., 2011). Temporal and spatial changes in the land uses, vacant land, and built-up spaces increase the land surface temperature and escalate environmental hazards associated with climate change.

Climate change is an alarming current trend in urban dynamics in developing countries challenging the energy consumption in transport-land uses (Grazi & van den Bergh, 2008). The exploration of oil sands and shale oil is a relevant option to fossil fuel scarcity and not necessarily climate change risks and therefore creates environmental damage (Rogner, 1997).

The capacity of urban spaces to address climate-related risks is affected by changes on land uses. A gradual decline in urban land uses is reflected in the challenges of the declining economic activities related to the slowing of development activities, other socio-economic factors, and the rising of a set of environmental issues of urban adaptive capacity in the context of climate change. This could increase the risks linked to climate change hazards such as increase in the level of evapotranspiration of land cover, heatwaves, and flooding in urbanized spaces lacking green infrastructure.

A systematic conceptual approach of the relationship between spatial planning, structure and transport, and climate change relates to the spatial transportation planning influencing climate change control (Greene & Schafer, 2003; IEA, 2006). Climate change analysis is dominated by models of local spatial organization of urban dynamics using aggregate approaches of urban planning and development, aimed to design climate policy and the implications in urban sustainability (Newman & Kenworthy, 1999). Climate change hazards are projected to emerge more strongly and become more severe with higher influence on growth and development of urban areas.

The urban-local dimension of climate policy identifies planning options in direct and indirect transport policies and non-transport policies. Spatial local planning and climate policy contribute to effective and efficient mitigation of greenhouse gas emissions. Urban climate hazard change impacts are projected to intensify (IPCC, 2014). Urban green spaces provide several changes among which is the climate change adaptation (Rosenthal et al., 2008). There are several applications based on scenario approaches to analyze changes in the urban climate as an impact of land use change although there is a gap in the research on the relationship between urban land use and climate change adaptation (Storch & Downes, 2011; Tong et al., 2012).

An adaptive capacity is related to a set of enabling conditions supporting the adaptation to climate change impacts (EEA, 2012: 63). There are many elements of adaptive capacity that are detrimental to climate change, including, among others, the lack of governance structures and social and urban community cohesion. The loss of adaptive capacity linked to green infrastructure functions increases the level of risks associated with climate change hazards.

Adaptive capacity is the capacity for preventing and managing the impacts of climate change (Swart et al., 2012: 11). Adaptive capacity interrelates a wide range of issues like the awareness of climate change, urban land use and urban green infrastructure coverage, institutional factors such as strategies and governance (Haddad, 2005; Smit & Wandel, 2006; Swart et al., 2012; Yohe & Tol, 2002). Urban green infrastructure enhances adaptive capacity to respond to global warming and climate change impacts (Swart et al., 2012).

The adaptive capacity that may exist within cities in the context of the urban green land use areas and infrastructure spaces and networks has an important role in the adaptation to climate change increasingly recognized by research communities and policy making (European Commission, 2013; Gill et al., 2007). The adaptive capacity supports activities to address the gap by reducing the urban climate risks. An urban spatial organization analysis of climate change policy for effective and efficient mitigation of greenhouse gas emissions allows assessing mechanisms of spatial urban policy, planning, and development in relation to climate change policies.

Climate change adaptation in declining and expanding cities requires to be supported by strategic planning of green infrastructure. Cities must develop urban strategies to grow and expand by enhancing urban green infrastructure to moderate climate change risks. Designing and implementing climate change adaptation strategies support decision-making to respond to the implications of shrinking and declining urban spaces and land uses, and enhance the potential value of adaptive capacity to foster urban green innovation.

3.8 Discussion

The changes in land surface temperatures of the land uses lead to an increase in surface temperature of urban vacant land, urban built-up spaces, and streams. Poor urban planning of uncontrolled land uses and inadequate environmental management lead to socio-ecological and environmental problems that alter urban climate (UNCHS, 1996; Zhau et al., 2006; Kalnay & Cai, 2003). The consequences of urban climate change calls for immediate action to avert negative environmental implications.

Revitalization of vacant land can be improved through the design and implementation of strategies addressing tax incentive structures, low rates to infill development of land forest structures, and rehabilitation abatement to increase the value.

Adaptive capacity is needed to design and implement urban planning-related strategies to adapt and deliver benefits linked to climate hazard change including land use banking and the use of vacant sites to create urban green spaces and develop green infrastructure projects (Schilling & Logan, 2008). Urban areas must maintain and enhance the adaptive capacity to the changing climate to determine the vulnerability to the severity of climate hazards and related risks. High adaptive capacity in urban areas tends to moderate the vulnerability and severity of climate-related risks.

Urban planning may incorporate vacant lots in its urban management strategy to promote cost-effective greening vacant lots, as the case of establishing the use of vacant lots in the storm-water management strategy to reduce the drains of the rainfalls, protecting the supply of drinking water, improving the safety of waterways, and mitigating the impact of hazards such as flooding.

References

Abdollahi, K. K., Ning, Z. H., & Appeaning, A. (2000). Gulf coast regional climate change council. In *Global climate change & the urban Forest*. Franklin Press.

Accordino, J., & Johnson, G. T. (2000). Addressing the vacant and abandoned property problem. *Journal of Urban Affairs, 22*, 301–315.

Ackerman, K. (2012). *The potential for urban agriculture in new York City: Growing capacity, food security, & green infrastructure*. Urban Design Lab at the Columbia Earth Institute.

Alexander, F. S. (2005). *Land Bank Authorities: A guide for the creation and operation of local land banks*. Local Initiatives Support Corporation. Available online: http://www.nhc.org/media/files/Land_Bank_Authorities-Alexander.pdf. Accessed 18 Dec 2015.

Angel, S., Sheppard, S., & Civco, D. (2005). *The dynamics of global urban expansion*. Transport and Urban Development Department, World Bank.

Baró, F., Chaparro, L., Gómez-Baggethun, E., Langemeyer, J., Nowak, D. J., & Terradas, J. (2014). Contribution of ecosystem services to air quality and climate change mitigation policies: The case of urban forests in Barcelona, Spain. *Ambio, 43*, 466–479.

Blakeman, C., Brown, B., Fitzpatrick, B. T., Shaw, I., & Williamson, A. (2008). *City-wide brownfield redevelopment plan*; report prepared for the City of Roanoke, VA, USA. City Council.

Blanco, H., McCarney, P., Parnell, S., Schmidt, M., & Seto, K. (2011). The role of urban land in climate change. In C. Rozensweig, W. Solecki, & S. Hammer (Vol. Eds.), *First Urban Climate Change Research Network (UCCRN) assessment report on climate change in cities*. Cambridge University Press.

Bonham, J. B., Jr., & Spilka, G. (1995). *Urban vacant land: Issues and recommendations*. Pennsylvania Horticultural Society.

Bonham, J. B., Spilka, G., & Rastorfer, D. (2002). *Old cities/green cities: Planning advisory service report number 506/507*. American Planning Association.

Bowman, A. O. M., & Pagano, M. A. (2004). *Terra incognita: Vacant land and urban strategies*. Georgetown University Press.

Branas, C. C., Cheney, R. A., MacDonald, J. M., Tam, V. W., Jackson, T. D., & Ten Have, T. R. (2011). A difference in differences analysis of health, safety, and greening vacant urban space. *American Journal of Epidemiology, 174*, 1296–1306.

Brandes, D. (1995). The flora of old town centers in Europe. In H. Sukoop, M. Numata, & A. Huber (Eds.), *Urban ecology as the basis for urban planning*. SPB Academic Publishing.

Burkholder, S. (2012). The new ecology of vacancy: Rethinking land use in shrinking cities. *Sustainability, 4*, 1154–1172.

Carter, J. G., Cavan, G., Connelly, A., Guy, S., Handley, J., & Kazmierczak, A. (2015). Climate change and the city: Building capacity for urban adaptation. *Progress in Planning, 95*, 1–66.

City of Pittsburgh. Greening Resources. (2012). Available online: http://www.pittsburghpa.gov/neighborhoodinitiatives/greenup/resources.htm. Accessed 19 June 2015.

Coleman, A. (1982). Dead space in the dying inner city. *International Journal of Environmental Studies, 19*, 103–107.
Crauderueff, R., Margolis, S., & Tanikawa, S. (2012). *Greening vacant lots: Planning and implementation strategies*. The Nature Conservancy.
Darlington, A. (1969). *Ecology of refuse tips*. Heinemann Educational.
Dietersdorfer, L., Efremova, V., Fernandez Agueda, B., Fleschurz, R., Mangialardi, G., Piscitelli, C., … Wolff, M. (2012). *Urban shrinkage and chances for adaptation to climate change*. Final report EU-COST action: TU0803.
European Commission. (2013). *Building a green infrastructure for Europe*. European Commission.
European Environment Agency (EEA). (2012). *Urban adaptation to climate change in Europe: Challenges and opportunities for cities together with supportive national and European policies*. Office for Official Publications of the European Union.
Fohrer, N., Haverkamp, S., Eckhardt, K., & Frede, H. G. (2001). Hydrologic response to land use changes on the watershed scale. *Hydrology, Oceans and Atmosphere, 26*(7–8), 577–582.
Gill, S., Handley, J., Ennos, A., & Pauleit, S. (2007). Adapting cities for climate change: The role of the green infrastructure. *Built Environment, 33*(1), 115–133.
Grazi, F., & van den Bergh, J. C. J. M. (2008). Spatial organization, transport, and climate change: Comparing instruments of spatial planning and policy. *Ecological Economics, 67*(4), 630–639. https://doi.org/10.1016/j.ecolecon.2008.01.014 (https://www.sciencedirect.com/science/article/pii/S0921800908000426).
Greene, D. L., & Schafer, A. (2003). *Reducing greenhouse gas emissions from U.S transportation*. Pew Center on Global Climate Change.
Grewal, S. S., Cheng, Z., Masih, S., Wolboldt, M., Huda, N., Knight, A., & Grewal, P. S. (2011). An assessment of soil nematode food webs and nutrient pools in community gardens and vacant lands in two post-industrial American cities. *Urban Ecosystem, 14*, 181–194.
Haddad, B. (2005). Ranking the adaptive capacity of nations to climate change when socio-political goals are explicit. *Global Environmental Change, 15*(2), 165–176.
Hall, P. (1993). Forces shaping urban Europe. *Urban Studies, 30*(6), 883–898.
Hodgson, K., Campbell, M. C., & Bailkey, M. (2011). *Urban agriculture: Growing healthy, sustainable places*. American Planning Association.
Hope, D., Gries, C., Zhu, W., Fagan, W. F., Redman, C. L., Grimm, N. B., & Kinzig, A. (2003). Socioeconomics drive urban plant diversity. *Proceedings of the National Academy of Sciences USA, 100*, 8788–8792.
IEA. (2006). *Energy technology perspectives: Scenario and strategies to 2050*. International Energy Agency, IEA.
Ifatimehin, O. O. (2007). An assessment of urban Heat Island of Lokoja town and surroundings using Landsat ETM data. *FUTY Journal of the Environment, 2*(1), 100–108.
IPCC. (2001). *Climate change 2001: The scientific basis*. D.L. Albritton, et al. (Eds.). Cambridge University Press.
IPCC. (2014). *Climate change 2014: Impacts, adaptation, and vulnerability. Part A: Global and sectoral aspects.* Intergovernmental Panel on Climate Change. Contribution of working group II to the fifth assessment report of the intergovernmental panel on climate change. Cambridge University Press.
Jakle, J. A., & Wilson, D. (1992). *Derelict landscapes: The wasting of America's built environment*. Rowman & Littlefield.
Jehlík, V. (1986). *The vegetation of railways in northern Bohemia* (eastern part). Academia.
Kalnay, E., & Cai, M. (2003). Impact of urbanization and land use change on climate. *Nature, 423*, 528–531. https://doi.org/10.1038/nature01675
Kamvasinou, K. (2011). The public value of urban vacant land. *Proceedings of ICE Municipal Engineer, 2011*(164), 157–166.
Kim, G. (2016). The public value of urban vacant land: Social responses and ecological value. *Sustainability, 8*(5), 486. https://doi.org/10.3390/su8050486

Kim, G., Miller, P. A., & Nowak, D. J. (2015). Assessing urban vacant land ecosystem services: Urban vacant land as green infrastructure in the City of Roanoke, Virginia. *Urban Forestry & Urban Greening, 14*, 519–526.

Kim, G., Miller, P. A., & Nowak, D. J. (2016). The value of green infrastructure on vacant and residential land in Roanoke, Virginia. *Sustainability, 8*, 296.

Klimeš, L. (1987). Succession in road bank vegetation. *Folia Geobot. Phytotaxon, 22*, 435–440.

Kremer, P., Hamstead, Z. A., & McPhearson, T. A. (2013). Social-ecological assessment of vacant lots in new York City. *Landscape and Urban Planning, 2013*(120), 218–233.

Lawson, L. J. (2005). *City Bountiful: A century of community gardening in America*. University of California Press.

Lennon, M., & Scott, M. (2014). Delivering ecosystems services via spatial planning: Reviewing the possibilities and implications of a green infrastructure approach. *Town Planning Review, 85*(5), 563–587.

Little, G. M. (2008). *629 Main street—A Revisioning of an urban void: An exploration into challenging cultural perceptions of an urban vacant lot*. The University of Manitoba.

Lynch, K. (1995). The openness of open space. In T. Banerjee & M. Southworth (Eds.), *City sense and city design: Writings and projects of Kevin Lynch* (pp. 396–412). The MIT Press.

Martinez-Fernandez, C., Audirac, I., Fol, S., & Cunningham-Sabot, E. (2012). Shrinking cities: Urban challenges of globalization. *International Journal of Urban and Regional Research, 36*(2), 213–225.

McKinsey and Company. (2016). *Urban world: Meeting the demographic challenge*. McKinsey and Company.

Millennium Ecosystem Assessment. (2003). *Ecosystems and human well-being: Synthesis*. World Health; Island Press.

Németh, J., & Langhorst, J. (2014). Rethinking urban transformation: Temporary uses for vacant land. *Cities, 40*, 143–150.

Neukrug, H. M. (2009). *Clean water, Green City: Blending interests of land and water*. Office of Watersheds.

Newman, P., & Kenworthy, J. (1999). *Sustainability and cities: Overcoming automobile dependence*. Island Press.

Nowak, D. J., Crane, D. E., & Dwyer, J. F. (2002). Compensatory value of urban trees in the United States. *Journal of Arboriculture, 28*, 194–199.

Pagano, M. A., & Bowman, A. O. M. (2000). *Vacant land in cities: An urban resource*. Brookings Institution, Center on Urban and Metropolitan Policy.

Pike, A., MacKinnon, D., Coombes, M., Champion, T., Bradley, D., Cumbers, A., et al. (2016). *Uneven growth: Tackling city decline*. Joseph Rowntree Foundation.

Robinson, S. L., & Lundholm, J. T. (2012). Ecosystem services provided by urban spontaneous vegetation. *Urban Ecosystem, 15*, 545–557.

Rogner, H. H. (1997). An assessment of world hydrocarbon resources. *Annual Review of Energy and Environment, 22*, 217–262.

Rosa, E. A., York, R., & Dietz, T. (2004). Tracking the Antropogenic drivers of ecological impacts. *Ambio: A Journal of the Human Environment, 33*(8), 509–512.

Rosenthal, J. K., Crauderueff, R., & Carter, M. (2008). *Urban Heat Island mitigation can improve new York City's environment*. Sustainable South Bronx.

Ruth, M., & Coelho, D. (2007). Understanding and managing the complexity of urban systems under climate change. *Climate Policy, 7*(4), 317–336.

Sackey, A. N. A. (2008). Environmental impacts of rapid urbanization of Ethiopia's cities. In M. P. van Dijk & J. Fransen (Eds.), *Managing Ethiopian cities in an era of rapid urbanization* (pp. 119–131). Eburon Delft.

Sassen, S. (2001). *The global city: New York, London, Tokyo*. Princeton University Press.

Schilling, J., & Logan, J. (2008). Greening the rust belt: A green infrastructure model for right sizing America's shrinking cities. *Journal of the American Planning Association, 74*(4), 451–466.

Schilling, J. M., & Mallach, A. (2012). *Cities in transition: A guide for practicing planners*. American Planning Association.

Shah, A. (2015). *Why Is Biodiversity Important? Who Cares?* Available online: http://www.globalissues.org/article/170/why-is-biodiversity-important-who-cares. Accessed 12 Sept 2015.

Shuster, W. D., Dadio, S., Drohan, P., Losco, R., & Shaffer, J. (2014). Residential demolition and its impact on vacant lot hydrology: Implications for the management of stormwater and sewer system overflows. *Landscape and Urban Planning, 125*, 48–59.

Smit, B., & Wandel, J. (2006). Adaptation, adaptive capacity, and vulnerability. *Global Environmental Change, 16*, 282–292.

Storch, H., & Downes, N. (2011). A scenario-based approach to assess Ho Chi Minh City's urban development strategies against the impact of climate change. *Cities, 28*, 517–526.

Sukopp, H., Blume, H. P., & Kunick, W. (1979). The soil, flora, and vegetation of Berlin's waste lands. In *Nature in cities: The natural environment in the design and development of urban green space* (pp. 115–132). Wiley.

Sun, Z., Li, X., Fu, W., Li, Y., & Tang, D. (2013). Long-term effects of land use/land cover change on surface runoff in urban areas of Beijing, China. *Journal of Applied Remote Sensing, 8*(1), 084596. https://doi.org/10.1117/1.JRS.8.084596

Swart, R., Fons, J., Geertsema, W., van Hove, B., Gregor, M., Havranek, M., … Peltonen, L. (2012). *Urban vulnerability indicators: A joint report of ETC-CCA and ETC-SIA*. http://cca.eionet.europa.eu/docs/TP_3-2012. Accessed 24 Nov 2017.

The Greening of Detroit. Urban Agriculture and Open Space. (2012). Available online: http://detroitagriculture.met/open-space-stories/. Accessed 7 May 2016.

Tong, S., Sun, Y., Ranatunga, T., He, J., & Yang, Y. (2012). Predicting plausible impacts of sets of climate and land use change scenarios on water resources. *Applied Geography, 32*, 477–489.

U.S. Department of Housing and Urban Development. (2009). *Revitalizing foreclosed properties with land banks*. Diane Publishing.

UNCHS. (1996). *An urbanizing world*. World global report on human settlements. Oxford University Press.

Wachter, S., Scruggs, G., Voith, R., & Huang, L. (2010). *Redevelopment Authority of the City of Philadelphia land use and policy study*. Penn Institute for Urban Research, Econsult Corporation.

Weng, Q. (2001). A remote sensing-GIS evaluation of urban expansion and its impact on surface temperature in the Zhujiang Delta, China. *International Journal of Remote Sensing, 22*(10), 1999–2014.

Wubet, M. T. (2003). *Estimation of absolute surface temperature by satellite remote sensing*, unpublished M. Sc thesis, International Institute for Geoinformation Science and Earth Observation, Netherlands.

Yadav, P., Duckworth, K., & Grewal, P. S. (2012). Habitat structure influences below ground biocontrol services: A comparison between urban gardens and vacant lands. *Landscape and Urban Planning, 104*, 238–244.

Yohe, G., & Tol, R. (2002). Indicators for social and economic coping capacity: 'Moving towards a working definition of adaptive capacity'. *Global Environmental Change, 12*(1), 25–40.

Zhau, S., Da, L., Tang, Z., Fang, H., Song, K., & Fang, J. (2006). Ecological consequences of rapid urban expansion: Shanghai, China. *Frontiers in Ecological Environment; The Ecological Society of America, 4*(7), 341–346.

José G. Vargas-Hernández is research professor at Instituto Tecnológico Mario Molina Unidad Zapopan, and earlier at University Center for Economic and Managerial Sciences, University of Guadalajara. Member of the National System of Researchers of Mexico. Professor Vargas-Hernández has a Ph.D. in Public Administration and a Ph.D. in Organizational Economics. He has been visiting scholar at Carleton University Canada, University of California Berkeley and Laurentian University, Canada. He holds a Ph.D. in Economics, Keele University; Ph.D. in Public

Administration, Columbia University; studies in Organizational Behavior at Lancaster University and has a Master of Business Administration. He has Published 9 books and more than 300 papers in international journals and reviews (some translated to English, French, German, Portuguese, Farsi, Chinese, etc.) and more than 300 essays in national journals and reviews. He has obtained several international awards and recognitions. He has also experience in consultancy. His main research is in organizational economics and strategic management. He teaches for several doctoral programs.

M. C. Omar C. Vargas-González serves as professor and head of the Department of Systems and Computing at the Technological National of Mexico Campus Ciudad Guzmán and professor of Telematics Engineering at the University Center of the South of the University of Guadalajara, with a master's degree in Computer Systems. He has received training in Entrepreneurship and Multidisciplinary Innovation at Arizona State University (2018) as well as a diploma in Ecosystems of Innovation and Entrepreneurship from the Harvard University School of Business. He currently conducts research in diverse areas such as Entrepreneurship, Economics, Statistics, Mathematics and Information and Computer Sciences, has collaborated in the publication of more than 15 articles in magazines and directs innovation and technological development projects.

Chapter 4
Strategic Organizational Sustainability

Vargas-Hernández

Abstract This chapter aims to analyze the various features and elements related to strategic organizational sustainability. The analysis departs from the assumption that traditional organizations have to face a lot of challenges posed by the new global economy context confronting contradictory patterns of globalization and de-globalization processes. In order to create the new conditions for the organizations not only to survive but to succeed, it is proposed a new model based on designing and implementing strategic organizational sustainability, abandoning the narrow focus on economic growth and profits to embrace the social inclusion and equity as well as the environmental sustainability issues.

Keywords Economic growth · Organizational sustainability · Organizations · Social inclusion · Strategy · Sustainability

4.1 Introduction

The contemporary society is going through rapid dynamic changes in the new global economy context in which organizations are facing new opportunities and threats to attain sustainable organizational development at their own pace and speed. A more comprehensive presence at work and organization embedded in the society is affected by complex economic and societal factors, where besides the organizational policies and job descriptions, the individual workers involved in their interactions have their own backgrounds and personal histories. The global economy has penetrating and changing organizations and their work processes and practices

Vargas-Hernández (✉)
Department of Management, University Center for Economic and Managerial Sciences, University of Guadalajara, Zapopan, Mexico
e-mail: jose.vargas@zapopan.tecmm.edu.mx

© The Author(s), under exclusive license to Springer Nature Switzerland AG 2023

S. Dhiman (ed.), *Sustainable Development and Environmental Stewardship*,
https://doi.org/10.1007/978-3-031-28885-2_4

lagging behind the organizational structures, processes, practices, etc., which are requiring organizational solutions. In the new global contextual economic and technological changes, organizations have to develop under specific environmental conditions.

Sustainability has become a relevant issue around the world because organizations and consumers today are more concerned enough on both green ecological and social issues when consuming and using products and services and therefore, they are taking more responsibility for strategic organizational sustainability and environmental management. Organizations may have different approaches to achieve these means and goals due to differential in capabilities, resources, knowledge, expertise, etc. Organizations engage in operational relationships of interactions between individuals, groups, organizations, and institutions at the workplace making decisions channeled for sustainable organizational and personal development.

Organizational changes in culture, structure, and individual behaviors have influence on community development. The organizational structure states the responsibilities of all the workers; manages a communication and consultation system to facilitate the knowledge exchange, organizational changes and progress, cooperation and collective bargaining; and ensures the implementation of employment policies, agreements agreed upon procedures, as well as the settlement of disputes and grievances. Traditional organizational structures are struggling with new internal, external, and environmental demands.

Organizational sustainability is affected by global and local sustainability challenges, concerns, and issues which need to be assessed and analyzed for designing policies, strategies, and practices that may have broader economic, social, and environmental impact.

The study begins with the analysis of the notions of sustainability and organizations to continue linking both the conceptualization of organizational sustainability and its components: economic growth and efficiency, social justice, equity and inclusion, and environmental sustainability. Finally, this chapter relates all the above components explained and analyzed in the topic of strategic organizational development before elaborating the final conclusions.

4.2 Sustainability

Around the world, more organizations are taking responsibility for sustainability and environmental management (Rondinelli & Berry, 2000). Sustainability has become a relevant issue because consumers today are concerned enough on both green ecological and social issues when consuming and using products and services.

Sustainability originally was defined as "Meeting the needs of the present without compromising the ability of future generations to meet their own needs" (Commission on Environment and Development, 1987). Sustainability is a critical issue for the organizations concerned with meeting the current needs of society without compromising the ability to meet the needs of future generations. Sustainability in essence is concerned with meeting the needs of people today without compromising the ability of future generations to meet their own needs (World Business Council for Sustainable Development, 2005).

Sustainability is an ecological concept and environmental development concerned with environmental stewardship, economic growth, and social justice, which are essential factors to characterize the sustainable organizational development. Sustainable organizational development is a concept that leads to promote growth and development (Sev, 2009). Organizational development has some basic constituents such as recognition and identification in an organizational diagnosis of the current situation, issues and problems to be solved and improved. Then, a planned organizational intervention through actions and corrective efforts followed by a follow-up of sustainable maintenance updating the measures on results as a continuous feedback to accomplish sustainable organizational development.

The framework that leads toward the environmental sustainability (Hill & Bowen, 1997) revolves around the organizational development management of life cycle synergies, involving workers in commitment development and decision-making process. Environmental sustainability is feasible with the development of the internal competencies. The ability for organizations to create and manage sustainable core competencies enables to implement interventions and initiatives on biophysical, economic, social, environmental, and technical sustainability (Hill & Bowen, 1997).

Ethics and sustainability have to do the right things when caring about others such as responding to human justice and fairness and in favor of a culture of peace. That is, the three more relevant values are: global ethics, a culture of peace, and sustainable development. Organizational sustainable ethics analyzes the issues regarding moral obligations of organizations regarding the bio-socio-ecosystems, environment, and natural resources to future generations. Violation of organizational obligations should produce organizationally an intense and detrimental response on the expectations in terms of individual costs and rewards that may vary in different organizations.

4.3 Organizations

Individuals are engaged in a system of relationships in organizations which in turn are connected in units and workplaces with different contextual characteristics, space and time, etc., creating an interest in finding the best methods to organize them (Scott, 1992). To create and develop a sustainable organizational solution, it is required to invest in resources and capabilities to create a systemic thinking patterns at all organizational levels, that is in terms of Senge (1999), a systems thinking.

Individuals become integrated to organizations that meet their needs and demands, development of human relations, and provide satisfaction by rewarding the work performed and contributions to production of goods and services with the best working conditions (Androniceanu, 2009). An organization is a group of individuals working together toward specific goals through the development of values used to operation and processes. The new organizational context from the economy and technology perspectives is a complex, uncertain, and dynamic system. Since Taylor, organizational science has tried to define organizational solutions to the human dimension leading toward a gradual development of flexible, liberal, and democratic organizational work.

Individuals in organizations are encouraged to freely develop relationships and connections in the workplace to share their ideas more democratically, regardless of their position (Dombrowski et al., 2007). The work organization takes into account the physical, the psychological needs of workers, and the nature of operations and work. Organizational operations derive some internalities from connecting functional organizational structure to more centralized and controlled structure making difficult to build conscious trust and openness to match the management standards. Organizational communication enables the development of group's identity in a self-categorization process leading to the sense of organizational community.

Trust on the information and operational communication patterns between individuals in an organization is reflected on a higher motivation and sense of commitment reducing uncertainty of changes and outcomes (Elving, 2005; Mahal, 2009). A mutual trusted organization is one that improves its workplace and best performs its operations. Organizational loyalty toward its workers must be demonstrated supporting them by recognizing their energy consumed at the workplace as well as the experiences of organizational fairness. Changes and developments require information and communication to find solutions to these grievances and conflicts, which may result in psychological contracts between employers and employees related to alignment of interests and goals, agreements on organizational events and actions.

The psychological contract assumed between employer and worker has the expectations to be fulfilled, being met or likely to meet focusing on the driving energy to achieve personal development. The discrepancies between organizational setting and the worker aspirations in the work place result in tensions, stress, and burnout in the working life. Workers accept the changes of their work and the institutional setting to derive support, but if they do not have this support, they become dissatisfied and frustrated for a sense of imbalance between their work performance and the work emerging from the organization, having always the option to leave the organization, although they may consume their own personal human resources to bridge the gap. This sense of an imbalance leads to dysfunctional experiences of social and organizational stress.

It was assumed that organizations operated in a stable and in a predictable context, but now is not the case. Organizations are confronting a complex, uncertain, ambiguous, and discontinuous environment (Nonaka & Takeuchi, 1995; Weick, 1995; March, 2006). Adequate organizational environment makes the workers feel to become genuinely interested in their own personal development and offers to provide appropriate interventions and programs. Organizations that have the goals of economic growth and wealth creation must care for all the socio-ecological systems and living species. Organizations need to have a viable relationship between economic and environmental goals aimed to increase the economic growth and efficiency while at the same time improving the environment.

The organizations create opportunities and possibilities for the individuals and organizations to develop capabilities to create, manage, and use resources framed by power structures and distribution of authority within an organization which enables the purposes and goals (Kanter, 2008; Leana, 1987).

The impact of economic globalization on organizations has shortened the cycles of change making faster to improve increasingly faster information, knowledge, processes, practices, uses of technologies, services, etc., leading to potential continuous need-oriented improvement, a holistic management system, and business models. The organizational work activity is already changing while the organizational change is not always taking place. The organizational structure of the economy is related with the development, qualitative growth, and efficiency of the economic system and the global economy.

Sustainable organizations are created by groups of individuals working together (Postmes, 2003). Organizations focused on sustainable development aimed at created common value need to be beyond profit-making. The objective of sustainable organization initiative must be aligned with the workforce motivation and awareness for the sustainable organizational development.

Organizations can be analyzed using the core competence in-out and out-in analysis (Han et al., 2010) placing relevance to the knowledge usage to identify courses of action supported by diversification strategies to develop the workforce and enable the new organizational position and the sustainable organizational growth.

Interaction among organizational stakeholders increases the potential of networked structures to achieve sustainable organizational development initiatives (Flyvbjerg et al., 2003). Organizational networking that includes collaboration among academic and research institutions, government agencies, and business make up the triple helix (Etzkowitz & Leydesdorff, 2000). A sustainable work system is one that meets the needs of the well-being of all the stakeholders involved and develops their creative potential in a continuous individual and organizational development.

Through continuous improvement of organizational processes, organizations create an organizational climate and implement a sustainable growth and development initiative in response to the environmental changes and challenges. The development of low-carbon products was achieved through environmental training aimed to climate change mitigation (Saturnino Neto et al., 2014). Organizational development requires specialized environmental training (Perron et al., 2006) to achieve a higher organizational performance of an environmental management system (Sarkis et al., 2010).

Organizations must develop strategic leadership centered on strategic sustainability practices pervading all the functional areas across the organization and embedding them for sustainable development systems and processes, and requiring greater involvement and commitment of all the stakeholders. Macke and Genari (2019) analyze the state-of-the-art and the elements of sustainable leadership and human resources management and environmental sustainability embedded in principles, organizational values, processes, and practices.

Holistic green organizations integrate and implement practices of green human resources management in facilitating the responsibilities to ensure policies and strategies, systems and processes to achieve the goals of profit-making and to allocate resources for an environmental sustainability. The holistic green organization meets its finances objective with practices of making profits and sets a sustainable environment through the creation of common good and value.

Green organizations may limit the usage of paper and avoid printing by accessing an online portal of the organization to most of the important documents, bills, procedures, etc. Paper shreds from the organization can help waste management and be sold as a waste paper recycle. A green cafeteria may try to make customers consciously serve themselves from the buffet with minimum use of plastic, while making them aware that the food wasted and left is collected for making fertilizer. Jute bags can be used for fruits and vegetables. Recyclable plastic dustbin can be used to dispose water cups etc. Green organizations are responsible toward profit-making, responding to the good and wellness of all its stakeholders and for the impact of its policies, practices, and activities on the environmental sustainability and the creation of common value.

Green organizations should be green corporate social responsibility creating value for the common good measured in terms of the impact in economic growth, environmental sustainability, and social development by benefitting the different sustainable involved and engage stakeholders (Freeman, 1984). A team or task force can have assignments to analyze the organizational environment issues, concerns, and hazards; promote green behavior; implement environmental interventions; and monitor within the organization. Some organizational green behaviors are socially responsible behaviors such as efficient use of water, energy, and lights, less printing, etc., which may require organizational learning and training.

Organizations need to focus on development to overcome any challenges that arise from the environmental changes (Han et al., 2010). Organizational management plays a critical role and responsibility in promoting innovation sustainability by creating the right behaviors, values, and attitudes engaged toward the organizational visions. Organizations set out long-term strategic planning aiming at specific goals of increased quality and profitability and reducing the risks, costs, and time (Garrido & Martos, 2016).

Organizations are implementing management systems based on international principles and standards, organization strategies and objectives, business processes and optimization, resource management, which are implemented in a systematic and structured way. The continual sustainable organizational development needs internal structure and systems (Hill & Bowen, 1997).

Organizations design, develop, and implement an incentive plan, motivation and communication system, and leadership guidance. Organization needs to create support for the workers together with managers and peer to reconfine and make sense of their own work at the workplace on priorities, tasks, roles, to make choices, personal judgments, and communicate to others.

Organizational structures, norms, rules, regulations, communication processes, values, etc. are related to the logic of organizational bureaucratic model. Organizational structure defined as the organizational system of relationships between authorities and tasks has an impact on organizational sustainable development through the use of resources and coordination of people to achieve the goals (Martin, 2014).

Sustainable organizational development needs an internal structure to constantly motivate, balance power, and opportunities to perform positive behaviors (Kanter, 2008). The organizational structure is a hierarchical arrangement of communications and authorities depending on the organizational objectives and strategies. Coordination of activities is a structural factor needed among the organizational abilities in which communication develops the social identities within the organization (Postmes, 2003).

Organization and communication are dependently intertwined in corporate or organizational communication developing in competitive advantage and ability to retain more proficient, motivated, and talented workers (Balmer & Gray, 1999). The communication channels of the organization must focus on empowering the workers to improve the level of development, motivation, and commitment of its individuals enabling sustainable organizational growth.

Organizational communication develops the structure and processes enabling the development of sustainable organizational development and growth. Organizational communication is a tool that helps to be sensible for all the stakeholders with the concept of organizational proudness, purpose, goal, and most important with the sense of organizational community ties. Organizational communication is the foundation for the interactions among individuals and sets the group's norms and social identity, promotes social cohesion, improves communication and collaboration, and encourages individual commitment (Elving, 2005; Haslam, 1997; Postmes, 2003; Tucker et al., 1996).

Organizational communication enables development of the personnel in areas of commitment as the foundation for motivation, involvement, and quality of life (Elving, 2005; Mahal, 2009). Organizational communication enables the interpersonal relationships of all the agents and stakeholders involved in the attainment of the strategic goals of the organization (Postmes, 2003). Organizational communication results in creating, expanding, and transferring knowledge in a continuous process of the workforce while developing a feeling of community of individuals with ties with the society enabling the improvement of the local and regional communities.

The organizational models of Barley and Tolbert (1997) and Crossan and Berdrow (2011) are in contradiction between the organizational setting and work to emerge with societal changes which have an impact in the work processes and challenge the institutions and organizations to adopt new post-bureaucratic organizational designs. Regenerative work emerges through collectively organizing the transitional development of work processes from bureaucratic structures to more post-bureaucracy challenging the implications on the work organizations and the working life.

Models for organizations across different contexts may duplicate organizational structures and processes and may be presented as autonomous analysis are based on similar decisions and norms for how to organize (Meyer & Rowan, 1977) in what is well known as organizational isomorphism.

The organizational post-bureaucratic work is emerging and changing in human considerations, an approach that should support work despite the classes between work without boundaries and the bureaucratic type of traditional work with strict rules and norms. Creating collective and regenerative work is specific to the type or organization but may be generalized to organizations with more post-bureaucratic working quality of life. Post-bureaucratic working quality of life is influenced by economic, ideological, social, technological, and political issues and developments leading to more sustainable organizational work systems characterized by more elusive and subtle nature of control subject to internal and external factors. There is a mismatch between the organizational post-bureaucratic design and work. This current situation leads to the need to design new organizational approach to work.

The post-bureaucratic characteristics and problems of work systems can be related to traditional bureaucratic workplaces and occupations subject to the internal and external factors as well as the type of organizations. Some elements of the organizational post-bureaucratic structures can coexist beyond the formal structures in the functions within the bureaucratic structures making it more stable. The nature of organizational work is changing rapidly and drastically with the transitional emergence of post-bureaucratic structures and posing new challenges to traditional organizational structures, although certain elements of post-bureaucratic organizations may be still found in more traditional ones with bureaucratic work (Howard, 1995).

Self-organizing structures in horizontal interaction have a synergetic effect to manage the large-scale economic system. The organizational synergic effects of resources and capabilities lead to innovativeness and productivity. Self-organization system aiming at synergetic effect is open and provides feedback of processes to the organizational structure based on objectives, resources, capabilities, policies, rules, values, etc. to strengthen the organization.

Human resources motivation and job satisfaction are relevant behavioral factors supporting sustainable organizational development and main priorities of long-term management (Deal & Jurkins, 1994; Erez & Early, 1993). Human resources motivation, job satisfaction, and wage policies are critical factors for economic and social activities (Ciocoiu, 2011; Colesca, 2010; Sims & Veres, 2007), supporting sustainable organizational development (Deal & Jurkins, 1994; Erez & Early, 1993). Workers hope that the wage policy of the organization can meet their personal needs by requiring a salary, duties, function, benefits, and cash rewards (Tyson, 2006) in exchange of work performance and contribution to production.

The development of traits and human relations at work is related to the existing work conditions and welfare state benefits (Androniceanu, 2009). The wage policy of the organization is linked to work satisfaction and to organizational climate which results in the level of organizational stress influencing the organizational development. There is a need to introduce an integrated differential motivation system including incentives to encourage performance of human resources and contribute to a sustainable organizational development. In the execution of environmental management systems, organizations have to confirm that the organizational performance appraisal have to develop, adopt, adapt, implement, and receive feedback on pro-environmental policies and criteria through the involvement of human resources and the stakeholders.

Organizational motivation is created by an enabling organizational climate that maintains communication and improves performance among the involved members (Patterson et al., 2005). Organizational communication has a relevant role on the attainment of the workforce commitment (Elving, 2005) to develop a sustainable competitive advantage within the organizational strategic core competencies (Tucker et al., 1996). Organizational communication allows members to interact and participate in solving problems and making decisions empowering them for innovation.

Empowerment has a significant positive impact on organizational sustainable development (Mumford et al., 2017). The empowerment awards powers to employees to get involved in making decisions and performing them within the organization and has an impact on organizational sustainable development while creating citizenship behavior by giving moral, ethical, psychological, social, and financial support (Bandura, 2002).

Trust and accuracy on information lead to make organizational decisions across the hierarchical structure with an impact on the overall performance (Roberts & O'Reilly, 1974). Many decision-making issues and details decided at the management group vanish in a black hole at the top management of the organization, one of the deteriorated symptoms of bureaucracy (Heckscher, 1994). Top management is responsible for making the organizational work system more comprehensible and manageable at all the levels of the organization.

The quality of information transmitted between individuals hinges on their level of trust, organizational behavior, and satisfaction (Roberts & O'Reilly, 1974). Trust development among individuals emerges from interpersonal relationships and interactions through communication which are tied to organizational behavior and structure that create a cohesive work environment enabling organizations to facilitate the transferal of information (Roberts & O'Reilly, 1974).

Organizational structures may mismatch work activities consuming more resources that challenge for a more balanced situation demanding more rational consumption and a more comprehensive presence of resources at the workplace. One possible solution is that work activities may be predefined in many workplace situations. Organizations adopting structures, practices, processes, etc. similar to other organizations is the result of organizational isomorphism and may be dysfunctional when the local factors are not taken into account. An organizational structure emerging from the internal interaction patterns and responsibilities may be more directive top-down decision-making while from the external factors may emerge as functional organizational structure with the expert departments and supporting functions.

The organizational structure correlates with the organizational behavior, motivation, communication, and the overall cohesion through the organizational climate. Organizational climate is the representation of the events that affects the level of motivation and performance at individual and organizational levels within an organization (Patterson et al., 2005; James et al., 2008).

Organizational climate is the symbol of organizational ambitions that refers to the collective opinions on regulations, processes, protocols, etc., and the method of attained goals (Elving, 2005). Organizational climate allows all the stakeholders and actors to reach a consensus regarding the methods to be used to achieve the goals of the organization. Actions consistent with the goals enable the improvement of relationships leading to a higher level of performance (James et al., 2008). The organizational climate has effects within the organization with a high degree of validated impact on the workforce.

The organizational climate enhances positive motivation, balances the structure of power, and improves the performance (Patterson et al., 2005) adjusted to the needs and satisfaction of all the stakeholders that leads to consistent and continual sustainable organizational development (Cummings & Worley, 2014). Determinants of organizational behaviors refer to the structure of proportions between the alignment of individuals and organizations in different situations, as a tool to determine the social type. This difference in proportions is referred to individual attitudes and organizational needs (Kanter, 2008).

Top management must leverage efforts to improve sustainable organizational practices by creating an environment where the talent retained works in harmony with the strategic vision of leadership. The organization top management must be confident about support for the promotion of initiatives for green behaviors in favor of greening and sustainability of the environmental development of the organization with green products, protecting natural resources, combating climate change. The environmental management system as an independent variable has challenges and offers opportunities mediated by the teamwork empirical-environmental training and employee involvement and the organization's environmental performance (Daily et al., 2012).

The organization's performance is measured in terms of a balance between economic, social, and environmental outcomes on attaining organizational growth and protecting, conserving, and maintaining natural resources for the use of future generations (Daily & Huang, 2001; Jennings & Zandbergen, 1995; Ramus, 2002).

Organizations have to report on sustainability initiatives implemented in conformity with regulations to show their commitment with their results. Organizations designed with agile sustainability structures are more responsive to internal and external environmental changes.

The future of organizations presents several challenges, risks, and opportunities that require the development of abilities and capacities to cope with these concerns to make decisions on planning organizational change and improvement for sustainable organizational growth and development.

4.4 Organizational Sustainability

Mindful organizations invest in sustainability which is one of the key drivers of economic, social, and environmental sustainability, with a positive impact on brand image, improving morale, increasing return on investments, etc. Sustainable organizational practices are a mindset that affects all levels of the organization through effective change management that enable economic growth and customer satisfaction. Organizational transformation structures and processes are subject to the innovative development and maturity of the economic system, able to ensure economic growth, and social and sustainable organizational development of the economic system.

A holistic approach to organizational sustainability enhances reputation, reduces costs, and gives a leadership position to the organizations. Holistic green organizations should utilize used and recycled components for designing new green products and green packaging. Naudé (2012) has proposed a tridimensional approach to sustainable development to include economic, social, and environmental dimensions combined with initiatives and strategies. The organizational sustainable development has the economic, social, and environmental dimension which can be managed by strategies aimed to face the socioeconomic development and environmental protection and maintenance.

Organizational sustainability has benefits among which it builds brand reputation, improves confidence, attractiveness, and environmental consciousness to talent, investors, suppliers, customers, confidence, better performance, etc. Organizational sustainability encompasses the need for motivation and cohesiveness of the teamwork toward setting guidelines and establishing the processes for the achievement of organizational goals. However, human resources management practices focused on profits and cost reduction are in tension with long-term social dimension of organizational sustainability, relationships with stakeholders, and performance.

Sustainability is the organizational ability to achieve its goals improving long-term shareholder value by integrating economic, environmental, and social opportunities into its organizational sustainable strategies (Symposium on Sustainability, 2001). The organizational sustainability framework (Hill & Bowen's, 1997) that analyzes as determinants are the biophysical, economic technical, and social sustainability.

Sustainable development in organizations is guided by some principles such as the active ownership, cooperation among actors, and learning through ongoing evaluation. The principles of organizational sustainability provide the foundations for the managerial strategies and policies aimed to attain sustainable organizational growth and development (Sev, 2009). Principles of mimic natural ecosystems affect the environmental sustainability of supply chains leading toward sustainable design and managed supply chains. Some of these ecological principles are the dialogue and synergies between industrial ecology, supply chain management, and organizational ethics.

Emotional energy could be a source of individual and organizational sustainability by continuously changing the internal and external factors of the working environment and well-being issues. Developing proactive initiatives of sustainable organizational practices is dependent on the talent and leadership that are conscious of the value-creation drive through the delivery of sustainability. Leadership committed to developing more sustainable organizational model must practice the values attached to sustainability by example and encourage workers to follow while assessing accurately the economic, social, and environmental risks and opportunities derived.

The facilitator of organizational sustainability processes facilitates the formulation and implementation of strategic change taking care that the organization responds to internal, external, and environmental challenges, remaining committed to the economic efficiency and growth, social justice and inclusion, and environmental sustainability.

Organizational sustainability values must have to be inculcated at all levels in the system of the organization to engage the initiatives while performing the tasks. Workforce diversity initiatives contributing to recruit, develop, and retain talent responds to sustainability challenges with economic, social, and environmental awareness. Organizational values must be consistent with sustainability in the workplace supported by the leadership of top management in charge of designing and implementing the organizational sustainability strategy for sustainable organizational practices.

Critical knowledge and skills regarding organizational sustainability have to be integrated to the agenda into programs, activities, and practices linked with economic and social betterment. A sustainability organizational agenda drives to design and implement sustainability strategic and operational planning at all levels to device the initiatives, practices, and activities led by top management with a proactive role and with the strategic support of professionals playing the roles of facilitators and agency. Sustainability strategic imperatives are in the most of the organizational agendas (Hee-Jae & Pucik, 2005).

Sustainable organizations must be able to attract, retain, and align the talent with the sustainability agenda and organizational practices. Organizational communication and information flows between the leaders, managers, and workers engage them in the commitment to an organizational sustainability agenda. Informal organizational work structures and organizational communication have to be adaptive to more functional structure evolving into a more matching organizational structure supportive of a more sustainable organizational development.

Corporate and organizational communication refers to the stakeholders developing organizational reputation and identity (Balmer & Gray, 1999). The development of organizational sustainability leading to corporate communication requires investments to receive the benefits (Balmer & Gray, 1999). The quality of information transmitted between individuals hinges on their level of trust, organizational behavior, and satisfaction (Roberts & O'Reilly, 1974). Trust development among individuals emerges from interpersonal relationships and interactions through communication which are tied to organizational behavior and structure that create a cohesive work environment enabling organizations to facilitate the transfer of information (Roberts & O'Reilly, 1974).

Top management is responsible to fulfill the organizational agenda of expectations on sustainability development, from the search of new environmental, internal, and external development design of a suitable structure, recruiting and developing the right talent to drive the sustainability initiatives to allocating efficient resources and implement the sustainable practices using the knowledge and skills. Organizational clusters, units, and groups should undergo changes and modifications toward sustainable organizational development, health, and efficiency.

Organizational practices that have an impact on the sustainability of the environment and community have the obligation for organizations to undertake sustainable reporting. Organizations and corporations have the obligation to show that they are transparent issuing sustainability reports.

Organizational sustainability is productive as the result of a congruent ideology of economic, societal, and ecological concerns that have influence on human resources management. Organizational growth is the natural expansion of size and development and is the ability to improve taking advantage of opportunities (Daly, 1996). Sustainable organizational development has been considered a pervasive philosophy, but environmental sustainability makes possible to protect, maintain, and manage natural resources, have economic growth, achieve more social equality, and respond to the global outcry of climate warming. Sustainable regenerative work systems build development plans from individual and organizational sustainable organizations (Docherty et al., 2002) contextualized in economic growth, social equity and justice, and environmental sustainability.

Organizational ethics has a direct link with the economic growth, social justice, and environmental sustainability in relationship with practices of sustainable organizational development. The relationship between organizational ethics and sustainable development in terms of the organizational economic and social goals is the equitability, as for example in ensuring a balance between pricing and quality and health of products. The ethical principles are in requirement in the analysis of sustainable organizational development and growth (Bryman et al., 2011) focusing on issues such as economic growth, social inclusion and equity, environmental sustainability, technological innovation, policies and regulations, and branding. All these being important factors that affect the ability for sustainable growth have to be considered when identifying the strategic focus of an organization.

Organizational performance management supported by organizational ethics and governance foster sustainable organizational development, conceived as radical change in organizational practices and are economically, socially, and environmentally sustainable. That is, sustainable organizational development must be efficient, equitable, and green. Sustainable development of the organizational environment is concerned with the so-called triple bottom-line sustainable activities and measures on natural resources and environment, the social impact on the organizational welfare, health and safety, protection of human rights, the economic well-being, labor productivity, investments in training, and economic efficiency and growth.

Economic Growth and Efficiency

The organizational sustainable objectives of economic development are achieved through industrial development based on the industrial harmony relationship between the employer and the workers for sustainable personal development. Sustainable organizational performance focus on the economic performance more than in the narrow financial performance, incorporating the social and ecological effects of production processes while maintaining image and sustainable outcomes. Organizational development interventions give meaning to practices, activities, duties, and commitments of personnel and facilitate accomplishments on efficiency and health performance.

The core competencies of organizations are the foundations for sustainable growth and development such as the capabilities to create organization tangible and intangible (Knowledge, technology, etc.) resources (Egbu et al., 2004; Spence & Mulligan, 1995) focusing on the economic benefit and encompassing environmental sustainability as the drivers for sustainable development (Vanegas et al., 1996; Ortiz et al., 2009; Presley & Meade, 2010). Human motivation considers the effective organizational wage policies to stimulate sustainable economic efficiency (Ciocoiu, 2011; Sims & Veres, 2007).

Organizational development is a regulated, integrated, and planned approach designed to cope with challenges and difficulties to boost efficiency in organizations. Organizational development is a response to the contextual and environmental changes through planned and implemented effort engaging the whole structural organization to revolutionize values, attitudes, beliefs, etc., aimed to improve the efficiency and health (Bennis, 1994). The prevalent attitude within the organization is related to the motivation and the output (Mahal, 2009). The efficiency management principles are based on the equilibrium and the ability of well-balancing working and labor time, full devotion-involving activities and flexibility.

Organizational development indexes are used to diagnose and identify the organizational problems, steer the interventions, and plan the actions toward improving the organizational development through the implementation of the corrective measures by units and groups aimed to achieve organizational efficiency and health. The worker is more efficient and cohesive in a stable workplace environment that enables to maintain standards and code of conduct in a more cohesive team leading to more initiatives and better performance (Mahal, 2009). Personal development of any worker is provided by the organization through behavioral and training interventions achieved through harmonious and cordial relationships aimed to result in higher performance, efficiency, and productivity.

Organizational democracy is the sound relation between management and the workforce based on motivation and consultation, to contribute with their best to the organizational performance, health, and efficiency. Organizational democracy motivates management and workers to mutual cooperation and agreement, to increase efficiency, productivity, and other tangible benefits, to contribute with their best efforts to economic growth, development, organizational health, and prosperity.

The sustainable organizational development requires high ability for changes in the new context that cannot be avoided, taking advantage of the environmental opportunities, and achieved by the capabilities of an efficient resource management.

The institutional environment development of any economic system has an impact on the economic growth, social justice, and environmental sustainability through the organizational management and property structure, distribution of income, etc. in such a way that with any increase in the level of institutional development, the economic growth also increases (IMF, 2017 World Economic Outlook Database). High institutional quality is related with negentropic processes in the economic systems of developed economies, while low institutional quality creates entropic processes in less developed economies where the environment is dissipative in economic growth and human capital (The Human Development Report, 2014). The dissipative system has certain order in the organization determined by the order of the economic system and characterized as a no equilibrium open system that maximizes the efficiency with the minimum entropy and energy dissipation. The organizational entropy is measured by its orderliness based on the structure and its impact on the economic system in terms of synergies, syncretic, and entropic (Prigozhin & Stengers, 1986).

Organizational development has an impact on the qualitative economic growth in national and global economic systems. Regional economic growth, competition, and technological innovation networks are some external factors that can be standardized for organizational growth and development (Miyatake, 1996; Tan et al., 2011).

Social Justice, Equity, and Inclusion

Social sustainability in organizations refers to sustainable human resources management practices that have an impact on the well-being of workers (Pfeffer, 2010). Social sustainability as a system may have institutionalized factors that provide some inertia to change. Organizational social systems are always in continuous change process in which people are able to be aware and participate to grow and develop in a never ending path (French & Bell Jr., 1973). Organizational actors have different personal contacts and informal and formal social relationships through interconnected communication networks that facilitate their presence and their roles at the workplace as a source of well-being and development.

A sustainable organization holds a group of individuals with different characteristics and backgrounds which lead to grievances, normal strife, and conflicts due to concerns and issues for justice and equity. Organizational conflict may be inevitable as a fact where people compete for scarce resources and power and may arise as intra- and interpersonal, inter- and intraorganizational, inter- and intragroup, etc.

Organizational power is shown in the hierarchical domination where individuals have capabilities to structure resources and get the desired results (Kanter, 2008). When only a group of individuals have the monopoly of power, they make decisions regarding the management of resources and methods. When inevitable conflict emerges in the organization, it should be managed and minimized ensuring organizational harmony. Measurement of human development takes into account physical, psychological, and social variables such as quality of life, living standards, health, and education levels.

A sustainable organizational development has some features such as awareness, open communication, and self-actualization; meets the needs for growth and development; implements open, inclusive, and transparent policies for all the stakeholders; creates an organizational environment of trust; and supports commitment for all the members.

Environmental Sustainability

Environmental sustainability and sustainable organizational growth can be maintained through the development of organizational knowledge and the community development with high standards in cultural diversity, ethics, social responsibility, safety issues, etc. Organizational structure and behavior continuously evolve to cope with environmental changes and need to develop and manage the core values. Sustainable organizational growth focuses on environmental sustainability, diversity, social responsibility, code of conduct, and worksite safety, supported by a clear technological and business approaches, leadership, innovation, creativity and responsible management, and alignment of individual involvement with organizational goals, all aimed at the development of core competencies to achieve a sustainable competitive advantage.

Organizations achieve environmental performance and sustainable development by developing an empirical environmental sustainability management system through the moderate and direct influence of human resources training and development. The intrinsic organizational competencies supporting the organizational development and focusing on environmental sustainability lead to sustainable organizational development and growth encompassing environmental sustainability.

4.5 Strategic Organizational Sustainability

Organizational sustainability is a critical concern in designing and implementing strategies and practices and leveraging management vision and leadership aimed at organizational sustainable development. Organizational sustainability is a competitive strategy to enhance the ability to develop sustainable change (Moore & Manring, 2009). Organizations adopt different approaches and initiatives for achieving environmental sustainability and protecting the environmental growth avoiding disruptions by managing internal factors such as managing the causal human behavior and the role of human behavior (Davis & Challenger, 2013; Ones & Dilchert, 2012;

Oskamp, 1995), and undertaking other environmental development initiatives outside the organizational boundaries (Uzzell & Moser, 2009).

Organizational development planning and implementing the interventions for change of practices, processes, procedures, and actions in relation to the environmental and contextual factors to design, formulate, and execute organizational structures, strategies, and policies are aimed to achieve the goals (Cummings & Worley, 2014). The planning and control approach for sustainable organizational development projects and programs can be supplemented with an open process and learning to promote long-term sustainability (Lindkvist & Söderlund, 2002; Morris et al., 2011; Brulin & Svensson, 2012; Svensson & Brulin, 2013). Strategic organizational planning for the formulation, implementation, and set-up long-term sustainability strategies provides a guide to all the stakeholders for theory involvement in sustainable development initiatives and practices across all levels of the organization to reap the best benefits.

Sustainable organization is the enabler of sustainable human resources management strategy (Fairfield et al., 2011). Environmental and internal analysis of the organization is needed to link environmental management and human resources management to formulate sustainable strategies to contribute toward greater environmental sustainability of the organization.

Development and management of sustainable core competencies focusing on environmental sustainability can lead to design and implementation of strategies aimed at sustainable organizational development and growth. An organizational strategy designed on management and communication of the workforce is crucial to enable sustainable organizational development and growth, more oriented toward environmental sustainability.

Design and implementation of a sustainability strategy throughout the organization need a structure to give support to operationalization of practices and activities at the workplace. Sustainable organizational strategies from the perspective of the whole sustainable core competencies must fit the core business model within the environment in order to implement the sustainable organizational practices, taking into account training and development of talent to fulfill the economic, social, and environmental sustainability. Global, local, and internal social networks can be used to get sustainability awareness, examining and analyzing specific issues related to the industry in order to design the sustainability strategy.

Change facilitation and agency must be aware of developing structures and systems to support sustainability initiatives in accordance with the agenda and aligned with the organizational culture and environmental strategy development. Organizational strategies should be aligned with human resource system to link workforce capabilities with environmental sustainability growth management systems and sustainable development (Ichniowski et al., 1997; Mendelson & Pillai, 1999; Collins & Clark, 2003; Boselie et al., 2001; Paauwe & Boselie, 2003; Jiang et al., 2012; Daily & Huang, 2001; Renwick et al., 2013).

Organizational leaders must recognize that the human resources function is in charge of implementing the sustainability strategy (Harmon et al., 2010). Organizational sustainability strategies must ensure to meet the short-term needs and obligations while sustaining for the long term as well focusing on profits and economic growth, social justice and inclusion, and environmental sustainability.

An organizational strategy to integrate workers can be that the organization has to develop the right organizational values and offer incentives, and fulfillment of personal perspectives of the workforce framed by a common organizational vision on diversity must reflect the society to create new perspectives on benefits. The formulation and implementation of an organizational sustainability strategy is beyond public and investor relations or environmental compliance and must be mandatory integrated into the fabric of the organizational making and practices to reap the value proposition from all the economic, social, and environmental benefits.

Social and environmental sustainability assessment of the impact of organizational decisions results in incorporating social responsibility practices in sustainable social strategies. All the organizational stakeholders need more clarity on the importance and meaning of green corporate social responsibility initiatives for the organization and the benefits in terms of the environmental sustainability. Green corporate social responsibility is related to social issues and environmental development aimed to improve the impacts of organizational activities on sustainability, such as reducing carbon dioxide emissions, eco-friendly cafeterias and dining halls, planting of trees, etc.

Sustainable strategies of organizations have to be held accountable for their impacts on economic growth and profits, but also on society inclusion and environment sustainability.

4.6 Conclusions

The impact of economic globalization on sustainable organizations has shortened the cycles of change to keep the pace of environmental sustainability through the design and implementation of more holistic organizational and business models, as well as strategic management systems and policies aimed to improve the use of organizational and technological knowledge, processes, and practices. However, the organizational change is not always taking place at the level of the economic structure, the organizational work activities and practices are already changing more toward the development, qualitative growth, and efficiency of the economic system and the global economy.

Organizations facing sustainability challenges need to adopt integral and holistic approaches and develop the talent by recruiting, educating, and training human resources centered on sustainability value based as the source of a sustainable competitive advantage which can leverage sustainability performance targets.

The development of a sustainable organizational vision and mission is facilitated by the incorporation of a sustainable agenda in core values. Shaping the organizational sustainability aligned on the vision and values of the organization must assess the impact on the stakeholders including local communities and consider issues of cultural diversity, intercultural dialogue, and organizational climate and environment to support improved performance. Sustainable organizational development dynamics implement innovation in knowledge and processes in human resource management and development practices aimed to develop infrastructure and increase in knowledge and innovativeness.

The individual abilities in the organization to the need to adapt to environmental changes is the task of developing organizational leadership and managerial effectiveness through adaptive processes in continuous developing stages gathering information, diagnosing, designing interventions, and implementing practices. Organizational leadership must be aligned with the sustainability initiatives besides that playing a modeling role and getting contacted with the external environmental communities and volunteering with nonprofit organizations for organizational and supervisory support. Organizational management is responsible for motivation and communication to set what is expected of individuals on the workforce and how to improve the functions of the organization that encompass the community and society as a whole. Communication is an intrinsic organizational competency and ability to retain the workforce.

Organizational sustainable development supports and concentrates more on regenerative and collaborative working processes to develop embedded organizational roles, practices, and tasks in the organization. To create collaborative and regenerative work enables sustainable individual and organizational development to be characterized in the post-bureaucratic working practices and their influences on the organizational structures, behaviors, and technological processes, during a long period of time.

Implementing sustainable organizational practices must take into account the current economic, social, and environmental needs as well as the future needs of all the stakeholders involved in sustainability to develop the organizational strategy, able to transform and make it happen.

4.7 Cross-References

- Agent-Based Change in Facilitating Sustainability Transitions: A Literature Review and a Call for Action
- Bio-Economy at the Crossroads of Sustainable Development
- Collaboration for Regional Sustainable Circular Economy Innovation
- Contributions of Urban Agro Ecological Agriculture to Ecosystem Services
- Ecopreneurship for Sustainable Development: The Bricolage Solution
- Environmental Intrapreneurship for Engaged Sustainability: Challenges and Pitfalls
- Environmental Stewardship: Pathways to Community Cohesion and Cultivating Meaningful Engagement
- From Environmental Awareness to Sustainable Practices: A Case of Packaging-Free Shopping
- Global Warming Impact on Quality of Life

- Green Economic Growth based on Urban Ecology and Biodiversity
- Oceans and Impasses of "Sustainable Development"
- People, Planet, Profit: Training Sustainable Entrepreneurs at the University Level
- Responsible Investing and Corporate Social Responsibility for Engaged Sustainability: Managing Pitfalls of Economics without Equity
- Responsible Investing and Environmental Economics: Green Finance and the Transition to a Green Economy
- Smart Cities: New Urbanism and New Agrarianism as a Path to Sustainability
- Strategic Management Innovation of Urban Green Spaces for Sustainable Community Development
- Strategic Organizational Sustainability
- Sustainable Decision-Making: Moving Beyond People, Planets, and Profits
- Sustainable Living in the City: The Case of an Urban Ecovillage
- Teaching Circular Economy: Overcoming the Challenge of Green-washing
- The Spirit of Sustainability
- The Sustainability Summit: Using Appreciative Inquiry to Engage the Whole System
- The Theology of Sustainability Practice: How Cities Create Hope
- To Be or Not to Be (Green): Does Communication Experts' Environmental Sensitivity Affect their Marketing Communication Plans?
- Transformative Solutions for Sustainable Well-Being
- Urban Green Spaces as a Component of an Ecosystem

References

Androniceanu, A. (2009). New public management model based on an integrated system using the informational and communication technologies. *Administration and Public Management Review, 13*, 13–25. https://doi.org/10.1080/12294659.2009.10805137

Balmer, J. M., & Gray, E. R. (1999). Corporate identity and corporate communications: Creating a competitive advantage. *Corporate Communications: An International Journal, 4*(4), 171–177. https://doi.org/10.1108/EUM0000000007299

Bandura, A. (2002). Social cognitive theory in cultural context. *Applied Psychology. An International Review, 51*(2), 269–290. https://doi.org/10.1111/1464-0597.00092

Barley, S. R., & Tolbert, P. S. (1997). Institutionalization and structuration: Studying the links between action and institution. *Organization Studies, 18*(1), 93–117. https://doi.org/10.1177/017084069701800106

Bennis, W. (1994). *On becoming a leader*. Addsion-Wesley.

Boselie, P., Paauwe, J., & Jansen, P. (2001). Human resource management and performance: Lessons from the Netherlands. *International Journal of Human Resource Management, 12*(7), 1107–1125. https://doi.org/10.1080/09585190110068331

Brulin, G., & Svensson, L. (2012). *Managing sustainable development programmes: A learning approach to change*. Gower Publishing.

Bryman, A., Collinson, D. L., Grint, K., Jackson, B., & Uhl-Bien, M. (2011). *The Sage handbook of leadership*. Sage.

Ciocoiu, N. C. (2011). Integrating digital economy and green economy: Opportunities for sustainable development. *Theoretical and Empirical Researches in Urban Management, 6*(1), 33–43.

Colesca, S. (2010). Online consumer. Theories of human relativism. *Amfiteatru Economic Journal, 12*(28), 694–699.
Collins, C. J., & Clark, K. D. (2003). Strategic human resource practices, top management team social networks, and firm performance: The role of human resource practices in creating organizational competitive advantage. *Academy of Management Journal, 46*(6), 740–751.
Commission on Environment and Development. (1987). World Commission on Environment and Development (Brundtland Report). [Cited 2010 May 14]. Available from: http://www.ace.mmu.ac.uk/eae/Sustainability/Older/Brundtland_Report.html
Crossan, M. M., & Berdrow, I. (2011). Organizational learning and strategic renewal. *Strategic Management Journal, 24*(11), 1087–1105. https://doi.org/10.1002/smj.342
Cummings, T. G., & Worley, C. G. (2014). *Organization development and change*. Cengage Learning.
Daily, B. F., & Huang, S. C. (2001). Achieving sustainability through attention to human resource factors in environmental management. *International Journal of Operations & Production Management, 21*(12), 1539–1552. https://doi.org/10.1108/01443570110410892
Daily, B. F., Bishop, J. W., & Massoud, J. A. (2012). The role of training and empowerment in environmental performance: A study of the Mexican maquiladora industry. *International Journal of Operations & Production Management, 32*(5), 631–647. https://doi.org/10.1108/01443571211226524
Daly, H. E. (1996). 14 sustainable growth: An impossibility theorem. In *Valuing the earth: Economics, ecology, ethics*. Beacon Press.
Davis, M. C., & Challenger, R. (2013). Environmentally sustainable work behaviors. In P. C. Flood & Y. Freeney (Eds.), *Wiley encyclopedia of management: Organizational behavior* (Vol. 11, 3rd ed.). Wiley.
Deal, T., & Jurkins, W. (1994). *Managing the hidden organization strategies for empowering your employee* (pp. 24–38). Oxford University Press.
Docherty, P., Forslin, J., Shani, A. B., & Kira, M. (2002). Emerging worksystems: From intensive to sustainable? In P. Docherty, J. Forslin, & A. B. Shani (Eds.), *Creating sustainable work systems: Perspectives and practices* (pp. 3–14). Routledge.
Dombrowski, C., Kim, J. Y., Desouza, K. C., Braganza, A., Papagari, S., Baloh, P., & Jha, S. (2007). Elements of innovative cultures. *Knowledge and Process Management, 14*(3), 190–202. https://doi.org/10.1002/kpm.279
Egbu, C.O., Hari, S., & Kumar, B. (2004). Knowledge capture in small and medium enterprises in the construction industry: Challenges and opportunities. In *20th AAnnual ARCOM Conference, Edinburgh*, 1–3 Sept. 847–855.
Elving, W. J. (2005). The role of communication in organisational change. *Corporate Communications: An International Journal, 10*(2), 129–138. https://doi.org/10.1108/13563280510596943
Erez, M., & Early, C. (1993). *Culture, self-identity and work* (pp. 310–316). Oxford University Press. https://doi.org/10.1093/acprof:oso/9780195075809.001.0001
Etzkowitz, H., & Leydesdorff, L. (2000). The dynamics of innovation: From national systems and 'mode 2' to a triple helix of university–industry–government relations. *Research Policy, 29*(2), 109–123. https://doi.org/10.1016/S0048-7333(99)00055-4
Fairfield, K., Harmon, J., & Behson, S. (2011). Implementing sustainability strategies: An integrative model. *Organization Management Journal, 8*(1), 1–17. https://doi.org/10.1057/omj.2011.3
Flyvbjerg, B., Bruzelius, N., & Rothengatter, W. (2003). *Megaprojects and risk: An anatomy of ambition*. Cambridge University Press. https://doi.org/10.1017/CBO9781107050891
Freeman, R. E. (1984). *Strategic management: A stakeholder approach*. Pitman.
French, W., & Bell, C. H., Jr. (1973, 1984). *Organization development: Behavioral science interventions for organization improvement*. Prentice-Hall.
Garrido, A., & Martos, M. (2016). Influence of customer quality perception on the effectiveness of commercial stimuli for electronic products. *Frontiers in Psychology, 7*, 336–342.

Han, S. H., Kim, D. Y., Jang, H. S., & Choi, S. (2010). Strategies for contractors to sustain growth in the global construction market. *Habitat International, 34*(1), 1–10. https://doi.org/10.1016/j.habitatint.2009.04.003

Harmon, J., Fairfield, K. D., & Wirtenberg, J. (2010). Missing an opportunity: HR leadership in sustainability. *People and Strategy, 33*(1), 16–21.

Haslam, S. A. (1997). Stereotyping and social influence: Foundations of stereotype consensus. In R. Spears, P. J. Oakes, N. Ellemers, & S. A. Haslam (Eds.), *The social psychology of stereotyping and group life* (pp. 119–143). Blackwell.

Heckscher, C. (1994). Defining the post-bureaucratic type. In C. Heckscher & A. Donnellon (Eds.), *The post bureaucratic organization: New perspectives on organizational change*. Sage. https://doi.org/10.1002/smj.461

Hee-Jae, C., & Pucik, V. (2005). Relationship between innovativeness, quality, growth, profitability, and market value. *Strategic Management Journal, 26*(6), 555–575. https://doi.org/10.1002/smj.461

Hill, R. C., & Bowen, P. A. (1997). Sustainable construction: Principles and a framework for attainment. *Construction Management and Economics, 15*(3), 223–239. https://doi.org/10.1080/014461997372971

Howard, G. (1995). *Between justice and beauty: Race, planning, and the failure of urban policy in Washington, D.C.* Johns Hopkins University Press.

Ichniowski, C., Shaw, K., & Prennushi, G. (1997). The effects of human resource management practices on productivity: A study of steel finishing lines. *The American Economic Review, 87*(3), 291–313.

IMF. (2017). *World economic outlook database*. International Monetary Fund. April 2017. Retrieved from: http://www.imf.org/external/pubs/ft/weo/2017/01/weodata/weorept.aspx. Accessed 29 March 2018.

James, L. R., Choi, C. C., Ko, C. H. E., McNeil, P. K., Minton, M. K., Wright, M. A., & Kim, K. I. (2008). Organizational and psychological climate: A review of theory and research. *European Journal of Work and Organizational Psychology, 17*(1), 5–32. https://doi.org/10.1080/13594320701662550

Jennings, P. D., & Zandbergen, P. A. (1995). Ecologically sustainable organizations: An institutional approach. *Academy of Management Review, 20*(4), 1015–1052. https://doi.org/10.5465/amr.1995.9512280034

Jiang, K., Lepak, D. P., Hu, J., & Baer, J. C. (2012). How does human resource management influence organizational outcomes? Ameta-analytic investigation of mediating mechanisms. *Academy of Management Journal, 55*(6), 1264–1294. https://doi.org/10.5465/amj.2011.0088

Kanter, R. M. (2008). *Men and women of the corporation*. New edition. Basic books.

Leana, C. R. (1987). Power relinquishment versus power sharing: Theoretical clarification and empirical comparison of delegation and participation. *Journal of Applied Psychology, 72*(2), 228. https://doi.org/10.1037/0021-9010.72.2.228

Lindkvist, L., & Söderlund, J. (2002). What goes on in projects? On goal- directed learning processes. In K. Sahlin-Andersson & A. Söderholm (Eds.), *Beyond project management: New perspectives on the temporary- permanent dilemma* (pp. 278–291). Liber.

Macke, J., & Genari, D. (2019). Systematic literature review on sustainable human resource management. *Journal of Cleaner Production, 208*, 806–815. https://doi.org/10.1016/j.jclepro.2018.10.091

Mahal, P. K. (2009). Organizational culture and organizational climate as a determinant of motivation. *IUP Journal of Management Research, 8*(10), 38.

March, J. G. (2006). Rationality, foolishness, and adaptive intelligence. *Strategic Management Journal, 27*(3), 201–214. https://doi.org/10.1002/smj.515

Martin, G. C. (2014). The effects of cultural diversity in the workplace. *Journal of Diversity Management, 9*(2), 89–95.

Mendelson, H., & Pillai, R. R. (1999). Information age organizations, dynamics and performance. *Journal of Economic Behavior and Organization, 38*(3), 253–281. https://doi.org/10.1016/S0167-2681(99)00010-4

Meyer, J. W., & Rowan, B. (1977). Institutionalized organizations: Formal structure as myth and ceremony. *American Journal of Sociology, 83*(2), 340–363. https://doi.org/10.1086/226550

Miyatake, Y. (1996). Technology development and sustainable construction. *Journal of Management in Engineering, 12*(4), 23–27. https://doi.org/10.1061/(ASCE)0742-597X(1996)12%3A4(23)

Moore, S. B., & Manring, S. L. (2009). Strategy development in small and medium sized enterprises for sustainability and increased value creation. *Journal of Cleaner Production, 17*(2), 276–282. https://doi.org/10.1016/j.jclepro.2008.06.004

Morris, P. W. G., Pinto, J. K., & Söderlund, J. (Eds.). (2011). *The Oxford handbook of project management*. Oxford University Press.

Mumford, M. D., Michelle, E., Higgs, T. C., & McIntosh, T. (2017). Cognitive skills and leadership performance: The nine critical skills. *The Leadership Quarterly, 28*(1), 24–39. https://doi.org/10.1016/j.leaqua.2016.10.012

Naudé, M. (2012). Sustainable organizational development and reflection: A good combination? *Corporate Ownership and Control, 9*(2), 364–375.

Nonaka, I., & Takeuchi, H. (1995). *The knowledge- creating company: How Japanese companies create the dynamics of innovation*. Oxford University Press.

Ones, D. S., & Dilchert, S. (2012). Environmental sustainability at work: A call to action. *Industrial and Organizational Psychology, 5*(4), 444–466. https://doi.org/10.1111/j.1754-9434.2012.01478.x

Ortiz, O., Castells, F., & Sonnemann, G. (2009). Sustainability in the construction industry: A review of recent developments based on LCA. *Construction and Building Materials, 23*(1), 28–39. https://doi.org/10.1016/j.conbuildmat.2007.11.012

Oskamp, S. (1995). Resource conservation and recycling: Behavior and policy. *Journal of Social Issues, 51*(4), 157–177. https://doi.org/10.1111/j.1540-4560.1995.tb01353.x

Paauwe, J., & Boselie, P. (2003). Challenging 'strategic HRM' and the relevance of the institutional setting. *Human Resource Management Journal, 13*(3), 56–70. https://doi.org/10.1111/j.1748-8583.2003.tb00098.x

Patterson, M. G., West, M. A., Shackleton, V. J., Dawson, J. F., Lawthom, R., Maitlis, S., Robinson, D. L., & Wallace, A. M. (2005). Validating the organizational climate measure: Links to managerial practices, productivity and innovation. *Journal of Organizational Behavior, 26*(4), 379–408. https://doi.org/10.1002/job.312

Perron, G. M., Côté, R. P., & Duffy, J. F. (2006). Improving environmental awareness training in business. *Journal of Cleaner Production, 14*(6), 551–562. https://doi.org/10.1016/j.jclepro.2005.07.006

Pfeffer, J. (2010). Building sustainable organizations: The human factor. *Academy of Management Perspectives, 24*(1), 34–45.

Postmes, T. (2003). A social identity approach to communication in organizations. Social identity at work. *Developing Theory for Organizational Practice, 81*, 191Á203.

Presley, A., & Meade, L. (2010). Benchmarking for sustainability: An application to the sustainable construction industry. *Benchmarking: An International Journal, 17*(3), 435451. https://doi.org/10.1108/14635771011049380

Prigozhin, I., & Stengers, I. (1986). *Order out of chaos: Man's new dialogue with nature*. Translation from English/Under the general editorship of Arshinov, V. I., Klimontovich, Yu. L., Sachkov, Yu. V. (p. 432). Progress.

Ramus, C. A. (2002). Encouraging innovative environmental actions: What companies and managers must do. *Journal of World Business, 37*(2), 151–164. https://doi.org/10.1016/S1090-9516(02)00074-3

Renwick, D., Redman, T., & Maguire, S. (2013). Green human resource management: A review, process model, and research agenda. *International Journal of Management Reviews, 15*(1), 1–14. https://doi.org/10.1111/j.1468-2370.2011.00328.x

Roberts, K. H., & O'Reilly, C. A. (1974). Failures in upward communication in organizations: Three possible culprits. *Academy of Management Journal, 17*(2), 205–215.

Rondinelli, D. A., & Berry, M. A. (2000). Environmental citizenship in multinational corporations: Social responsibility and sustainable development. *European Management Journal, 18*(1), 70–84. https://doi.org/10.1016/S0263-2373(99)00070-5

Sarkis, J., Gonzalez-Torre, P., & Adenso-Diaz, B. (2010). Stakeholder pressure and the adoption of environmental practices: The mediating effect of training. *Journal of Operations Management, 28*(2), 163–176. https://doi.org/10.1016/j.jom.2009.10.001

Saturnino Neto, A., Chiappetta Jabbour, J. C., & Lopes de Sousa Jabbour, B. A. (2014). Green training supporting eco-innovation in three Brazilian companies: Practices and levels of integration. *Industrial and Commercial Training, 46*(7), 387–392. https://doi.org/10.1108/ICT-02-2014-0010

Scott, W. R. (1992). *Organizations: Rational, natural and open systems* (3rd ed.). Prentice Hall.

Senge, P. (1999). *Changing dance, spellbacks for sustainable change in learner organizations*, translators: Mashayekhi, Ali Naghi et al., Ariana Industrial Research Group, First Edition.

Sev, A. (2009). How can the construction industry contribute to sustainable development? A conceptual framework. *Sustainable Development, 17*(3), 161–173. https://doi.org/10.1002/sd.373

Sims, R., & Veres, J. (2007). *Keys to employee success in coming decades* (pp. 179–264). Quorum Books, Greenwood Publishing Group.

Spence, R., & Mulligan, H. (1995). Sustainable development and the construction industry. *Habitat International, 19*(3), 279–292. https://doi.org/10.1016/0197-3975(94)00071-9

Svensson, L., & Brulin, G. (2013). Effects of programmes and projects. In L. Svensson, G. Brulin, S. Jansson, & K. Sjöberg (Eds.), *Capturing effects of projects and Programmes* (pp. 15–33). Studentlitteratur.

Symposium on Sustainability. (2001). *Profiles in leadership*. Prentice Hall.

Tan, Y., Shen, L., & Yao, H. (2011). Sustainable construction practice and contractors' competitiveness: A preliminary study. *Habitat International, 35*(2), 225–230. https://doi.org/10.1016/j.habitatint.2010.09.008

The human development report. (2014). *Ensuring a sustainable progress of mankind: Decreasing vulnerability and forming resilience*. The entire world.

Tucker, M. L., Meyer, G. D., & Westerman, J. W. (1996). Organizational communication: Development of internal strategic competitive advantage. *The Journal of Business Communication, 33*(1), 51–69. https://doi.org/10.1177/002194369603300106

Tyson, S. (2006). *Essential of human resource management* (5th ed.). Butterworth – Heinemann.

Uzzell, D., & Moser, G. (2009). Introduction: Environmental psychology on the move. *Journal of Environmental Psychology, 29*(3), 307–308. https://doi.org/10.1016/j.jenvp.2009.08.003

Vanegas, J. A., DuBose, J. R., & Pearce, A. R. (1996). Sustainable technologies for the building construction industry. In *Proceedings, symposium on design for the global environment, Atlanta*.

Weick, K. E. (1995). *Sensemaking in organizations*. SAGE Publications.

World Business Council for Sustainable Development. (2005). *Driving success: Human resources and sustainable development*. University of Cambridge Programme for Industry.

Vargas-Hernández is research professor at Instituto Tecnológico Mario Molina Unidad Zapopan, and earlier at University Center for Economic and Managerial Sciences, University of Guadalajara. Member of the National System of Researchers of Mexico. Professor Vargas-Hernández has a Ph.D. in Public Administration and a Ph.D. in Organizational Economics. He has been visiting scholar at Carleton University Canada, University of California Berkeley and Laurentian University, Canada. He holds a Ph.D. in Economics, Keele University; Ph.D. in Public Administration, Columbia University; studies in Organizational Behavior at Lancaster University and has a Master of Business Administration. He has Published 9 books and more than 300 papers in international journals and reviews (some translated to English, French, German, Portuguese, Farsi, Chinese, etc.) and more than 300 essays in national journals and reviews. He has obtained several international awards and recognitions. He has also experience in consultancy. His main research is in organizational economics and strategic management. He teaches for several doctoral programs.

Chapter 5
The Role of Indian Culture in Enlightening the Notions of Environmental Stewardship and Sustainable Living Mechanism

Nidhi Kaushal

Abstract Environmental stewardship is not a contemporary practice but a human obligation, which has been explained through significant poetic works, mythological epics, folklore, and the framework of ecological-spiritual feelings as the duty of humans toward nature for centuries. Ancient sages have exquisitely classified the empirical knowledge of making harmony with life by caring for flora, fauna, and *Prakriti* (Mother Nature) in their treatises and ethnic methodologies and have preserved the wisdom of moral practices toward maintaining the Earth's sustainability through the cultural rituals, folktales, songs, poems, festivals, and devotional observances. This work is to identify the classical facts of environmental stewardship and sustainable living and asserts the feasibility of Indigenous techniques as well as their applicability for saving nature in the diversely changing modern era. It also gives an implication of noble approaches of the seers or intellectuals as ecologists through the ideology of reviewing the old concepts. Moreover, it advances an attitude of reverence for nature-oriented cultural reforms, epic endeavors, and narrative endowments.

Keywords Cultural reforms · Ecology · Flora and Fauna · Panchagavya · Poetry · Prakriti · Reverence · Stewardship · The Cow · Yajna

5.1 Introduction

Nature's legislation is growth and development (Bharti, 1995), and its diversity has fascinated humans from ancient times (Sheshadri, 2016). Human beings are predominantly related to the sustainable development of the environment (Krishnamoorthy, 2017), because their prosperity, wellbeing, and development

N. Kaushal (✉)
Leadership Practitioner/Researcher, Yamunanagar, India

© The Author(s), under exclusive license to Springer Nature Switzerland AG 2023

S. Dhiman (ed.), *Sustainable Development and Environmental Stewardship*,
https://doi.org/10.1007/978-3-031-28885-2_5

are contingent upon natural resources to provide the goods and services, which directly or indirectly generate socio-economic benefit (Sasidharan, 1998). Earth's stewardship has emerged from a praxis, which is ancient, contemporary, collective, and individual (Rozzi et al., 2015). The concept of Earth as Mother Nature for sustainment of the human race has been deeply rooted in ancient Indian thought and culture (Misra, 1995), and the Indian concept of environment stewardship refers to following the ideology of *Vasudhaiva Kutumbakam*, which means a person should consider the entire universe as his or her extended family, and all the living organisms or creatures are members of its family (Dwivedi, 2016), and its notion provides the perspective of the ancient tradition of reverence and harmony with nature. The primary goal of environmental learning is to develop nature-oriented knowledgeable citizens who are involved in working towards a more liveable future (O'Brien & Stoner, 1987).

Nature's invaluable wealth is a gift to human life, and its flora and fauna are the frameworks of the existence of humanity. Humans have always been indebted to plants and trees, which fulfil all their requirements in return for a little maintenance, care, and attention without any monetary gains. Sensibility and awareness to preserve nature have defined sustainable living, and the measures which have been taken to protect it are collectively known as environmental stewardship. It is one such social and personal aspect that plays a crucial role in inflicting the sensibility of the obligation for protecting the nature of all beings. It is a practice related to the conservation and use of natural resources with responsibility. Legendary litterateurs and ancient sages, through their writings and knowledge, have considered the conservation and enhancement of the environment the most, and described in various ancient *Puranas* and *Nitishastra* through education rendering and awareness. They have considered the cow as an integral part of environmental sustainability because the use of *Panchagavya* (a kind of product made up of five items) obtained from it represents the perspective of purifying nature and a unique example of connecting its preservation along with reverence and religious sentiment towards the cow.

Scholarly poets devoted to nature have drawn special attention to environmental safety, value, and awareness through their work. The universalism of their poems is beneficial in advancing the trend of people towards the environment in the changing scenario of present environmental conditions and also presents a holistic approach. Likewise, legends or stories presenting the importance of ecology have a subject matter of interest and curiosity enduringly, which act as a guide for ecologists or common naturalists to understand the importance of plants and trees, their benefits, and the need for existence. Due to rising concerns about environmental hazards, there is a requisite for redeveloping and identifying the ancient old traditional knowledge of preserving nature and reviving its glory. Identically, some key aspects of the earth's stewardship based on ethnic approaches, and primordial wisdom of Indian treatise and folk culture, are recognized through this chapter and demarcated into five important categories: (1) Relevance and

Significance of Indian Poetic Literature to the Environment; (2) The Consecrated Aspect of Earth or Prakriti (Mother Nature) and Its Importance in Ancient Indian Civilization; (3) Role of the Cow and Yajna-Ritual in Conserving the Environmental Sustainability Cycle; (4) Contribution of Stories to Awaken the Spirit of Stewardship Along with an Emotional Connection to Ecology; and (5) The Primacy of Environmental Conservation in Ancient Indian Folks and Their Reverence for Plants and Trees.

5.2 Literature Review

Every aspect of nature is sacred to Indic religions, such as forests, grooves, gardens, rivers, plants and seeds, animals, mountains, and other water bodies (Krishna, 2017). The term natural is often equated with "eco-friendly," which implies a rigid divide between humans and the rest of nature, thereby obscuring the many interconnections between all living things, humans included (Crowley, 2010), and the word environmental stewardship has defined various activities like planting trees, developing community gardens, and reducing waste (Bennett et al., 2018). The nature conservation conception views restoration and preservation of habitats and ecosystems as ends in themselves, irrespective of economic and social consequences, while the sustainable development conception views the protection of habitats and ecosystems as a means towards the fulfilment of economic and social objectives laid down by society (Gray & Hatchard, 2007). Leopold (1949) observes that when people see land as a community to which they belong, they use it with love and respect because this kind of perspective develops reverence for it. They get spiritual happiness by interacting and mixing life with nature because this practice has developed and strengthened their personalities (Bharti, 1995).

Literature examines accepted values to identify a distinct perspective (Misra, 2003), and Rudra Dhar Mahamahopadhyay (a Sanskrit scholar) has stated in his text Sraddha Vivek (1872) that the collection of items of equivalent quality is called literature (Awasthi, 1998). The representation of Indian poetry includes the study of Indian poetics along with dramaturgy and prosody as closed disciplines (Winternitz, 1985). Kavi (the poet) is an illustrator of life who draws inspiration from the world (Premshankar, 2008) because he makes use of irony as a mode of perception to expose the hollowness of rituals and religious practices (Hivale, 2019). Kālidāsa has been considered a contemporary poet of King Vikramaditya of the first century and counted among the Navaratnas of his court (Joshī, 2005). He has been recognized as a man of sound and rather extensive education and preserves his intellectual balance and spiritual initiative in his works (Ryder, 1920). He endeavoured propriety in his play's character's utterances (Keith, 1992), and his *Meghadūta* is the most beautiful and poignant short poem in all Sanskrit literature (Majumdar and

Chauhan, 1984). He visualized the natural form of the famous six seasons in India in his first work *Ṛtusaṃhāra* (Upadhyay, 2015).

Sir William Jones (1746–1794), who worked as the Chief Justice of the court of Bengal state of India in 1973, after reading the works of Kālidāsa declared him as Shakespeare of India (Ārasu & Kumaran, 1995). Kālidāsa has more fancy and imagination than the classical style of poetry, while Bhavbhuti has a more sentimental and passionate approach in his creative works (Bhavabhūti & Kāle, 1997). Indian poet Suryakant Tripathi Nirala has an easy alliance of compassion and prowess (Sharma, 2009), while Gangadhar has been known as the poet of nature (Prakriti Kavi) (Nayak, 2010) and represented Prakriti as a friend, philosopher, and guide of man. In sharp contrast to Rousseau's naturalism, Gangadhar has a socialized approach to nature (Panda, 2008). Jayant Mahapatra's poetry acknowledged several landscapes and represented Puri (a city in the Orissa state of India) on the sea as a real-life character (Padihari, 2007).

The *Purānic* culture is a composite picture of the Hindu religion, which has also modified its philosophy (Saran & Pandey, 1992). Similarly, ancient Sanskrit scholar Panini (fourth century BC) described several beautiful trees in *Ashṭādhyāyī*, Shudraka (100 BC) gave an account of gardens and flowers in *Mrichhakatikam*, and Sarangadhara (AD 1300) has described the art of gardening in his texts *Upavana Vinoda* and *Sarangadhara Paddhati* (Valsalakumari, 2008). The place of horticulture in ecology, both at the home garden and the field scale, has been well explained in *Vrikshayurveda* (Sadhale & Nene, 2009), and it also includes instructions for the organization, cultivation, and preservation of gardens (Ramachanran, 1984). Prithvi (the Mother Earth) sustains plants' life in all its multitudinous variety (Vatsyayan, 1992), and in many ancient Indian texts, trees have been considered a symbol of hospitality for visiting guests of the hermitage (a place of penance) (Sheshadri, 2015). Yajna performed through cow's ghee nurtures the environment, removes dangerous pollutants, and makes the Earth's elements pure. The action of the Sun evaporates the waters into the clouds that bring the rain that causes to sprout the flora, which sustains animals and men (Chaitanya, 1992). For rains, the fascinating perspective of the climate given in the ancient Vedic texts like the cloud is also sculpted as a cow by the seers because the cloud in the form of a cow gives birth to a calf in the form of lightning (Shastri, 2014).

Vedic scriptures teach that trees must be protected, as must all species which live on the land (Prime, 1996), and Yoga Sutra attributes direct moral standing to animals and plants (Framarin, 2014). The *Śrīmad Bhagavad Gītā* has been summarized as the expression of metaphysical knowledge and also analysed as an approach to associating the individual, not with the end physical world, but with its relation to the metaphysical (Dangwal, 1999), and it is presumably already plain enough that the *Upanishads* teach the fictitious and unreal nature of the world. The fictitious character of the world of semblances has everywhere implied in the doctrine of the sole existence of the impersonal self (Gough, 2013). The *Upanishads* record the philosophical speculations regarding the beginning and the end of the world, the transmigration of the soul, and the ultimate unity of the individual with the World Soul (Sharma, 1985, p. 1).

The ecosystem is a fundamental ecological unit that refers to associated species of living organisms in a non-living physical environment and the structural and functional relationships among them (Debnath, 2003), and an ecologist studies forms of life not in isolation but as parts of a system, an economy that sustains them and that they constitute (Kerridge, 2001). Ecocriticism studies the relationship between nature and literature, and its fundamental premise is that human beings are deeply influenced by their environment, which consists of both natural and cultural aspects (Baindur, 2015). Ethics and motivation of normative actions are treated in terms of rationalized moral principles, attitudes, preferences, and traits (Enqvist et al., 2018). Environmental ethics is the code of human behaviour regarding nature, which ensures progress without jeopardizing the ecological balance (Misra, 1995), and these ethics, as propounded by ancient Hindu scriptures and the seers, have been practised by common men, rulers, and kings, and they observed these fundamentals as religious duties, rules of administration, or obligation for law and order (Dwivedi & Tiwari, 1999). The crisis mind sees the forest and its trees as weeds, valued commercially, and converts even afforestation into deforestation and desertification (Shiva, 1999), and environmental degradation has been identified as a major impediment to agricultural growth (Sasidharan, 1998). Indigenous societies have an understanding of the fleeting and transient nature of human life on the evolutionary and geological scale (Strong, 1992). Ethical values, traditions, metaphors, poetic creativity, contemporary and vernacular ecological knowledge, ecology, institutional networks, local communities, and exemplary lives complement the scientific perspectives of activities towards Earth stewardship (Rozzi et al., 2015).

The organization's perspective of environmental stewardship fosters eco-friendly material for production, which is relevant to technical strategies, business models, consumer preferences, and policy instruments (Worrell et al., 2016). Bamboo has great advantages of yield to the environment, and through its plantation, the rate of deforestation has been reduced because it grows instinctively without much care (Atanda, 2015). It is a multipurpose material (Srivastava, 2008), and its shoots have been used in the treatment of many diseases, as recommended in Ayurveda (Sarkara et al., 2015); due to its fair resistance to diseases (Rosulu et al., 2019), it is a substitute for plastics and structural other materials (Gupta & Kumar, 2008). Tulsi and Neem plants have medicinal properties as well. Tulsi plant has healing properties and auspicious values (Cohen, 2014), while Neem has a scope in reforestation and hygienic uses (Tomar & Singh, 2009).

Nature is perceived by sensory awareness, and rituals liturgically express people's relationship with environmental sustainability (Singh, 2012b). The Hindu and Buddhist myths are cosmic myths based on the cyclic view of time and belong to cosmic revelation (Saigal, 2008). Environmental awareness develops learning opportunities for people to make connections between environmental knowledge, socio-scientific issues, and their life, and it includes raising the issue of aesthetic awareness along with the ethical values associated with such awareness (Hadzigeorgiou & Skoumios, 2013). Environmental initiatives and practices must follow precautionary movements (Mawdsley, 2006). Facts and ideas about the natural environment and their interrelationship with human life embedded in the plot of the story are very effective for making people aware of nature's value and respect (Hadzigeorgiou & Judson, 2017). The law of Deuteronomy for protecting the environmental world advocates that it must require justice for the environment, for all creation, as it does for the human beings who live within the environment, and it demands a life lived in daily mindfulness of the gifts of creation (Hilbert, 2001).

The effectiveness of any religion in protecting the environment depends upon the faith of its believers in its precepts and injunctions (Dwivedi, 1993). In India, most of the tribal communities are nature worshippers. They protect the environment considering it as their responsibility towards their ancestors because they believe that their ancestors lived in nature and resided in natural objects (Bain, 2017), and its example is the Chipko movement by the local people of India, a non-violent protest to oppose the felling of timber from outside contractors (Crowley, 2013). Affinity or cultural ecofeminism theorizes the psychological perspective that women have a caring attitude toward nature's nourishment and nurturance (Curry, 2013). Indigenous traditions acknowledge the need to conserve and preserve their respective environments (Fergusson et al., 2017). The fabric of the cultural landscape of the Garhwal Himalaya has been woven with the multiracial elements of culture, and deities of both genders are worshiped in these landscapes (Pala et al., 2014), and these mountains have dense forests and a rich ecosystem (Manjrekar, 2017).

5.3 Relevance and Significance of Indian Poetic Literature to the Environment

The role of poets in the objective of making people environmentally conscious has a remarkable approach because the views of the unfathomable beauty of nature have always been the inspiration of the poets, in which they have preached the preservation and protection of the environment, stating the importance of the environment. Caube (1994) observes that nature is a reflection of human emotions. The language of myth and poetry, of the concrete imagination, engages the senses, the feelings, the affections, and the will as well as the reason, and so leads to the transformation (Saigal, 2008). India has indicated its respect for the environment, as represented in the Vedic and post-Vedic texts, including additional literary writings of Kālidāsa (first century BCE), Bāṇabhaṭṭa (seventh century), Bhavbhuti (eighth century) that have recognized the significance of nature (Chapple, 1998). Their poetry shows a deep understanding of the environment, and these poets have made nature a partner in sensing humans in their way (Misra, 2003).

Sanskrit is a highly inflected language (Rajan & Kalidasa, 2005), and Sanskrit poetry known as Kavya is not artificial, but it includes a learned approach because the real poet has studied most of the different sciences (Winternitz, 1985), which includes the portrayal of flora and fauna (Sheshadri, 2017). Kālidāsa's knowledge of nature is not only sympathetic, but it is also minutely accurate. Not only are the snows and the windy music of the Himalayas, but the mighty current of the sacred Ganges; his too are smaller streams and trees and every littlest flower (Ryder, 1914, p. 19). He has been called the best poet of simile due to his deep experience of life and rites of melody, compassion, and glee, and his simile method is not just decking (Prasāda, 1995), and glimpses of the life of India and nature are enduring features of his creations (Ārasu & Kumaran, 1995). He personifies rivers, mountains, and tress, and to him, they have a conscious individuality as truly and as certainly as animals or men or gods (Ryder, 1920). His unique creative work *Meghadūta* is a dynamic transformation of the Earth's aspiration (Misra, 2003), which represents the graphical descriptions of nature, along with topographical details of ancient India (Chaudhari, 2017), and Himalayan grandeur and sublimity attracted him very deeply (Ryder, 1920).

Kālidāsa has presented the latest form of illustration of the change of seasons as well as the transitional form of nature in his poetry. He has described six seasons in *Ṛtusaṃhāra*, and this description provides evidence of his subtle observation and love of nature (Majumdar & Chauhan, 1984). The rhythmic presentation of the behaviour of all fauna and humans according to the change in the season of nature has been given in ancient Indian literature (Sheshadri, 2014). The epigraphy has described the six seasons starting from the Grisham (summer) to Vasant (spring) Ritu, which include Grisham, Pavas, Sharad, Hemant, Shishir, and Vasant Ritu, respectively (Upadhyay, 2015). Folk songs are used in three ways, first as Ritu Mangal like spring festival (Vasantotsav), Rain festival ((Varshotsav), and as a season song (Ritu Geet) on the occasion of autumn (Sharad Ritu); second as in the form of Mangal Geet on the occasion of ceremonies; and third as the form of song of the work (Shram Geet) (Mishra, 2009). He is the only Sanskrit poet who has firstly described and introduced the 'saffron flower' that grows in the Kashmir valley of India (Ryder, 1920). His works have a relevant and intellectual approach to the study of ecology and sustainable living (Sahasrabuddhe, 2017).

A Sanskrit play is known as *drsya-kavya*, a spectacle poem (Vidyākara et al., 1968), and in folk literature, birds and beast are implied in the novel *Kadambari* to play human roles, which is one of the creative works of ancient Indian poet Bāṇabhaṭṭa that includes parrot's talk to make the content impressive (Krishnamoorthy, 1982). Like Kālidāsa, Bhavbhuti has also admired nature because he has roamed in the *Dandakaranya* (Dandaka forest), near to his birthplace, and described its hills and dwells, the streams and cataracts graphically in the *Uttararāmacarita* (a Sanskrit play) (Mirashi, 1996), and this play has signified his clarity of vision and intellectuality (Keith, 1992). The name of Bhavbhuti in

Sanskrit literature is the highest in the eloquence of expression and sublimity of imagination (Borooah, 1878). As a poet of nature and a describer of the picturesque, he has represented various kinds of ancient trees like Bakula (*Mimusops elengi*), Kadamba (*Neolamarckia cadamba*), Matulunga (*Citrus medica*), and others in his writings, along with their universal significance and diligence (Bhavabhūti & Kāle, 1997).

Poetry survives the poets because of its timeless and intrinsic value (Kumar, 2018). Scientific knowledge of nature and an awareness of ecology notify humans through poetry, which emphasizes nature's impressive strength rather than its beauty (Daruwalla, 1999). Several other Indian contemporary poets like Rabindranath Tagore, Suryakant Tripathi Nirala, Gangadhar Meher, Jayant Mahapatra, and Sarojini Naidu have also created awareness regarding sustainable living and performed their duty of environmental stewardship through the poetical works. Aesthetes Kālidāsa and Rabindranath Tagore believed in the absolute power of nature, which they have reflected through their writings (Sharma, 2008). Rabindranath Tagore (1861–1941) exercised a remarkable influence on the style and imagery of the new poetry. His works like *Gitanjali*, *The Gardener*, *Stray Birds*, and *The Crescent Moon* popularized certain modes of utterances like the devotional lyric, epigram, and poetic drama. As a poet, philosopher, dramatist, essayist, short story writer, novelist, and critic, he has given literary expression to the renaissance in all its variety, phases, and opulence (Gokak, 1960).

Suryakant Tripathi Nirala (1896–1961), an Indian poet, novelist, essayist, and story writer, has reflected his love for nature and its importance in his various compositions like *Bela*, *Naye Patte*, *Juhi Ki Kali*, and several other works. He has revived the philosophical thinking of the *Rigveda* through his compositions, and his creations are based on emotions, and this sentiment is related to his ideology (Sharma, 2009). Oriya poetry has a tradition of robust realism (Gokak, 1960), and the poet from the Orissa region of India, Gangadhar Meher (1862–1924), has considered being a staunch naturalist and describes nature as a living spirit like all other human characters of his literature (Panda, 2008). Nature imbibes its comely, gracious, fierce, tranquil, and auspicious forms in various contexts. With his poetic insight, he sees human feelings, conscious life, and internal beauty in nature, and vividly and exhaustively delineates its beautiful facets (Nayak, 2010).

Sarojini Naidu (1879–1949) has drawn inspiration for her creative works from the world of flora and fauna (Hivale, 2019), and has felt a heavenly delight in describing nature (Nagendra & Reddy, 2018). Her poem 'Harvest Hymn' from the book *The Golden Threshold* (1905) presents a group of men and women signing a hymn of adoration and thanksgiving to the four great deities responsible for their rich harvest—Surya, the Sun God; Varuna, the God of rain; Prithvi, the Goddess of Earth; and Brahma, the creator of all life in the universe (Kaur, 2003, p. 126). This poem is about expressing gratitude to the Sun by the farmers during the harvesting season (Naravane, 1996). Flowers are everywhere in her poetry, and the poem 'Spring' from the book *The Bird of Time: Songs of Life, Death & the Spring* (1912) depicts beautiful pictures of the season (Nagendra & Reddy, 2018) and also provides a vivid description of songs of life, nature, especially springtime. Her approach to environmental stewardship through her literary works has significantly influenced the individuals for sustainable living.

Contemporary poet Jayant Mahapatra is a true poet of the soil and a gifted poet fascinated by the Indian heritage, myths, and legends (Padihari, 2007). Nature does not have a voice to speak like humans, but it communicates by showing the changes in its creatures, seasons, flora, and fauna, which get affected through human negligence. He has reflected the turmoil created by people's harsh activities in the environmental domain through his poetry, and his poems such as 'Bone of time' from the Book *A Whiteness of Bone: Poems* (1992), and 'Summer' and 'Dawn at Puri' (Orrisa landscape) in the book *A Rain of Rites* (1976) have depicted the worth of nature. These poetic works are very effective in awakening the consciousness of the masses because of their simple language and meaning, the sensation of people about nature has been evoked, and the importance of nature has been identified through ages.

5.4 The Consecrated Aspect of Earth or Prakriti (Mother Nature) and Its Importance in Ancient Indian Civilization

The responsibility for the protection of nature is not only that of an individual, but of all those who live in a natural environment, and this fact is not related to contemporary issues but has been given in the *Vedas* and *Puranas* as a guideline. Nature is our life, and it protects us like a mother and gives all the essential commodities that people require for sustainable living. Every living organism of nature consists of five major elements, which are described as a doctrine of *Pancha Maha-Bhuta* in Vedic texts and recognized as the constant element (Prithvi), volatile element (Vaayu), cold element (Jal), warm element (Agni), and comprehensive plus united element (Aakash). Dwivedi and Tiwari (1999) identify that the traditional attitude of Hindu culture towards the ecosystem has been kind and respectful, and animals and birds are not only given respect, but they also received reverence, and worship of trees and plants has been based partially on utility, but mostly as a religious duty. Ghosh (2003) finds that the religious beliefs and values, as enshrined in the *Vedas* and *Upanishads*, constitute the ethics of Hinduism, and this religion has been noted for its deep respect for all forms of nature and the unique role that each life form plays in the ecology of the earth (Krishna, 2017).

Ancient Indian culture has the commingling of cultural elements, from the north to south, and from the east to west. The Aryan elements of the ancient period have been equated with the *Vedic* and *Purānic* culture of the north and the pre-Aryan with the Dravidian and Tamil culture of the south, and each culture has its orientation towards environmental sustainability and growth (Sharma, 2006). The *Puranas* are Indian myths, and their stories are classical legends narrating Indian religion, philosophy, Sanskrit, and social life (Narsimhachari, 2014), because they have recorded the measures to understand the world (Misra, 2003), and notions of dharma are communicated and narrated through relevant epical fictions since ancient period (Narayanan, 1997). From the *Vedic* age through the *Upanishadic* to the *Purānic* age, there developed a chain of mythologies that have been preserved in various forms of religious symbols. While claiming no direct link with modern theories, these symbols define the concept of human experience and wisdom (Singh, 2012a).

The tradition of depicting nature in nature has been going on for a long time (Caube, 1994), and in almost all the texts of Sanskrit literature, there is an inscrutable precept of environmental protection and reflects a commitment to it (Shukla, 2018). The *Upanishads, Vedas,* and *Puranas* abound in moral lessons that the ultimate responsibility for preserving the earth's resources rests on humankind (Ghosh, 2003). *Puranas* are not manuscripts for ritual activities, but they have contained the ocean of wisdom to live life with morality. For saving life on Earth, at the time of the cataclysm, environment sustainability has given prominence, which has indicated in the ancient Indian treatise *Matsya Purana*, that Lord Vishnu has instructed King Satyavrata to preserve the sampling of some vegetation, grain seeds, and other flora, concerning the rehabilitation of human civilization and development of nature.

Earth is known to all civilizations and cultures as the great Mother Goddess, and the Vedas dedicate many hymns to Bhumi, Earth or Prithvi, and the *Prithvi-Sukta* is one of the greatest hymns (Vatsyayan, 1992). Its seer is the Vedic sage Atharva, and he has explained the eulogy of both the physical and spiritual forms of the Earth and prayed for the best blessing of nature's Excellency heartily. It deals with the Earth as our benefactress from the beginning of history and has its social and cultural roots grounded in the earth (Chamola, 2007). One of the shlokas of the *Prithvi-Sukta* has been given here:

> *Vishvasvam Maataram-Ossadhiinaam Dhruvaam Bhuumim Prthiviim Dharmannaa Dhrtaam* |
> *Shivaam Syonaam-Anu Carema Vishvahaa* ||*17*||

'We all worship the strong, extended, wellbeing nature of the goddess Earth, including its all-pervasive form of wealth, which generates all grains and pulses, and has the religious observance.' *Suktas* or *Samhitas* explicit that human beings pray all natural objects, to bless them with prosperity by being sweet and kind, and *Vedas* provide many shreds of evidence to culminate that how nature remains a constant companion of human beings, and they have offered their homage to nature as a deity (Patil, 2019). The natural and human aspects of the Sun as God has been elaborated in the *Vedas*, and saints and sages have maintained the Sun worship mandatory for human flourishing, which has been given in the mythological epics (Saran

& Pandey, 1992). Vedic traditions teach that Surya the Sun is a manifestation of ancient yet always young God Agni (Fire), who is the energy of the universe, including life energy and associated with water for the creation and maintenance of life (Vannucci, 1992). Ancient Indian mythological text *Brahma Purana* has declared the Sun god as the cause of all activities of the world because, according to its textual composition, he is the creator of day, night, months, seasons, fruits, vegetables, food, flowers, and medicines (Saran & Pandey, 1992). It is the only source of energy for the flourishing of plants in the natural environment.

The Vedic seers have not only been concerned and conscious of the purity of the natural environment, including earth, water, air, light, and space, but also of the human mind and spirit. In the *Yajurveda*, a prayer for the Goddess Earth is mentioned, which includes the message for its protection, sustainability, and control of land pollution – O' Mother Earth, you should be nurtured well. We should never misuse your patience, and not malnourish you with excessive use of chemical fertilizers and pesticides. Rather sow the crop by variation, and nourish you with cow dung, water, etc. Plant trees to prevent corrosion because our nutrition depends on your nutrition (Maheshawari, 2019). Ancient Hindu traditional texts provide the basis for environmental ethics and ascribe the intrinsic value and direct moral standing to all living beings and their virtues (Framarin, 2011). There are vastly different indigenous traditions and languages throughout the world which have operated for thousands of years in distinct and unique ecologies, each having its own set of specialized principles and practices, and each having a valuable view of the environment (Fergusson et al., 2017). The facts about plants, as known to cultural traditions, have been based on the framework of shabda (authority), pratyaksha (perception), anumana (inference), and upamana (analogy) because indigenous people have used these means to attain information and discover new insights regarding environmental sustainability (Sadhale & Nene, 2009).

Indian philosophy establishes communion in Rishi and Kavi (the poet), and according to Rigveda, they are creators of the divine forms (Premshankar, 2008). The Vedic Indians show great love for nature and animal life and its preservation. Kālidāsa (one of the greatest poets and dramatists) reflected the same sensitivity and concern for nature in his literary work *Shakuntalam* (a Sanskrit play) (Biju, 2006). It is all because of the natural environment that the air is clean, the water is pious, and the Earth congenial to human existence (Dangwal, 1999). Some exceptional uses of special grasses have been given in various mythological texts, and this tendency describes the vast knowledge and intellectual level of ancient sages. Grass grows automatically without any special care or weather conditions. Bamboo is the most advanced and ever-growing forest grass on this Earth. The flute, which is the oldest Indian Instrument, is also called *vamsa vadya* (bansuri) due to made of bamboo (vamsa) (Mehta, 1995).

Nature, in all its forms and countless modes, is nothing but an appearance of God (Banwari & Vohra, 1992), and humans have adored the nature in its various manifestations, through the persona of a divinity (Daruwalla, 1999). A description of the great tree plantation called *Utsav-Tree* festival has been given in the epics, which also affirmed that the value of planting one tree is equivalent to begetting ten sons (Singh, 2016). There is a reference to Aranyani, the Goddess of forests in the *Rig Veda* (Ramachanran, 1984). Evidence of various deities in trees and forests has been confirmed in the Vedic texts, and provision for the punishment has been made for cutting them or doing any other kind of damage. The realization of the divinity of nature helps to get rid of the thought of doing any harm to any of its forms (Banwari & Vohra, 1992), and felling of plants without permission has been considered as a penal offense in ancient India, and it is linked with calamities, disorders, and sin so that people would fear to destroy a tree (Dwivedi, 2017).

Spiritual ethics exhibit the worship of nature. Human society has the responsibility to see to the welfare of all and to destroy the ignorance of the people; Lord Krishna has stated in the *Śrīmad Bhāgavatam* that their environment should be their utmost concern, and during the occasion of Govardhan Puja (a ritual to worship the sacred mountain), he advised to all beings that it's better to worship hills, cows, and trees, rather than to perform a ritual for the demigods (Prime, 1996). All sentient entities have direct moral standing, and the text *Manusmriti* has attributed sentient to both animals and plants (Framarin, 2014). In ancient times, large trees were being planted to determine the extent of the village. Like, this treatise has indicated that Vat (Banyan), Peepal (*Ficus religiosa*), Semar, Sakhuva (Sal), Tal (*Borassus flabellifer*), and other trees should be planted on the boundary of the residential area to mark the boundary. And the person who destroys the fruits, flowers, and leaves of trees should be punished in the same way. Nature (*Prakriti*), made of Earth, water, fire, sky, and the air, is the basis of human civilization of the world. So, all individuals have to protect it and think about the damage that happens to the environment before doing any activity.

5.5 Role of the Cow and Yajna-Ritual in Conserving the Environmental Sustainability Cycle

Vedas provide the wisdom for the duty of humans, and they have instructed them to breathe in pure air, drink pure water, eat pure grain, perform agricultural activities in pure soil, and remain to prosper. Observation and exploration of ancient facts related to nature in various ritualistic cultures are very important in order to discover new dimensions of environmental protection and conservation. The cow is not only the basis of Indian culture but also the basis of all human civilizations that have a sense of love for all creatures, like animals and birds. The meaning of the word 'Gau' in Sanskrit is Gaaye (Cow) and Prithvi (Earth). The word 'cow' has been used many times in the meaning of the earth, and it also means senses. Thus, the idea of the cow-element is considered through the relationship between the Earth and the senses (Shastri, 2014). Cow and earth complement each other. Earth is the basis of organisms, and the cow is for the sustenance of organisms. In this way, both are identical (Tripathi, 2014).

The cow is not only an animal, but it is also a land-goddess who is inhabited by all the deities, and it is called *Kamadhenu* because it fulfils all desires. *Kamadhenu* is the most common, virtually generic name of the cow that symbolizes the source of all prosperity in India, and she is the cow of plenty (Biardeau, 1993). Dasji (2014) identifies that the four motives of *Puruṣārtha*, such as Dharma, Artha, Kama, and Moksha, have been achieved through revering and caring of the cow because ancient seers have recognized the holy aspects and applications of cow's products in human's chores and culture. (i) Dharma—not any religious work is complete without the cow. Its milk, ghee, curd, dung, and urine are used in all religious works due to their auspiciousness; (ii) Artha—raising the cow, and trading of its ghee and milk, leads to wealth; (iii) Kama—the milk and ghee of the cow are also used, especially in the Vedic rituals performed for the accomplishment of wish; (iv) Moksha is attained by serving the cow in a spiritless way because the conscience gets purified by serving it.

The Vedas have explained the concept of environmental stewardship and the cycle of its preservation and conservation through the cow itself and the preformation ritual activity yajna through cow's ghee only because the cow has been considered as the living form of the Earth. Tripathi (2014) observes that the productiveness of the Earth is always protected by cow dung manure because the cow dung and urine enhance the fertility of the Earth, and this growth process is natural due to its stability and continuity; accordingly, cow dung is very dear to the Earth. Cow dung manure is the food of the earth, and this also makes the environment pure, clean, and healthy. Farms and cows have a close relationship because cows get nourishment by eating the grass of the field, and their dung and urine make the field productive in return. They retain the fertility of the soil, and food products produced by using the cow dung and urine are very pure in the quality (Dasji, 2014). Being a part of agriculture, food, medicine, and enterprises, the cow has a big contribution to the prosperity of the environment. Ancient texts state that the *Suryaketu* nerve present on cow's back makes the environment clean by preventing harmful radiations (Sharma, 2014). The cow is called Vrishabh (Taurus) in the *Rig Veda*, and the golden rays of dawn have been depicted as a cow (Shastri, 2014). *Panchagavya* is the product made from cow's milk, curd, ghee, urine, and dung. It is a pure and holy substance, a dietary, provides bliss to the heart, force, and wisdom. It has a significant value from the therapeutic, ritualistic, and natural-puristic perspectives. Nowadays, for increasing the awareness of organic farming and for maintaining the productiveness of the soil, and discarding the harmful fertilizers, the cow dung manure is being given more prominence.

The cow is the mother of the world (*Gavo Vishvasya Matar:*). While performing yajna to the Sun (God of Fire) and Varuna (God of Water), the confirmation is given by the offering, and is obtained from the ghee of the cow itself. Dasji (2014) explains that in the ritual worship procedure, only cow's ghee is considered as auspicious and offered to the God through yajna, that reinforces the sun's rays and brings plentiful rainfall, and this rain produces all kinds of grains, plants, grasses, etc., which nourishes all real, movable, variable, constant creatures. Yajna-Agni spreads the aroma of the offered material a hundred times in the whole environment, which makes the environment pollution free and the rain is confirmatory as well as nourishing. Food grains and vegetation fed by this rain are also nutritious and healthy.

On earth, the cow is like fire in its glory and body. The interpretation of the superpower, which represents idolatrous cow on the Earth, is not practicable even in the *Vedas*. The presence of the cow at any place on the Earth itself is an important contribution to the purification of the environment. When people use ghee in the Sanskar Karmas like Yajna, the ozone layer gets strong through this practice, and the Earth is protected from harmful solar radiation (Sharma, 2014). Cows conduct the yajna and are its main form, and Vedas have explained the concept and cycle of environmental protection through them. It is an integral part of India's customs and has been a part of the religious activities of the people of India for centuries. It is not only a divine act but also an amicable ritual since ancient times, and for the fulfilment of this activity with perfection, only cow's ghee should be used, as declared in the *Vedas*.

Likewise, Lord Krishna has vividly defined the ecology system and environment sustainability cycle in the *Śrīmad Bhagavad Gītā*, which derives Satya, the integrity of man, the wholesomeness of his behaviour in the world from Satt, reality, the nature of the world, and its working. This reality is an orchestrated system, such as every component and process in nature works for the totality. This work distinguishes between two aspects of nature or two levels in its operation, like atomistic and holistic, segmental and integral, mechanistically causal, and overarching teleological (Chaitanya, 1992). The best way to teach environmental concern is through Lord Krishna's life. His practice of religion includes living in harmony with nature, showing love to all creatures, never harming any living being, and rejoicing in the beauty of the natural life of simplicity (Prime, 1996). The style of this holy text is terse and packs rich meanings with brevity (Chaitanya, 1992), and its related phrases are as follows:

aha-yajñāḥ prajāḥ sṛiṣhṭvā purovācha prajāpatiḥ
anena prasaviṣhyadhvam eṣha vo 'stviṣhṭa-kāma-dhuk [3:10]

At the beginning of creature, the Lord of all beings, created humans along with the sacrifice, saying: 'By this shall you prosper; this shall be the cow of plenty granting all your desires.'

When God created the human being, he puts the qualities of charity, austerity, and service inside of each human being. If people have this attitude, it will be a blessing to themselves. They will become like the wish-fulfilling cow, *Kamadhenu*, so they don't even need to ask anything.

devān bhāvayatānena te devā bhāvayantu vaḥ
parasparaṁ bhāvayantaḥ śhreyaḥ param avāpsyatha [3:11]

By this, may you nurture the gods, and the gods will nurture you in return. Thus, nurturing one another, you will obtain the highest good. Each person is a manifestation of the divine. Through the service which you render, selflessly, with a pure heart to humanity, you will receive a ten times greater blessing (Vishwananda, 2017).

Yajna is originated from the beginning of creation, and it is a means of both our prosperity advent (cosmic advancement) and the ultimate welfare. It is the ancient Indian practice to achieve happiness and prosperity through pleasing the deities who operate the world-cycle. These gods nourish our flora and our lives through them, by arranging heat (sun), moonlight, rain, etc. on time for us. They provide us the earth to live, water to quench the thirst, fire to cook the food, wind to breathe, and vacation to roam around (Goenka, 2014). Indian culture is karma predominantly, and yajna is also a form of karma. Just as Yajna-chakra is not possible without the cow, similarly, the cow is needed to make the Karma-chakra beautiful, pleasant, and favourable (Shastri, 2014).

The cow is the main reason for getting the best grain. It gives (*Havishya*) the best food through yajna as an offering to all the deities (Yangalwar, 2015). The practice of performing yajna is known as *Agnihotra* in ancient Hindu texts, and it is a holistic concept of growing plants in a healthy atmosphere and maintain ecological balance by performing *Agnihotra* in the middle of the farm and using the Yajna-ash as fertilizer, which is a safe, productive, and holistic approach towards environmental sustainability (Acharya & Joshi, 2001). It constitutes the custom of dry mango's wood and Hawan Samagri (a mixture of certain herbs) combustion along with cow's ghee which is offered in the fire with rhythmic chanting of mantras particular to that yajna (Saxena et al., 2018). In ancient times, this ritual has been continuously performed by the seers to make the environment clean and indicates that ancient Vedas are the foremost texts defining the green movement for environmental conservation and sustainability.

5.6 Contribution of Inspirational Stories to Awaken the Spirit of Stewardship Along with an Emotional Connection to Ecology

Ecology is the science which deals with the relationships between all living organisms and their total environment (Sasidharan, 1998), and includes the influence and change of environmental factors on organisms (Baarschers, 2013). The eco-pedagogical potential of ecology-oriented literature has a significant contribution in recognizing the need for awareness about the environment (Curry, 2013). Ancient literature has a wealth of knowledge related to various aspects of flora, fauna, and environmental sustainability (Awasthi, 1998), and the modern form of environmental texts is known as eco-fiction, which provides an understanding of the critical aspects of nature's variances. Environmentalism literature includes a new and differentiated approach from the conservative tradition of writing (Kerridge. 2001). Novels or fiction works have a significant perspective to identify the role of humans embedded in the Earth's ecology, along with technological and scientific nexus (Curry, 2013).

Human understands nature only based on his consciousness (Caube, 1994), and ancient narratives have projected inextricable and intimate bonding between man and nature (Patil, 2019). For teaching the lessons of moral behavior and ethical treatment to all living creatures, animal symbolism and imagery are used in the stories (Hivale, 2019) because the art of storytelling has the power to develop and engage emotions, which help people to understand the reality of facts through fiction, and also exhibit their relation with the natural environment (Hadzigeorgiou & Judson, 2017). The lessons of social stories develop a profound and comprehensive understanding of complex environmental issues (Mahasneh et al., 2017).

Stories have a vital aspect in creating awareness because they have the power to awaken the conscience and signify an influential approach to the ethical learning of individuals. Many references in the literature show an awareness of protecting trees and their sustainability. Trees also have life and feel like humans, and pieces of evidence of this phenomenon have been found from time to time through various incidents, which make people aware of their existence, and give the message of not cutting trees, and loss, consequences, and punishment for cutting trees. A related incident as a didactic narrative is given here:

Saint Namdev (1270–1350) was a famous devotee of Maharashtra, India. He was a very observing and mindful person. There was a harmony of devotion and knowledge in his personality. One day, when he returned to his house in the evening after cutting the wood from the forest, then his dhoti (an Indian ethnic dress) was bloodied. Her mother gets horrified when she saw the blood.

Immediately, she asked, "What happened to him, and where did he fall"?

He replied that he did not fall but had chipped off his foot with an axe. His mother observed that skin of his foot had peeled at one point.

She scolded him and said that he had performed a foolish act of peeling the skin with an axe because if his foot would be broken or the wound rots, then there could be a chance of getting a leg amputated.

Namdev replied that if this could happen to human beings, then the trees would also get hurt because, at her instance on that previous day, he had driven an axe on the *Palash* tree (Bastard teak) and brought the tree bark. So abruptly, by considering this aspect and observing that the tree could feel the same pain as he felt, he had removed his skin.

After understanding his views, his mother cried with joy and blessed him that he would be a monk one day due to his sacred values. Trees also have a life, like humans, and he has the power to understand their feelings. As we suffer due to injuries, so do they. Now she would never ask him to do such a wrong thing.

This incident from Saint Namdev's life has given the meaningful message of protecting nature, which is an important requirement of today's era. The existence of human life rests only on nature, which provides air, water, and food, so everybody should have an emotional understanding and approach to plants and ecology (Sahay, 2018). This narrative encourages the sense of conserving and adoring nature's valuable gifts because if the sensation towards flora or fauna can be developed among individuals, then no one will try to harm or exploit the natural resources. Life evolves in various forms on this beautiful earth, and humans must give respect to all species and be protective and sensible while dealing with them.

5.7 The Primacy of Environmental Conservation in Ancient Indian Folks and Their Reverence for Plants and Trees

The plants and trees found in the present environment system have botanical as well as mythological importance because these plants have existed as a unique gift of God on the Earth before the development of human civilization, and their agricultural, herbal, and medicinal importance have been collected and recognized as Samhita by seers and sages. Varāhamihira of the fifth century is known for centuries as an enlightened farsighted, wise, and learned botanist of unmatched excellence, who had known the cause and types of disease caused to trees and suggested their remedies as well (Dwivedi, 2005). He has compiled the text *Brihat Samhita* (505 AD–581 AD), an ancient text of plant pathology, which includes a chapter on plants and trees (Borkar, 2017). Likewise, Surapala's *Vrikshayurveda* (tenth-century treatise) includes many recommendations concerning the treatment of soil, seeds, and plants with various and often seemingly bizarre mixtures of animal and plant products (Willis, 2007). It has elaborated on the importance and glory of trees and signified the common people's beliefs, religious attitudes, and folk knowledge to persuade individuals to plant trees (Valsalakumari, 2008).

Knowledge of herbs is developed by seers, sages, travelers, and tribes through observation and intuition, and they have discovered the many properties of plants and their products (Menon, 2002). Sheshadri (2016) finds that ancient Sage Mātaṅga has explained various flowers, which have never been plucked, and even if plucked, never fade and remain fresh always. The *Rig Vedic* seers symbolized trees as attributes of several deities and considered them animate beings who feel happiness and sorrow (Dwivedi, 2008). Plants and trees feel hunger and thirst and sleep too at night (Ramachanran, 1984). They use the senses of smell and taste for communication and need each other for growth (Flannery, 2016). Each plant has a particular season of maximum responsiveness (Bose, 1902).

Religion has the ancient ways of protecting and nurturing nature. This is evident in the tradition of worshiping of Sun, wind, land, trees, plants, and water, which forms the very basis of human survival (Bain, 2017). The ancient culture of Hindus defines their careful observation of moral teachings regarding the treatment of nature (Dwivedi, 1993) because many deities have an association with animals and plants, which depict their love for nature (Mawdsley, 2006). Rituals are highly structured actions and express one of the most pervasive forms of religious behaviour in tribal as well as other culture (Debnath, 2003), like, there is a ritual of reciting the particular mantra related to the tree before its transplantation (Sadhale & Nene, 2009). Trees are the future of humanity. Various writers have mentioned the attainment of heavenly pleasure, material happiness, and fulfilment of other wishes from the sacred trees in their compositions. For example, planting *Triveni* (a combination of Peepal, Vat, and Neem tree), and other sacred trees, and watering, worshiping, and protecting them confirms the way to salvation. A man is capable of attaining the highest conceivable perfection by planting trees (Ramachanran, 1984), and this activity leads to the attainment of *Puruṣārtha* such as Dharma, Artha, Kama, and Moksha.

Menon (2002) observes that plants' spiritual factor has an association with floras, such as the Tulsi plant with Lakshmi (Goddess of wealth), Peepal to Lord Vishnu, Bel tree signifies the Lord Shiva, Amla tree to God of wealth, and Lord Krishna with the Kadamba tree. The usefulness for rituals of certain species has necessitated the practice of statutory domestication, such as holy Tulsi (*Ocimum sanctum*), which is grown at each Hindu homestead (Deb & Malhotra, 2001). Varāhamihira, in his text *Brihat Samhita* advocates that one can guess the happiness of mankind from the Rasaal (Mango), danger from the Bhallataka (*Semecarpus anacardium*), health from the Peelu (*Salvadora persica*), famine from the Khadira (*Acacia catechu*), and Samee (*Prosopis cineraria*), and good rain from the Arjuna (*Terminalia arjuna*) trees, by observing their condition of growth and development (Shastri et al., 1946)

From the early times, plants and people have a strong relationship in the form of a traditional system of sacred forests, groves, and trees (Mehta, 2017), and the traditional systems of knowledge are not just curiosities, but they are important for rediscovering new principles, for the more sustainable uses of the natural environment (Berkes et al., 1995). The life-enhancing paradigm emerged from India's ancient forest culture, in all its diversity, which creates a sustainable, renewable forest system, supporting and renewing food and water resources (Shiva, 1999). Sacred forests present an alternative view of conservation that is led by norms and taboos rather than formal legal frameworks (Ormsby, 2010), and sacred groves are areas that comprise forests or natural vegetation usually dedicated to local folk deities or tree spirits (*Vanadevatas*) (Kala & Sharma, 2010). Krishna (2014) defines that the sacred trees have a great significance in the ecological heritage and worshipped from time immemorial, and documented in all ancient societies all over the world. In India, the groups of sacred trees include *Kalpa vriksha* (a tree which fulfils the wishes), *Chaitya vriksha* (an open-air sanctuary where daily offerings are made) (Pal, 1997), and *Sthala vriksha* (official tree of the temple for pilgrims and devotees, which are planted by the local devotees) (Narayanan, 1997).

Similarly, forest area encompasses *Sreevanam* (immediate surroundings of the village), *Mahavanam* (a kind of reserved forest), and *Tapovanam* (densest forest for great saints to do meditation and austerity (Manjrekar, 2017), and *Latagrha* is the name given to the garden area or nursery (Ramachanran, 1984). The social, moral, and cognitive beliefs of any culture have a crucial aspect in human societies' adaptation to the natural environment, and their spiritual and moral ideas motivate them to care for the world around them (Saigal, 2008). Religious faith plays a key role in respecting the flora and fauna along with the honesty and respect for ancestors, and the observance of long-held community values (Kamga-Kamdem, 2010). Like, in Rajasthan (a desert state of India), the Khejri tree has been valued for its moisture-retaining properties, and people never cut it, even if it comes between the constructional works (Kala & Sharma, 2010). Trees are the abode of gods, beasts, birds, insects, human beings, and other creatures because all these take shelter under trees (Dwivedi, 2005), and their association with the spiritual power of deities presents a divine aspect of nature (Gupta, 1971).

The Chipko Movement in the 1970s of India has great importance to understand the concept of environmental stewardship in the modern context, which has always been memorialized by environmentalists and local people with great reverence in the process of saving the environment. It is a kind of feminist movement to protect nature from the greed of men (Dwivedi, 1993), and its narratives share much with a particular strain of friluftsliv (nature-loving) thinking that idealizes mountain culture and advocates a hands-off approach to environmentalism (Crowley, 2013). Plants are often associated with many myths and folklore, and certain social customs and traditions like Vratas (fasting) and festivals related to trees strengthened this ideology (Dwivedi, 2017), like the ritual of worshipping the Banana tree, Peepal tree, or Amla tree for seeking the blessings of Lord Vishnu and Goddess Lakshmi. The only motive of ancient people behind this practice is environmental conservation.

Brihat Samhita defines unique criteria of making demarcation about people and items in any area or town through trees or plant flourishing, such as traders are indicated by Suvarnapushpa (golden flower); brahmins by Lotuses; royal priests by Lillies; the commander of the army by Saugandhika flower; and the increase of gold by the Arka plants (Shastri et al., 1946). Valsalakumari (2008) observes that the plant's whole life is a worship of light (material symbol of the divine), and India has a rich heritage in ornamental gardening. For example, one of the auspicious tree Sindur (*Bixa Orellana*) is an ornamental plant, and its seeds are used in making Sindur (sacred traditional red or orange-coloured powder) and have a very religious purpose in Hindu dharma. Tucker and Grim (2001) find that traditional cultures have a persuasive vision for a sustainable future. The art of the folk or common people is rooted in ageless traditions drawn from community and culture native to different places within India. Aggarwal (2018) observes that *Madhubani*, *Kalamkari*, *Warli*, *Gond*, and *Kashmiri* art forms are especially well-known for their eye-catching representations of flora and fauna, and these artworks visualize a good understanding and harmony among humans and nature. Accordingly, there is no any special provision required for environmental conservation because this process gets completed fundamentally and indubitably through tree plantation only.

Stewardship of earth acknowledges that humans are members of the household of nature and that they bear the responsibility to care attentively for it (Chapin III et al., 2015), and understanding of the environmental stewardship must be developed in every individual to globally find solutions to prevailing issues of the environment (Rogayan, 2019). The exercise of self-restraint is essential to ensure the sustainable use of resources and realize their full potential (Ghosh, 2003). Entrepreneurs, leaders, or managers for working holistically have to acquire the ability to manage skilled teams for environment preservation rather than remaining in the safety of their own professional expertise (Turner & Evans, 2004). Plants and trees are not resources only, but they are lively objects, which feel and respond, and their development and taking care is the responsibility of everybody, especially of those people who make use of their by-products in commodities. Nature should not be treated as a property of humans for the fulfilment of their desires and needs, but it must be respected for its kindness.

5.8 Conclusion

Sustainable living and relative goals and efforts of environmental stewardship have been a part of human life for eras. Every living organism has comprised the *Pancha Maha-bhuta*, and the virtue of these five elements of nature is to complement each other because human life has no existence without them, and also worldly creation has no meaning, and accordingly, it is the moral responsibility of every person to strive for the purification of these elements. Trees have not just made the Earth elegant and rich, but they have been fulfilling human interests with their spirit of loyalty and self-sacrifice. The ancient perspective of the purpose of environmental stewardship to preserve nature is very advantageous in making today's leaders and managers awake in the sense of their responsibility to ecology and represents an integral relationship, psychological understanding, objectivity, sensitivity, and reverence of it. Environmental stewardship is a movement that inspires people to save trees, and its practice has included the aspect of awakening for plantations and the management of the Earth's valuable resources, where trees teach philanthropy from the conception of selfless service to all beings.

5.9 Reflection Questions

1. How nature-oriented poetic works are helpful in understanding and developing emotions about environmental stewardship in human consciousness?
2. How do mythological texts relevantly define the vital aspects of sacred trees to achieve prosperity and growth in the life-work of humans?
3. What is the importance and contribution of the cow in protecting nature's harmony and maintaining the cycle of ecology?
4. How do stories present a coherent way of generating poignant sensations among individuals for understanding the concept of environmental protection?
5. Why is it imperative for naturalists or ecologists to have a holistic approach to sustain the sanctity of plants?

5.10 Relevant Lessons

- Poets have always shown respect for nature through their conscious ideology, which identifies their attitude to environmental stewardship and gives inspiration for affection and protection for it.
- The ancient wisdom of Indian *Puranas* present the essence of the antiquity of the environmental stewardship and reinforces the feeling of reverence for nature.
- The Cow and *Panchagavya* are particularly useful in maintaining the fertility of the Earth, and contaminated elements of the environment can be eliminated by using cow's ghee in Yajna and ritual activities.
- Stories can develop a sense of responsibility for nature in individuals by presenting the emotional facts of thoughts within the human sensation.
- Universal and abstracted information related to the conservation of flora and fauna and ecology has been provided in texts like *Brihat Samhita* and *Vrikshayurveda*, which can pave the way to knowledge of environmental sustainability in the modern era.

References

Acharya, P. S. S., & Joshi, R. (2001). *The integrated science of Yagna*. Shantikunj.

Aggarwal, M. (2018). Depictions of nature in Indian art, themes and imagery drawn from the environment. *Terragreen, 10*(10), 24–28.

Ārasu, & Kumaran, V. T. (1995). *Sāhityānuvāda: samvāda aura samvedanā*. Vani Prakashan.

Atanda, J. (2015). Environmental impacts of bamboo as a substitute constructional material in Nigeria. *Case Studies in Construction Materials, 3*, 33–39. https://doi.org/10.1016/j.cscm.2015.06.002

Awasthi, D. S. (1998). *Sahitya Vidhaon Ki Prakriti*. Radhakrishna Prakashan.

Baarschers, W. H. (2013). *Eco-facts and eco-fiction: Understanding the environmental debate*. Routledge.
Bain, W. K. (2017). Conservation of environment through traditional knowledge and wisdom with special reference to beliefs and practices in tribal India: An overview. *Heritage: Journal of Multidisciplinary Studies in Archaeology, 5*, 224–243.
Baindur, M. (2015). *Nature in Indian philosophy and cultural traditions, Volume 12 of Sophia studies in cross-cultural philosophy of traditions and cultures*. Springer India.
Banwari, & Vohra, A. (1992). *Panchvati: An Indian approach to environment*. Shree Vinayaka Publication.
Bennett, N. J., Whitty, T. S., Finkbeiner, E., Pittman, J., Bassett, H., Gelcich, S., & Allison, E. H. (2018). Environmental stewardship: A conceptual review and analytical framework. *Environmental Management, 61*, 597–614. https://doi.org/10.1007/s00267-017-0993-2
Berkes, F., Folke, C., & Gadgil, M. (1995). Traditional ecological knowledge, biodiversity, resilience and sustainability. In C. A. Perrings, K. G. Mäler, C. Folke, C. S. Holling, & B. O. Jansson (Eds.), *Biodiversity conservation. Ecology, Economy & Environment* (Vol. 4, pp. 281–299). Springer. https://doi.org/10.1007/978-94-011-0277-3_15
Bharti, A. (1995). *Prasangatah*. Bhartiya Jnanpith.
Bhavabhūti, & Kāle, M. R. (1997). Bhavabhūti' and Kalidasa. In M. R. Kāle (Ed.), *Bhavabhūti's Mālatīmādhava: With the commentary of Jagaddhara*. Motilal Banarsidass Publications.
Biardeau, M. (1993). Kamadhenu: The religious cow, symbol of prosperity. In Y. Bonnefoy (Ed.), *Asian mythologies* (pp. 99–116). University of Chicago Press.
Biju, M. (2006). *Sustainable dimensions of tourism management*. Mittal Publications.
Borkar, S. (2017). *History of plant pathology*. CRC Press.
Borooah, A. (1878). *Bhavabhuti and his place in Sanskrit literature*. Khetromohan Mukherjea.
Bose, J. C. (1902). *Response in the living and non-living*. Longmans, Green, and Company.
Caube, P. (1994). *Loka-Cetanā Ke Rāshṭrīya Kavi Śivamaṅgala Siṃha "Sumana"*. Vani Prakashan.
Chaitanya, K. (1992). The earth as sacred environs. In G. Sen (Ed.), *Indigenous vision: Peoples of India attitudes to the environment, Vol.19, Issues 1–2 of India International Centre quarterly* (pp. 34–48). Sage Publications.
Chamola, S. D. (2007). *Kautilya Arthshastra and the science of management: Relevance for the contemporary society*. Hope India Publications.
Chapin, F. S., III, Pickett, S. T. A., Power, M. E., Collins, S. L., Baron, J. S., Inouye, D. W., & Turner, M. G. (2015). Earth stewardship: An initiative by the Ecological Society of America to Foster Engagement to Sustain Planet Earth. In R. Rozzi, F. S. Chapin III, J. B. Callicott, S. T. A. Pickett, M. E. Power, J. J. Armesto, & R. H. May (Eds.), *Earth stewardship linking ecology and ethics in theory and practice* (Vol. 2, pp. 173–194). Springer.
Chapple, C. K. (1998). Toward an indigenous Indian environmentalism. In L. E. Nelson (Ed.), *Purifying the earthly body of god: Religion and ecology in Hindu India* (pp. 13–38). SUNY Press.
Chaudhari, B. H. (2017). The portrayal of nature and language in Kalidasa's Meghaduta. *International Journal of Advanced Education and Research, 2*(4), 75–77.
Cohen, M. M. (2014). Tulsi - Ocimum Sanctum: A herb for all reasons. *Journal of Ayurveda and integrative medicine, 5*(4), 251–259. https://doi.org/10.4103/0975-9476.146554
Crowley, T. (2010). From "Natural" to "Ecosocial flourishing": Evaluating evaluative frameworks. *Ethics & the Environment, 15*(1), 1–36.
Crowley, T. (2013). Climbing Mountains, hugging trees: A cross-cultural examination of love for nature. *Emotion, Space and Society, 6*, 44–53.
Curry, A. (2013). *Environmental crisis in young adult fiction: A poetics of earth*. Springer.
Dangwal, S. (1999). *Hinduism in T.S. Eliot's writings*. Deep and Deep Publications.
Daruwalla, K. N. (1999). Poetry and nature. *India International Centre Quarterly, 26*(3), 128–140.
Dasji, S. R. (2014). *Gaaye ki Mehtta aur Avshyakta. GO SEVA ANK* (pp. 130–135). Gita Press.
Deb, D., & Malhotra, K. C. (2001). Conservation ethos in local traditions: The West Bengal heritage. *Society and Natural Resources, 14*, 711–724.

Debnath, D. (2003). *Ecology and rituals in tribal areas*. Sarup & Sons.
Dwivedi, O. P. (1993). Human responsibility and the environment: A Hindu perspective. *Journal of Hindu Christian Studies, 6*, Article 8, 19–26. https://doi.org/10.7825/2164-6279.1077
Dwivedi, B. (2005). *Environmental Vaastu*. Diamond Pocket Books.
Dwivedi, O. P. (2008). Classical India. In D. Jamieson (Ed.), *A companion to environmental philosophy* (pp. 37–51). John Wiley & Sons.
Dwivedi, O. P. (2016). *India's environmental policies, Programmes and stewardship*. Macmillan Publishers Ltd.
Dwivedi, D. V. (2017). Importance of plants as depicted in Purānas. *Indian Journal of History of Science, 52*(3), 251–274. https://doi.org/10.16943/ijhs/2017/v52i3/49157
Dwivedi, O. P., & Tiwari, B. N. (1999). Environmental protection in the Hindu religion. In G. A. James (Ed.), *Ethical perspectives on environmental issues in India* (pp. 161–188). APH Publishing.
Enqvist, J. P., Westb, S., Mastersonb, V. A., Haiderb, L. J., Svedinb, U., & Tengö, M. (2018). Stewardship as a boundary object for sustainability research: Linking care, knowledge and agency. *Landscape and Urban Planning, 179*, 17–37. https://doi.org/10.1016/j.landurbplan.2018.07.005
Fergusson, L., Kettle, D., & Wells, G. (2017). Indigenous accounts of environmental stewardship in light of the theory and language of Maharishi Vedic Science. *International Journal of Society, Culture & Language, 5*(1), 69–81.
Flannery, T. (2016). *Forward. The hidden life of trees: What they feel, how they communicate—Discoveries from a secret world* (Vol. 1, pp. vii–ix). Penguin UK.
Framarin, C. G. (2011). Ātman, identity, and emanation: Arguments for a Hindu environmental ethic. *Comparative Philosophy, 2*(1), 3–24.
Framarin, C. G. (2014). *Hinduism and environmental ethics: Law, literature, and philosophy*. Routledge.
Ghosh, A. (2003). *Natural resource conservation and environment management*. APH Publishing.
Goenka, J. D. (2014). *Go-Mahima aur Goraksha ki Avshaykta. GO SEVA ANK* (pp. 91–94). Gita Press.
Gokak, V. K. (1960). Tagore's influence on modern Indian poetry. *Indian Literature, 4*(1), 99–115.
Gough, A. E. (2013). *The philosophy of the Upanishads and ancient Indian metaphysics*. Routledge.
Gray, T. S., & Hatchard, J. (2007). Environmental stewardship as a new form of fisheries governance. *ICES Journal of Marine Science, 64*(4), 786–792. https://doi.org/10.1093/icesjms/fsl041
Gupta, S. M. (1971). *Plant myths and traditions in India*. Brill Archive.
Gupta, A., & Kumar, A. (2008). Potential of bamboo in sustainable development. *Asia-Pacific Journal of Management Research and Innovation, 4*(3), 100–107. https://doi.org/10.1177/097324700800400312
Hadzigeorgiou, Y., & Judson, G. (2017). Toward more effective storytelling for raising environmental awareness in young students. *Journal of Advances in Education Research, 2*(1), 12–18. https://doi.org/10.22606/jaer.2017.21002
Hadzigeorgiou, Y., & Skoumios, M. (2013). The development of environmental awareness through school science: Problems and possibilities. *International Journal of Environmental & Science Education, 8*, 405–426. https://doi.org/10.12973/ijese.2013.212a
Hilbert, B. S. (2001). Beyond "Thou Shalt not": An eco critic reads deuteronomy. In K. Armbruster & K. R. Wallace (Eds.), *Beyond nature writing: Expanding the boundaries of ecocriticism* (pp. 29–40). University of Virginia Press.
Hivale, G. T. (2019). *Animals in Indian English poetry*. Lulu.com.
Joshī, D. (2005). *Bhāratiya Samskriti Ke Sarjaka*. Star Publications.
Kala, M., & Sharma, A. (2010). Traditional Indian beliefs: A key toward sustainable living. *Environmentalist, 30*, 85–89. https://doi.org/10.1007/s10669-009-9247-z

Kamga-Kamdem, S. L. (2010). Ancestral beliefs and conservation: The case of sacred sites in Bandjoun, West Cameroon. In B. Verschuuren, R. Wild, J. McNeely, & G. Oviedo (Eds.), *Sacred natural sites conserving nature and culture* (pp. 121–155). Earthscan.
Kaur, S. (2003). *Indian people. Sarojini Naidu's poetry melody of Indianness* (p. 126). Sarup & Sons.
Keith, A. B. (1992). *The Sanskrit Drama in its origin, development, Theory & Practice*. Motilal Banarsidass Publications.
Kerridge, R. (2001). Ecological hardy. In K. Armbruster & K. R. Wallace (Eds.), *Beyond nature writing: Expanding the boundaries of ecocriticism* (pp. 126–142). University of Virginia Press.
Krishna, N. (2014). *Sacred plants of India*. Penguin UK.
Krishna, N. (2017). *Hinduism and nature*. Penguin Random House India Private Limited.
Krishnamoorthy, K. (1982). *Banabhatta* (Sanskrit Writer). Sahitya Akademi.
Krishnamoorthy, B. (2017). *Environmental management: Text and cases*. PHI Learning Pvt. Ltd.
Kumar, A. (2018). *100 great Indian poems*. Bloomsbury Publishing.
Leopold, A. (1949). *A Sand County Almanac, and sketches here and there, environmental studies* (Vol. 263). Oxford University Press.
Mahasneh, R. A., Romanowski, M. H., & Dajani, R. B. (2017). Reading social stories in the community: A promising intervention for promoting children's environmental knowledge and behaviour in Jordan. *The Journal of Environmental Education*, 1–14. https://doi.org/10.1080/00958964.2017.1319789
Maheshawari, S. L. (2019). Bhartiya Sanskriti Main Pryavarna Sanrakshan. *Kalyan, 6*, 1–50.
Majumdar, R. C., & Chauhan, S. S. (1984). *Shrenya Yug Hindi translation of classical age*. Motilal Banarsidass Publishers.
Manjrekar, S. M. (2017). Study of sustainability practices ingrained in Indian culture. In *5th International Conference on Civil, Architecture, Environment and Waste Management (CAEWM-17)*, pp. 270–276.
Mawdsley, E. (2006). Hindu nationalism, neo-traditionalism and environmental discourses in India. *Geoforum, 37*, 380–390. https://doi.org/10.1016/j.geoforum.2005.06.004
Mehta, T. (1995). *Sanskrit play production in ancient India* (Vol. 5). Motilal Banarsidass Publishers.
Mehta, S. (2017). Role of traditional practices in conserving environment: A case of Manesar Village, Gurgaon. *IOSR Journal of Humanities and Social Science, 22*(11), Ver. 10, 10–16.
Menon, I. (2002). *The story of Ayurveda*. Publications Division Ministry of Information & Broadcasting.
Mirashi, V. V. (1996). *Bhavabhūti: His date, life and works*. Motilal Banarsidass Publishers.
Mishra, V. N. (2009). *Lok Aur Lok Ka Swar*. Prabhat Prakashan.
Misra, R. (1995). *Environmental ethics: A dialogue of cultures*. Concept Publishing Company.
Misra, V. N. (2003). *Anchhue Bindu*. Kitabghar Prakashan.
Nagendra, S., & Reddy, P. S. (2018). Sarojini Naidu as a poet of nature. *Language in India, 18*(7), 140–143.
Naravane, V. S. (1996). *Sarojini Naidu: An introduction to her life, work and poetry*. Orient Blackswan.
Narayanan, V. (1997). "One tree is equal to ten sons": Hindu responses to the problems of ecology, population, and consumption. *Journal of the American Academy of Religion, 65*(2), 291–332.
Narsimhachari, S. (2014). *Telugu Sahitya-Andarbh Aur Samiksha*. Vani Prakashan.
Nayak, R. K. (2010). Gangadhar Meher: A comparative figure to Sanskrit literature. *Orissa Review*, 30–33.
O'Brien, K., & Stoner, D. K. (1987). Increasing environmental awareness through children's literature. *The Reading Teacher, 41*(1), 14–19.
Ormsby, A. A. (2010). Sacred forests of India: A Strong tradition of community-based natural resource management. *Environmental Conservation, 37*(3), 320–326. https://doi.org/10.1017/S0376892910000561
Padihari, S. (2007). Jayanta Mahapatra: The poet of the soil. *Indian Literature, 51*(3 (239)), 168–176.

Pal, P. (1997). *Divine images, human visions: The Max Tanenbaum collection of South Asian and Himalayan Art in the National Gallery of Canada*. National Gallery of Canada.
Pala, N. A., Negi, A. K., & Todaria, N. P. (2014). The religious, social and cultural significance of Forest landscapes in Uttarakhand Himalaya, India. *International Journal of Conservation Science, 5*(2), 215–222.
Panda, J. (2008). Philosophical doctrines of poet Gangadhar Meher. *Orissa Review*, 51–53.
Patil, S. (2019). *Ecofeminism and the Indian novel*. Routledge.
Prasāda, K. (1995). *Kavitā tīre*. Vani Prakashan.
Premshankar. (2008). *Prasad Ka Kavya* (2nd ed.). Radhakrishna Prakashan.
Prime, R. (1996). *Hinduism and ecology: Seeds of truth*. Motilal Banarsidass Publishers.
Rajan, C., & Kalidasa. (2005). *The loom of time*. Penguin UK.
Ramachanran, C. (1984). VRIKSHAYURVEDA (arboreal medicine in ancient India). *Ancient Science of Life, IV*(2), 110–111.
Rogayan, D. V. (2019). I heart nature: Perspectives of university students on environmental stewardship. *International Journal on Engineering, Science and Technology, 1*(1), 10–16.
Rosulu, H., Balongun, E. O., & Hassan, T. I. (2019). Exploring the neglected potential of bamboo cultivation and utilization in Nigeria for sustainable development: An overview. *International Journal of Research and Scientific Innovation (IJRSI), VI*(XI), 229–237.
Rozzi, R., Chapin, F. S., Callicott, J. B., Pickett, S. T. A., Power, M. E., Armesto, J. J., & May, R. H. (2015). Introduction: Linking ecology and ethics for an interregional and intercultural earth stewardship. In R. Rozzi, F. S. Chapin III, J. B. Callicott, S. T. A. Pickett, M. E. Power, J. J. Armesto, & R. H. May (Eds.), *Earth stewardship linking ecology and ethics in theory and practice* (Vol. 2, pp. 1–14). Springer.
Ryder, A. W. (1914). *Kalidasa–His life and Writings. Kalidasa translations of Shakuntala, and other works* (p. 19). Prabhat Prakashan.
Ryder, A. W. (1920). *Kalidasa* (Vol. 1). Library of Alexandria.
Sadhale, N., & Nene, Y. L. (2009). Ancient Indian traditional and scientific thought on plants: Sir JC Bose and Vrikshayurveda. *Asian Agri-History, 13*(2), 101–111.
Sahasrabuddhe, S. (2017). *The Supreme Poet (Kalidasa)*. Retrieved from Ancientindianwisdom: http://ancientindianwisdom.com/intellectual-heritage/the-supreme-poet-kalidasa
Sahay, R. (2018). *Karma Hi Pooja Hai*. Prabhat Prakashan.
Saigal, K. (2008). *Sustainable development: The spiritual dimension*. Gyan Publishing House.
Saran, A. B., & Pandey, G. (1992). *Sun worship in India: A study of Deo Sun-Shrine*. Northern Book Centre.
Sarkara, P., Dh, L. K., Dhumal, C., Panigrahi, S. S., & Choudhary, R. (2015). Traditional and ayurvedic foods of Indian origin. *Journal of Ethnic Foods, 2*(3), 97–109.
Sasidharan, C. K. (1998). Ecologic-economic evaluation of renewable resources for sustainable development. In S. Chattopadhyay & H. S. Sharma (Eds.), *Sustainable development: Issues and case studies* (pp. 238–261). Concept Publishing Company.
Saxena, M., Kumar, B., & Matharu, S. (2018). Impact of Yagya on particulate matters. *Interdisciplinary Journal of Yagya Research, 1*(1), 1–8.
Sharma, S. (1985). *Introduction. Life in the Upanishads* (p. 1). Abhinav Publications.
Sharma, R. (2006). *India's ancient past*. Oxford University Press.
Sharma, R. (2008). *Bharat Ke Pracheen Bhasha Pariwar Aur Hindi Bhag-3*. Rajkamal Prakashan.
Sharma, R. (2009). *Nirala Ki Sahitya Sadhana* (Vol. 1–3). Rajkamal Prakashan.
Sharma, K. K. (2014). Go-Seva: Jan Seva. In P. Jha (Ed.), *Gauravshali Bharat* (pp. 307–314). Prabhat Prakashan.
Shastri, M. S. (2014). *Sanskriti ki Drishti se Gau ka Mehtav. GO SEVA ANK* (pp. 84–85). Gita Press.
Shastri, P. V. S. S., Bhat, V. M. R., & Mihira, V. (1946). *Brihat Samhita of Varaha Mihira*. V.B. Soobbiah and Sons.
Sheshadri, K. (2014). Seasonal behaviour and treatments of elephants in ancient India. *Gajah Journal of the Asian Elephant Specialist Group (AsESG), 41*, 20–24.

Sheshadri, K. (2015). Tree welfare as envisaged in ancient Indian literature. *Asian Agri-History, 19*(3), 231–239.
Sheshadri, K. (2016). Classification of flowers as gleaned from ancient Indian literature and culture. *Asian Agri-History, 20*(3), 181–199.
Sheshadri, K. G. (2017). Animals and animal products as reflected in Smṛti texts. *Prācyā, 9*(1), 1–21.
Shiva, V. (1999). Women in the Forest. In G. A. James (Ed.), *Ethical perspectives on environmental issues in India* (pp. 73–112). APH Publishing.
Shukla, N. K. (2018). Environment-protection in Kalidas-made season and identification Shakuntalam. *International Journal of Scientific Research in Science and Technology, 4*(7), 467–472.
Singh, N. N. (2012a). Puranic thought of planting and protecting trees: A lesson to the modern mankind. *Odisha Review*, 88–90.
Singh, R. P. B. (2012b). Eco spirituality in Indian tradition: Messages for global order. In D. Tiwari & A. Mishra (Eds.), *Environment ethics: Indian perspectives* (pp. 25–50). YS Books International.
Singh, P. K. (2016). *Samrath Bharat*. Diamond Pocket Books Pvt Ltd.
Srivastava, V. C. (2008). *History of agriculture in India, up to C. 1200 A.D.* Concept Publishing Company.
Strong, M. F. (1992). Only one earth. In G. Sen (Ed.), *Indigenous vision: Peoples of India attitudes to the environment, Vol. 19, Issues 1–2 of India International Centre quarterly* (pp. 49–52). Sage Publications.
Tomar, A., & Singh, K. K. (2009). Neem: An introduction. In K. K. Singh, S. Phogat, A. Tomar, & R. S. Dhillion (Eds.), *Neem, a treatise* (pp. 3–15). I. K. International Pvt Ltd..
Tripathi, B. K. (2014). *Gau Devi Sampda he. GO SEVA ANK* (pp. 168–169). Gita Press.
Tucker, M. E., & Grim, J. A. (2001). Introduction: The emerging alliance of world religions and ecology. *Daedalus, Religion and Ecology: Can the Climate Change? (Fall, 2001), 130*(4), 1–22.
Turner, M., & Evans, C. (2004). Users influencing the management of practice. In D. Statham (Ed.), *Managing front line practice in social care* (Vol. 40, pp. 47–62). Jessica Kingsley Publishers.
Upadhyay, R. (2015). *Rtusamhara of Kalidas: With Maltimaninanjusa Sanskrit to Hindi Commentary*. Chaukhamba Sanskrit Pratishthan Oriental Publishers & Distributors.
Valsalakumari, P. K. (2008). *Flowering trees* (Vol. 12). New India Publishing.
Vannucci, M. (1992). Tradition and change. In G. Sen (Ed.), *Indigenous vision: Peoples of India attitudes to the environment, Vol. 19, Issues 1–2 of India International Centre quarterly* (pp. 25–33). Sage Publications.
Vatsyayan, K. (1992). Ecology and Indian myth. In G. Sen (Ed.), *Indigenous vision: Peoples of India attitudes to the environment, Vol. 19, Issues 1–2 of India International Centre quarterly* (pp. 156–180). Sage Publications.
Vidyākara, Miśra, V., & Ingalls, D. H. H. (1968). Spring. In V. Miśra (Ed.), *Sanskrit Poetry, from Vidyākara's Treasury, Volume 44 of Harvard oriental series*. Harvard University Press.
Vishwananda, P. S. S. (2017). *Shreemad Bhagavad Gita: The song of love*. Bhakti Marga Publications.
Willis, R. (2007). *The history of allelopathy*. Springer Science & Business Media.
Winternitz, M. (1985). *History of Indian literature*. Motilal Banarsidass Publication.
Worrell, E., Allwood, J., & Gutowski, T. (2016). The role of material efficiency in environmental stewardship. *Annual Review of Environment and Resources, 41*, 575–598. https://doi.org/10.1146/annurev-environ-110615-085737
Yangalwar, Y. G. (2015). *Go Mahatmya Batanewale Go-Sukt*. Nachiket Prakashan.

Nidhi Kaushal is a scholar of Management Studies at Indian Institute of Technology Roorkee. She holds a master's degree in Business Administration and a bachelor's degree in Computer Science from Kurukshetra University, India. She has been interested in the research works related to Entrepreneurship, Leadership, Literature, Management and Indigenous Studies. During her Ph.D. studies, she has identified the indigenous studies of literature and folklore related to Leadership and Management and presented her work in various international conferences and research publications across the globe. She has worked at many managerial and academic positions. Being as a research fellow, she is also an event organizer and has organized conferences and other related events in the Institute. She is exploring Leadership with the study of creative writings, and this is her contribution to her academic research. This area will not only enrich management studies but also become immensely useful for entrepreneurs

Chapter 6
Organizational Citizenship Behaviors Toward the Individual, Organizational, Environmental, and Contextual Dimensions

José G. Vargas-Hernández and M. C. Omar C. Vargas-González

Abstract This study analyzes the individual, organizational, environmental, and contextual dimensions of organizational citizenship behaviors. It is supported under the assumption that organizational, individual, environmental, and contextual factors lead to the innovative behavior and organizational citizenship behavior as extra-role behaviors. The methods employed are the analytical-descriptive leading to the reflective analysis based on the theoretical and empirical review of the literature on these issues. It is concluded that the individual, environmental, contextual, and organizational factors facilitate more innovative organizational citizenship behaviors that contribute to individual and organizational efficiency.

Keywords Contextual dimension · Environmental dimension · Individual dimension · Organizational citizenship behaviors

6.1 Introduction

Organizational, individual, environmental, and contextual factors lead to the innovative behavior and organizational citizenship behavior as extra-role behaviors. Lack of organizational support has moderating effects on the relationship between the individual, organizational, environmental, and contextual factors, the innovative behavior, and organizational citizenship behaviors. Individual, environmental,

J. G. Vargas-Hernández (✉)
Postgraduate and Research Division, Instituto Tecnológico Mario Molina Pasquel y Henríquez, Unidad Académica Zapopan, Zapopan, Jalisco, Mexico
e-mail: jose.vargas@zapopan.tecmm.edu.mx

M. C. O. C. Vargas-González
Instituto Tecnológico de México, Campus Cd. Guzmán, Cd. Guzmán, Jalisco, Mexico
e-mail: omar.vg@cdguzman.tecnm.mx

© The Author(s), under exclusive license to Springer Nature Switzerland AG 2023

S. Dhiman (ed.), *Sustainable Development and Environmental Stewardship*,
https://doi.org/10.1007/978-3-031-28885-2_6

contextual, and organizational factors facilitate innovative behavior and organizational citizenship behaviors.

Organizational citizenship behavior is a discretionary individual behavior not directly and not explicitly recognized by formal reward system which promotes the organizational effective functioning (Organ, 1988). Organizational citizenship is defined as the discretionary individual behavior, directly and explicitly recognized by the formal reward system aimed to promote the functioning of the organizations (Organ et al., 2006, p. 3).

Organizational citizenship behavior (OCB) is an individual behavior that promotes the goals of the organization by contributing to its social and psychological environment (Organ, 1997; Rotundo & Sackett, 2002). Discretionary individual behaviors are that which contribute to environmental management organizations and that are not formally recognized and rewarded. Voluntary and discretionary organizational citizenship behaviors for the environment reduce consumption of resources and energy and promote all the green activities leading to lower the level of carbon foot printing (Boiral & Paillé, 2012). The outcomes of organizational citizenship behaviors are the job characteristic and job satisfaction, the fit person-organization, person-job, and organizational commitment (Sawalha & Kathawala, 2019; Farzaneh et al., 2014).

A theoretical framework of self-regulation theory focuses on intraindividual development in organizational citizenship behavior and intends to explain the affective, cognitive, and unconscious processes. The intraindividual development in organizational citizenship behavior may be explorative. The relational perspective of mobile usage develops organizational citizenship behavior among the individuals. The relational mobile usage has a mediating relationship between the personal values of individualism, collectivism, academic self-efficacy, and organizational citizenship behavior (Ahmad et al., 2021). Individuals with high level of self-efficacy are likely to get more involved and committed and able to face difficulties and accomplish more complex behaviors. Collectivism is related to relational mobile usage while individualism is not significantly related to relational mobile usage.

Research is focusing on the scope of factors causing the organization citizenship behaviors and effects on the individual workers and organization (ALHussain, 2011). Self-report measures of organizational citizenship behavior are used in meta-analysis and are unmonitored (Carpenter et al., 2014). From the meta-analysis approach, perceived organizational support relates to dimensions of organizational citizenship behaviors, among others helping coworkers (Rhoades & Eisenberger, 2002). The longitudinal research on dynamic construct and the relationships of organizational citizenship behaviors has not analyzed the intraindividual development of increasing or withholding behaviors at different times (Bergeron, 2007; Bolino et al., 2012; Sturman, 2007). The findings of the previous research on intra-individual development of organizational citizenship behavior and methodological issues are further analyzed.

Earlier empirical studies of OCB found that that the criteria of the factor analysis were assimilated in two forms. The OCB-I rendered to an individual coworker, and the OCB-O to measure the conscientiousubjective normess of organizational management. Organizational citizenship behavior has a variety of individual and organizational outcomes (Podsakoff et al., 2009) as well as environmental outcomes. Organizational citizenship behavior can be directed toward individuals, organizations, the environment, etc. (Temminck et al., 2015; Xerri & Brunetto, 2013).

The social exchange theory is the theoretical framework to explain the differences between persons in organizational citizenship behavior (Cropanzano & Mitchell, 2005), although there have been several alternative frameworks developed to explain the dynamic nature. Both approaches have different developments on organizational citizenship behaviors and on individuals.

Organizational citizenship behaviors are classified into organizational and individual dimensions (Williams & Anderson, 1991) defined as pro-social behavior or helping behaviors directed toward the coworkers. Grounded in the dimensions of Organ (1988) has been proposed a two-dimensional framework consisting of individual-oriented and organization-oriented organizational citizenship behaviors. Organizational citizenship behavior oriented toward the individual is different from the one oriented toward the organization related to different antecedents although they are supported by factor analysis (Turnley et al., 2003; Williams & Anderson, 1991).

Organizational citizenship behaviors are targeted toward individuals, organizations, contexts, and the environment. Counterproductive working behavior can be targeted toward the organization or individuals. Organizational citizenship behavior can be assessed with measuring behaviors directed to other individuals (Williams & Anderson, 1991). Organizational, contextual, and environmental- and individual-level factors affect the extra-role behaviors of innovative behavior and organizational citizenship behaviors.

6.2 Organizational Citizenship Behaviors Toward the Individual

The organizational citizenship behavior oriented toward the individual refers to beneficial behaviors directed toward other individuals, such as helping others and assisting supervisors. Organizational citizenship behaviors are defined as the discretionary individual behaviors not recognized by a formal reward system and promote the effective organizational function (Organ, 1988, p. 4). Organizational citizenship behavior is an individual extra-role behavior voluntarily carried out and not included in formal job requirements but able to increase the organizational effectiveness (Organ & Lingl, 1995).

Organizational citizenship behavior is the discretionary individual behavior not explicit or directly recognized by a formal reward system and promotes the efficient organization (Organ, 1988; Organ et al., 2006: 3). Organizational citizenship behavior is defined by Oikarinen et al. (2007) as individual behaviors at the workplace being productive but not recognized by a formal reward system.

There is an individual development in organizational citizenship behavior that differs among the individuals. The individual organizational citizenship behavior develops over time as a dynamic construct (Bolino et al., 2012). Organizational citizenship behaviors are different among the individuals. Individuals varied in their level of OCB at different points in time and that there is a substantial variability in individual growth trajectories. There are between-person differences in the individual average level of organizational citizenship behavior, although there is not a significant effect on the rate of change. Organizational citizenship behaviors among individuals differ leading to individual development over time.

Individuals vary in organizational citizenship behavior at different points in time, although there is not a common trend, it develops over time, and that there are differences in the individual development. Within-individual development predicts accurately organizational citizenship behaviors (Bolino et al., 2012; Mitchell & James, 2001; Ployhart & Vandenberg, 2010; Singer & Willett, 2003). The individual development of organizational citizenship behavior and individuals differ in this development over time.

The dynamic nature of organizational citizenship behavior leads to an intraindividual development in organizational citizenship behavior over time. An intraindividual development in organizational citizenship behavior and individuals differ in their development. Individuals regulate citizenship behaviors before, during, and after the potential promotion creating a discontinuous slope modeled by a temporal predictor leading to an increasing rate of change in organizational citizenship behaviors (Hui et al., 2000; Singer & Willett, 2003).

Organizational citizenship behavior at individual level factor has a positive relationship with the ability to achieve in society an organizational strategy and competitive advantage. The different levels of organizational citizenship behaviors are affected by the individual psychological, physical, and emotional resources (Beal et al., 2005; Binnewies et al., 2010; Minbashian & Luppino, 2014).

The more committed an individual is to the organization, the results of work and organizational citizenship behavior are more positive. Individual voluntary initiatives referred as organizational citizenship behaviors are not directly rewarded but contribute to improve organizational operations (Van Dyne et al., 1994; Niehoff, 2005). At the individual level, autonomy has a positive impact on innovative behavior and organizational citizenship behaviors.

Social exchange explains why individuals are willing to show organizational citizenship behavior (Vizano et al., 2020; Sena et al., 2020) based on the assumptions of reciprocity and equity. Organizational citizenship behavior is an individual choice and initiative out of formal reward system (Bies & Organ, 1989). Individuals that go beyond their formal duties indulge in discretionary behaviors that are called organizational citizenship behaviors (Bogler & Somech, 2004).

Factors influencing the development of organizational citizenship behavior are such as the individual motives, attitudes and dispositions, job satisfaction, group cohesiveness, organizational commitment, and organizational. Age-related changes leading to expressions for personality traits are beneficial for individuals and organizations in organizational citizenship behaviors and counterproductive working behaviors in the work context. Older organizational members become high on honesty-humility, conscientiousubjective normess, and openness to experience, and low on emotionality. Variation in OCB occasions indicate the variations within and across individuals (Dalal et al., 2009).

The specific dimensions of organizational citizenship behavior oriented toward the individuals include altruism, conscientiousubjective normess, courtesy, civic culture and sportsmanship, peacekeeping, and cheerleading (Organ, 1990; Podsakoff et al., 2009). Psychological property is the state of individual consciousubjective normess, thoughts, and beliefs (Ahmad & Zafar, 2018). Organizational citizenship behavior oriented to individual is categorized into courtesy and altruism while organizational citizenship behavior oriented to organizations is categorized into civic virtue, sportsmanship, and conscientiousubjective normess. A courtesy and respect organizational citizenship behaviors oriented toward organizations. Sportsmanship is the behavior of someone who does not file a protest or dissatisfaction.

The model of organizational citizenship behavior developed by Organ (1988) consists of five factors: altruism defined as helping behavior to other individuals, conscientiousubjective normess as the attendance beyond the limits, courtesy as prevention of work-related problems; sportsmanship as tolerance to inevitable inconveniences, and civic virtue as concerns about organizational life. A subsequent model removes conscientiousubjective normess and combines altruism and courtesy (Podsakoff & MacKenzie, 1994).

Human resource management has implications in organizational support for the individual efficacy and organizational citizenship behavior. Managerial support promotes the organizational citizenship behavior as the proactive and discretionary behavior displayed by an individual to contribute voluntarily to the organization (Lavelle, 2010). The organizational citizenship behavior has positive effects on the sustainability of the organization through the individual behaviors outside their job descriptions and carrying out their own work responsibilities. The individual practices of discretionary behavior engage in organizational citizenship behavior (Supriyanto et al., 2019).

Most affectionate and reliable individuals contribute to increase strong relationships between citizens. Individuals who have high emotional intelligence develop the ability to adapt and empathize, affecting the feelings in the workplace and increase the organizational citizenship behavior. A socially included individual behaves and acts to benefit the group by engaging in organizational citizenship behavior (Ellemers & Jetten, 2013).

Individuals experience their social inclusion in organizational teams and groups while being under negative gossip by others, leading to affect their organizational citizenship behavior. Negative gossip has causal effects on organizational citizenship behavior and social inclusion. Negative gossip effects on organizational citizenship behavior and social inclusion can be mitigated by strengthening the perceptions of individual to be accepted for social inclusion.

Negative gossip may have beneficial results and sustainable effects on organizational citizenship behaviors, contribute to individual and organizational reputational capital of targets. Negative gossip of targeted individuals has effects on the organizational citizenship behaviors of well-functioning groups that contribute voluntarily without monitoring (Nielsen et al., 2009). Self-monitoring of negative gossip targets may use organizational citizenship behavior as a tool and tactics to enhance reputation management (Xie et al., 2019).

Some individuals can increase time and effort expended in OCB when being under pressure, as well as the sense of entitlement which allows them to violate regulations. Time management skills support the individual core job requirements and expectations related to OCB (Rapp et al., 2013). Reduction of organizational citizenship behaviors leads to reduction in organizational trust reciprocated by the individual trust in coworkers (Turnley & Fieldman, 2000).

Counterproductive working behavior is determined by individual differences more than organizational citizenship behavior and has more variance in counterproductive working behavior (Lee et al., 2019).

Ethical leadership of committed organizational members is beyond their basic duties. Ethical leadership has a significant impact toward individual and organizational citizenship behaviors (Lu, 2014). The interrelationships of ethical leadership between the perceived organizational support, and individual and organizational citizenship behavior. Ethical leadership is positively related to perceived organizational support related to both dimensions of individual and organizational citizenship behaviors. Additionally, other researchers found that committed employees having ethical leadership left (Boehm & Dwertmann, 2015). There is empirical evidence about the mediating role of ethical leadership and leader-member exchange, individually and conjunctively between the relationship of prosocial motivation and organizational citizenship behavior.

6.3 Organizational Citizenship Behaviors Toward the Organization

Organizational citizenship behavior (OCB) is an extra-role discretionary behavior that is not direct and explicitly recognized by the formal system to function effectively and efficiently in organizations (Organ, 1988). Discretional helping behaviors foster the adoption of best practices between the organizational members to enhance abilities to become more efficient in the organization (Podsakoff et al., 1997). Organizational citizenship behavior is defined as the informal behavior outside of the proper job description and expected behaviors that contribute to the organizational effectiveness (Shenkar, 2005). Organizational citizenship behavior improves organizational effectiveness (Podsakoff et al., 1997).

Organizational citizenship behavior is voluntary aimed to support the organizational social system to accomplish the goals. Social inclusion based on social exchange theory sustains that negative gossip influences the engagement of targets in organizational citizenship behavior. Social inclusion is a mediating mechanism that explains the effects of negative gossip on organizational citizenship behavior. Low social inclusion of targets predicts negative gossip experiences and decreases organizational citizenship behavior.

Social inclusion is likely to be positively related to organizational citizenship behavior. Inclusion needs of members that are not met by the work team or group may manifest lower levels of organizational citizenship behavior, hostility, and lack of group loyalty (Ellemers & Jetten, 2013). Organizational work teams and groups require the contributions of their members with the proper organizational citizenship behaviors to develop a smooth functioning (Nielsen et al., 2009).

Organizational citizenship behavior is oriented toward the benefit of the organization in general such as adherence to informal rules and attendance beyond the norm (Williams & Anderson, 1991). Organizational citizenship behavior is a helping organizational tool that contributes to the individuals to become more self-driven and indulge in extra works. Organizational citizenship behavior is the set of behaviors supporting and enhancing the organizational cooperation system and is bound to receive special rewards.

Organizational citizenship behavior is based on the notion that the organization is a system of cooperation where individuals are willing to contribute through esprit de corps, solidarity, and loyalty (Barnard, 1937; Organ, 2016). Cooperative and discretionary and organizational citizenship behaviors lead to the organizational survival (LePine et al., 2002).

Organizational citizenship behavior is an extra-role that benefits the organization, not being recognized through the payment methods (Chughtai & Zafar, 2006; Triatmanto et al., 2019; Hartinah et al., 2020). The existence of organizational citizenship behavior in the organizations is difficult to establish. Organizational citizenship behavior is the helping behavior to coworkers, managers, customers, suppliers, etc. (Penner et al., 1997; Van Dyne et al., 1995).

Organizational citizenship behaviors are voluntary and active behaviors that improve organizational productivity, job effectiveness, and interactions (Dekas et al., 2013; Kataria et al., 2013). Managerial support has an impact on the relationship between prosocial motivation and citizenship behavior in the organizational context. Organizational management supports its workers who respond with organizational commitment and organizational citizenship behavior (Jehanzeb & Mohanty, 2020). Managerial support mediates the relationship between organizational citizenship behaviors and prosocial motivation and organizational commitment. Prosocial motivation organizational members support ethical leadership to achieve challenging tasks.

There is a relationship between prosocial motivations and organizational citizenship behavior (Lazauskaite–Zabielske et al., 2015). Prosocial motivation has an indirect effect on organizational commitment and organizational citizenship behavior with the managerial support. Prosocial motives include attention to others, group norms, prosocial worth, etc. for citizenship behaviors (Zbierowski, 2018). The relationship between prosocial motivation on organizational behavior has an impact on organizational commitment. Relationships between prosocial motivation, organizational commitment, and organizational citizenship behaviors are mediated by management support. Helping coworkers and organizations predicts organizational citizenship behavior concerning the impression management.

Organizational citizenship behavior responds to the need to strengthen the relationships between the organizational members and customers (Anaza, 2015). Organizational members have an impact on changes that can benefit the organization supported by investments related decisions (Neessena et al., 2020). The non-formal behaviors of organizational members are extra works that benefit them and the organization.

The outcomes of emotional intelligence have an impact of the organizational citizenship behavior contributing to more profitable organizations, leading to less stressful organizational members, counterproductive behaviors, and absenteeism (Di Paola & Tschannen-Moran, 2001).

Public organizations develop extra-role behaviors besides the innovative behaviors and the organizational citizenship behaviors among other determinants, organizational and individual factors such as the organizational cooperative culture and organizational support. Nonprofit organizations with conventional properties require substantial amounts of OCB including the dimension of voice (Graham, 1989, 1991). At these organizations, the voice may take various forms including constructive feedback, complaints, criticisms, suggestions, self-serving, etc. (Maynes & Podsakoff, 2014).

6.4 Organizational Citizenship Behaviors Toward the Environment

Organizational citizenship behavior for the environment refers to the involvement and engagement of organizational members in undemanding environmental practices, that are not formally recognized and awarded but complement the defensive attitude of the environment by social inhabitants and the strategic green development. Organizational environment supports the sustainable management (Felin et al., 2015). Organizational citizenship behaviors for the environment support the development of self-awareness of organizational members which is beneficial to proactive citizenship, environmental behaviors, organizational and societal contexts. Organizational citizenship behavior for the environment influences the organizational environmental practices (Paillé et al., 2013; Paillé & Mejía-Morelos, 2014; Lamm et al., 2015; Temminck et al., 2015).

Sustenance of environmental security contributes to increase the organizational citizenship behavior for the environment (Priyankara et al., 2018). Organizational citizenship behavior for the environment (OCBE) is related to formal organizational environmental practices. The higher level of organizational hierarchical position must implement the organizational citizenship behavior to sustained environment into concrete and specific individual behaviors in organizations (Boiral et al., 2015).

Despite that empirical evidence is limited, perceived organizational support has effects on self-efficacy and indirectly with organizational citizenship behaviors for environmental, health and safety issues. Self-efficacy is a mediator in the relationship between perceived organizational support and organizational citizenship behavior for environment, health, and safety. Organizational citizenship behavior addresses environment, health, safety as integrated practices in organizations. The determinants of organizational citizenship behaviors are related to health, safety, and environmental issues.

Rising perception of self-efficacy among the organizational members who utilize pro-environmental behaviors in organizational citizenship behavior for the environment. Organizations with a diffuse organizational citizenship behavior for the environment tend to reduce environmental impact (Paillé et al., 2014). Voluntary initiatives focusing on positive effects on organizations are relevant where organizational citizenship behavior is more diffuse although to elucidate the determinants are less appealing.

The organizational citizenship behavior for environment is the discretionary individual behavior not recognized by formal reward system and benefits the natural environment, contributes to the organization, and benefits individuals (Robertson & Barling, 2017: 58). Positive organizational behavior leads to proactive behavior toward the society and organization. The environmental benefits result in changes in the lifestyle of individual citizens. Organizational citizenship behavior for a sustainable environment at personal and organizational levels has several barriers, such as personal attitudes, lack of knowledge, personal barriers, lack of organizational support, lack of resources, etc. (Yuriev et al., 2018).

The organizational citizenship behaviors for environment of organizational members have an influence on environmental practices (Paillé et al., 2013; Paillé & Mejía-Morelos, 2014; Lamm et al., 2015; Temminck et al., 2015). Employees engage in organizational citizenship behavior for the environment. Organizational members may develop self-awareness and behaviors that benefit the organizational and societal contexts, environmental protection, and citizenship. Organizational members with a sense of ownership engage in proactive and positive behaviors for organizational citizenship behaviors for the environment.

The descriptive norms in the work environment act on predisposition to become internalized personal norm influencing the decision to adopt organizational citizenship behavior for environmental, health, and safety issues influencing the adoption of behaviors. The attitudes toward health, safety, and environment mediate the relationship between the social norms and the organizational citizenship behaviors.

Behaviors compromised in organizational citizenship behaviors for environment are inspiring and motivating organizations and coworkers to participate, energy conservation and renewable energies, waste reduction and recycling, and other discretionary environmental behaviors. The nomological networks of organizational citizenship behavior for the environment considered as predictor have consequences in spiritual leadership. Harmonious environmental passion is a contingent factor on the relationship between the spiritual leadership and organizational citizenship behavior for the environment. Spiritual leadership affects positively the organizational citizenship behavior for the environment.

Harmonious environmental passion has a moderated effect on the positive association between spiritual leadership and organizational citizenship behavior for the environment. The harmonious environmental passion combines with spiritual leadership to enhance organizational citizenship behavior for environment. High harmonious environmental passion leads to a stronger organizational citizenship behavior for environment.

The interaction between the formal environment management systems and organizational citizenship behavior for the environment enhances the reputation of environmental protection, addresses the regulatory systems, and reduces the environmental costs (Paillé et al., 2014; Alt & Spitzeck, 2016; Zhang et al., 2016). Among the managerial implications of the personal attitudes and self-efficacy, firms may exert direct influence on organizational citizenship behavior and organizational citizenship behaviors for environment, health, and safety. Attitudes are related to pre-existing values of the person (Organ, 1990).

Organizations adopt green human resources practices and strategies to foster pro-environmental behaviors such as organizational citizenship behavior for the environment (Renwick et al., 2013; Boiral & Paillé, 2012; Lamm et al., 2013; Lülfs & Hahn, 2013; Mesmer-Magnus et al., 2012; Ones & Dilchert, 2009, 2012; Paillé & Boiral, 2013; Prajogo et al., 2014). Green human resources can contribute to adopt organizational citizenship behaviors toward the environment and influence the organizational efficiency.

Implementation of organizational strategies for the environment supports sustainable management (Felin et al., 2015). Cultural context affects the different forms of organizational citizenship behavior. Organizational citizenship behaviors are affected by cultural differences. The emotional and psychological support and empowerment lead to committed organizational green behaviors and organizational members exert organizational citizenship behavior environment. However, emotionality has no correlation with organizational citizenship behavior, although that sentimentally correlates positively, but whereas fearfulness and anxiety correlate negatively (Pletzer et al., 2020).

Organizational citizenship behavior to sustained environment in relation to intrapreneurship has been studied by Chen et al. (2015), Fellnhofer et al. (2016), Moriano et al. (2014), Rigtering and Weitzel (2013), Temminck et al. (2015), Wang et al. (2013) for a deductive study (Saunders et al., 2019). Organizational citizenship behavior to sustainable environment is related to circular purchasing but not through intrapreneurship and the hierarchical level (Neessena et al., 2020). Organizational citizenship behavior to sustainable organizational environment and intrapreneurship are behaviors and intentions that can be stimulated by the organizations, and the mediating role of intrapreneurship between circular purchasing and organizational citizenship behavior to sustainable environment (Neessena et al., 2020).

There is a relationship between organizational citizenship behavior to sustainable environment that entails actions beyond the job requirements and entrepreneurship, such as commitment to activities beyond the requirement by the job description and taking risks (Temminck et al., 2015). There is a positive relationship between organizational citizenship behavior to sustainable environment and circular purchasing in a higher level hierarchy (Neessena et al., 2020). Organizational citizenship behavior to sustainable environment has no direct effect on high-level circular purchasing and not for low level. (Neessena et al., 2020).

The level of intrapreneurial behavior of the circular purchasing behavior has a mediation relationship between the level of organizational citizenship behavior to sustainable environment and circular purchasing. Circular purchasing is stimulated with high level of organizational citizenship behavior to sustainable environment. If the organizational citizenship behavior is low, it leads to low circular purchasing. (Neessena et al., 2020). The hierarchical position of the circular purchasing has an impact on the association between organizational citizenship behavior to sustained environment and intrapreneurship.

Intrapreneurship has a mediating role, and the hierarchical role has a moderating role on the relationship between circular economy and organizational citizenship behavior to sustainable environment. There is a relationship between organizational citizenship behavior to sustainable environment and intrapreneurship, but there is no relationship between intrapreneurship and circular purchasing (Neessena et al., 2020). The relationship between organizational citizenship behavior to sustained environment and intrapreneurial behavior is stronger in higher hierarchical positions.

Behavioral control perception is related to the association between organizational citizenship behavior to sustained environment and intrapreneurial behaviors. The hierarchical level of the circular purchasing has a moderation relationship between organizational citizenship behavior to sustainable environment and circular purchasing. Organizational control facilitates the association between circular purchasing and organizational citizenship behavior to sustainable environment.

The strength of the relationship between organizational citizenship behavior to sustained environment and circular purchasing is related to the level of hierarchy position, the higher the level the higher the level of strength. The hierarchical level of the circular purchasing behavior has a moderation relationship between organizational citizenship behavior to sustained environment and the level of intrapreneurial behavior. The circular purchasing at higher level leads to more discretion to implement organizational citizenship behavior for sustained environment into intrapreneurial behaviors.

Organizational citizenship behavior for the environment is measured by the scale developed by Boiral and Paillé (2012). The organizational citizenship behavior of individuals measured by the scale elaborated by Williams and Anderson (1991) refers to citizenship behaviors directed at individuals, colleagues, and supervisors in the organization. The relationship between organizational citizenship behavior to sustainable environment and circular purchasing requires further research to determine the influence of intentions according to the planned behavior theory.

A correlation analysis was conducted between organizational citizenship behavior for sustainable environment (OCBSE) and intrapreneurship, but it was not found to be multicollinear, and the control variables such as the type of organization, age, gender, and education were not correlated with the variables of organizational citizenship behavior for sustainable environment (OCBSE), intrapreneurship, and circular purchasing. Temminck et al. (2015) developed a scale to measure organizational citizenship behavior for the environment (OCBE).

6.5 Individual and Contextual Dimension

Individual and contextual dimensions are related to organizational citizenship behavior (Somech & Drach-Zahavy, 2004). An integrative theoretical framework and empirical perspective on sustainability management and organizational citizenship behavior explain the driving forces in relation to the individual and contextual factors influencing health, safety, and environmental issues. The theoretical perspective of organizational citizenship behavior has several managerial implications leading to the enhancement of environmental, health, and safety performance based on the discussions on antecedents of organizational citizenship behaviors (Boiral & Paillé, 2012; Fugas et al., 2013; Hofmann et al., 2003; Lülfs & Hahn, 2013; Paillé et al., 2013).

The analysis of antecedents and determinants of organizational citizenship behavior focuses on contextual and individual factors (Eisenberger et al., 1990; Fishbein, 1963; Ramus & Killmer, 2007; Temminck et al., 2015; Witt & Wilson, 1991).

In determining the factors leading to proactive behaviors are related to environmental, health, and safety issues combine theories on organizational and individual behavior that derive from a combination of contextual factors and individual predisposition (Eisenberger et al., 1990; Ramus & Killmer, 2007; Ajzen & Fishbein, 1980; Fishbein & Ajzen, 2010). The influence of individual attitudes on environmental and social dimensions is positive in relation to organizational citizenship behaviors for health, safety, and environment. Social norms play a crucial role on organizational citizenship behaviors for the environment, health, and safety, although they can be mediated by attitudes toward them (Testa et al., 2020; Lülfs & Hahn, 2014; Clarke, 2006) findings based on the reciprocity between individuals and organizations (Randall et al., 1999; Rhoades & Eisenberger, 2002).

Organizational citizenship behaviors (OCBs) related to the environment, health, and safety are shaped by voluntary individual initiatives as factor for efficient organizations. The organizational citizenship behavior for environment, health, and safety as a theoretical framework on individual behavior based on theory of planned. However, personal attitudes do not predict proactive safety behavior. Organizational and individual variables are motivators of safety citizenship behaviors (SCB) (Didla et al., 2009).

Organizational and individual perspectives on health, safety, and environment are combined to analyze the discretionary actions of organizational citizenship behavior that provide individual, organizational, and societal benefits. Individual attitudes, organizational support, self-efficacy, and social norms are factors influenced by perceived organizational support and social norms and have positive influence on organizational citizenship behaviors related to environmental and health and safety issues.

6.6 Conclusion

Organizational, individual, environmental, and contextual factors lead to the innovative behavior and organizational citizenship behavior as extra-role behaviors. Innovative behavior and organizational citizenship behavior explain both the organizational- and the individual-level dependent variables which have a positive influence. Organizational management must encourage members to participate in discretionary citizenship behaviors for the environment aimed to protect the natural environment and achieve green objectives.

The developments in organizational citizenship behaviors examine the individual engagements in organizational citizenship behavior over time. The dynamic nature of organizational citizenship behaviors over time is analyzed using longitudinal approach into the directions and forms of intraindividual development. Contextual and personal determinants must be explored from an organizational perspective.

Social inclusion explains variations in organizational citizenship behavior. Social inclusion is a mediation relationship between negative gossip and organizational citizenship behavior. In correlational studies, the perceived negative gossip is related to social inclusion and organizational citizenship behaviors. Social inclusion and social closeness may be difficult to appreciate in an experimental setting and experimental group as they would be different in an organizational group.

Management of organizations plays a relevant role to encourage the organizational citizenship behavior for the environment supported by a spiritual leadership in the mitigation of climate change through actions such as reduction of the carbon footprint. Organizational members with high harmonious environmental passion are influenced by the effects of spiritual leadership on organizational citizenship behavior for the environment.

Future research should design longitudinal studies on negative gossip and its effects on organizational citizenship behavior and social inclusion. Individuals participating on experimental settings have less opportunities to develop social inclusion and more opportunities to develop social closeness and predict organizational citizenship behavior. Further research should examine the occurrence of organizational citizenship behavior and counterproductive working behavior related to personality domains. Future research must address the implications of harmonious environmental passion for organizational citizenship behavior for the environment. Future research is needed on the linkage between contingency factors related to spiritual leadership and organizational citizenship behavior and environment.

References

Ahmad, I., & Zafar, M. A. (2018). Impact of psychological contract fulfillment on organizational citizenship behavior: Mediating role of perceived organizational support. *International Journal of Contemporary Hospitality Management, 30*(2), 1001–1015. https://doi.org/10.1108/IJCHM-12-2016-0659

Ahmad, B., Iqbal, S., Hai, M., & Latif, S. (2021). The interplay of personal values, relational mobile usage and organizational citizenship behavior. *Interactive Technology and Smart Education, 19*, 260. https://doi.org/10.1108/ITSE-01-2021-0016

Ajzen, I., & Fishbein, M. (1980). *Understanding attitudes and predicting social behaviour*. Prentice Hall.

ALHussain, A. Z. (2011). *Barriers to knowledge management in Saudi Arabia*. The George Washington University.

Alt, E., & Spitzeck, H. (2016). Improving environmental performance through unit-level organizational citizenship behaviors for the environment: A capability perspective. *Journal of Environmental Management, 182*, 48–58. https://doi.org/10.1016/j.jenvman.2016.07.034

Anaza, N. A. (2015). Relations of fit and organizational identification to employee-customer identification. *Journal of Managerial Psychology, 30*(8), 925–939. https://doi.org/10.1108/JMP-12-2012-0389

Barnard, C. I. (1937, December). *The functions of the executive*. Harvard University Press.

Beal, D. J., Weiss, H. M., Barros, E., & MacDermid, S. M. (2005). An episodic process model of affective influences on performance. *Journal of Applied Psychology, 90*(6), 1054–1068. https://doi.org/10.1037/0021-9010.90.6.1054

Bergeron, D. M. (2007). The potential paradox of organizational citizenship behavior: Goodcitizens at what cost? *Academy of Management Review, 32*(4), 1078–1095. https://doi.org/10.5465/AMR.2007.26585791

Bies, R. J., & Organ, D. W. (1989). Organizational citizenship behavior: The good soldier syndrome. *The Academy of Management Review, 14*(2), 385–400. https://doi.org/10.2307/258426

Binnewies, C., Sonnentag, S., & Mojza, E. J. (2010). Recovery during the weekend and fluctuations in weekly job performance: A week-level study examining intra-individual relationships. *Journal of Occupational and Organizational Psychology, 83*(2), 419–441. https://doi.org/10.1348/096317909X418049

Boehm, S. A., & Dwertmann, D. J. (2015). Forging a single-edged sword: Facilitating positive age and disability diversity effects in the workplace through leadership, positive climates, and HR practices. *Work Aging Retire, 1*(1), 41–63.

Bogler, R., & Somech, A. (2004). Influence of teacher's empowerment on teachers' organizational citizenship behaviour in schools. *Teaching and Teacher Education, 20*(3), 277–289.

Boiral, O., & Paillé, P. (2012). Organizational citizenship behaviour for the environment: Measurement and validation. *Journal of Business Ethics, 109*(4), 431–445.

Boiral, O., Talbot, D., & Paille, P. (2015). Leading by example: A model of organizationalcitizenship behavior for the environment. *Business Strategy and the Environment, 24*(6), 532e550. https://doi.org/10.1002/bse.1835

Bolino, M. C., Harvey, J., & Bachrach, D. G. (2012). A self-regulation approach to understanding citizenship behavior in organizations. *Organizational Behavior and Human Decision Processes, 119*(1), 126–139. https://doi.org/10.1016/j.obhdp.2012.05.006

Carpenter, N. C., Berry, C. M., & Houston, L. (2014). A meta-analytic comparison of self-reported and other-reported organizational citizenship behavior. *Journal of Organizational Behavior, 35*(4), 547–574. https://doi.org/10.1002/job.1909

Chen, M. H., Chang, Y. Y., & Chang, Y. C. (2015). Entrepreneurial orientation, social net-works, and creative performance: Middle managers as corporate entrepreneurs. *Creativity and Innovation Management, 24*(3), 493e507. https://doi.org/10.1111/caim.12108

Chughtai, A. A., & Zafar, S. (2006). Antecedents and consequences of organizational commitment among Pakistani university teachers. *Applied H.R.M Research, 11*(1), 39–64.

Clarke, S. (2006). Contrasting perceptual, attitudinal, and dispositional approaches to accident involvement in the workplace. *Safety Science, 44*(6), 537–550.

Cropanzano, R., & Mitchell, M. S. (2005). Social exchange theory: An interdisciplinary review. *Journal of Management, 31*(6), 874–900. https://doi.org/10.1177/0149206305279602

Dalal, R. S., Lam, H., Weiss, M., Welch, E. R., & Hulin, C. (2009). Within-person approach to work behavior and performance: Concurrent and lagged citizenship-counterproductivity associations, and dynamic relationships with affect and overall job performance. *The Academy of Management Journal, 52*(5), 1051–1066.

Dekas, K. H., Bauer, T. N., & Welle, B. (2013). A review and qualitative investigation of organizational citizenship behaviors for knowledge workers at Google and beyond. *Academy of Management Perspectives, 27*(3), 219–237. https://doi.org/10.5465/amp.2011.0097

Di Paola, M., & Tschannen-Moran, M. (2001). *Report on the 2001 Virginia Principals Study School of Education* the College of William and Mary in conjunction with VASSP and VAESP Supported by a grant from the Virginia Department of Education.

Didla, S., Mearns, K., & Flin, R. (2009). Safety citizenship behaviour: A proactive approach to risk management. *Journal of Risk Research, 12*(3–4), 475–483.

Eisenberger, R., Fasolo, P., & Davis-LaMastro, V. (1990). Perceived organizational support and employee diligence, commitment, and innovation. *Journal of Applied Psychology, 75*(1), 51.

Ellemers, N., & Jetten, J. (2013). The many ways to be marginal in a group. *Personality and Social Psychology Review, 17*(1), 3–21. https://doi.org/10.1177/1088868312453086

Farzaneh, J., Farashah, A. D., & Kazemi, M. (2014). The impact of person-job fit, and person organization fit on OCB The mediating and moderating effects. *Personnel Review, 43*(5), 672–691. https://doi.org/10.1108/PR-07-2013-0118

Felin, T., Foss, N. J., & Ployhart, R. E. (2015). The micro foundations movement in strategy and organization theory. *The Academy of Management Annals, 9*, 575–632. https://doi.org/10.1080/19416520.2015.1007651

Fellnhofer, K., Puumalainen, K., & Sjogren, H. (2016). Entrepreneurial orientation and performance, are sexes equal? *International Journal of Entrepreneurial Behavior and Research, 3*, 346–374. https://doi.org/10.1108/IJEBR-12-2015-0286.P.C.M

Fishbein, M. (1963). An investigation of the relationship between beliefs about an object and the attitude toward that object. *Human Relations, 16*, 233–239.

Fishbein, M., & Ajzen, I. (2010). *Predicting and changing behavior: The reasoned action approach*. Taylor and Francis.

Fugas, C. S., Silva, S. A., & Meliá, J. L. (2013). Profiling safety behaviors: Exploration of the sociocognitive variables that best discriminate between different behavioral patterns. *Risk Analysis, 33*(5), 838–850.

Graham, J. W. (1989). Organizational citizenship behavior: Construct redefinition, operationalization, and validation. Unpublished working paper, Loyola University of Chicago, Chicago, IL, 68.

Graham, J. W. (1991). An essay on organizational citizenship behavior. *Employee Responsibilities and Rights Journal, 4*, 249–270.

Hartinah, S., Suharso, P., Umam, R., Syazali, M., Lestari, B., Roslina, R., & Jermsittiparsert, K. (2020). Teacher's performance management: The role of principal's leadership, work environment and motivation in Tegal City, Indonesia. *Management Science Letters, 10*(1), 235–246.

Hofmann, D. A., Morgeson, F. P., & Gerras, S. J. (2003). Climate as a moderator of the relationship between leader-member exchange and content specific citizenship: Safety climate as an exemplar. *Journal of Applied Psychology, 88*(1), 170.

Hui, C., Lam, S. S., & Law, K. K. (2000). Instrumental values of organizational citizenship behavior for promotion: A field quasi-experiment. *Journal of Applied Psychology, 85*(5), 822–828. https://doi.org/10.1037/0021-9010.85.5.822

Jehanzeb, K., & Mohanty, J. (2020). The mediating role of organizational commitment between organizational justice and organizational citizenship behavior: Power distance as moderator. *Personnel Review, 49*(2), 445–468. https://doi.org/10.1108/PR-09-2018-0327

Kataria, A., Garg, P., & Rastogu, R. (2013). Employee engagement and organizational effectiveness: The role of organizational citizenship behavior. *International Journal of Business Insights and Transformation, 6*, 102–113.
Lamm, E., Tosti-Kharas, J., & Williams, E. G. (2013). Read this article, but don't print it: Organizational citizenship behavior toward the environment. *Group & Organization Management, 38*, 163–197. https://doi.org/10.1177/1059601112475210
Lamm, E., Tosti-Kharas, J., & King, C. E. (2015). Empowering employee sustainability: Perceived organizational support toward the environment. *Journal of Business Ethics, 128*, 207–220. https://doi.org/10.1007/s10551-014-2093-z
Lavelle, J. J. (2010). What motivates OCB? Insights from the volunteerism literature. *Journal of Organizational Behavior, 31*, 918–923. [CrossRef].
Lazauskaite-Zabielske, J., Urbanaviciute, I., & Bagdziuniene, D. (2015). The role of prosocial and intrinsic motivation in employees' citizenship behaviour. *Baltic Journal of Management, 10*(3), 345–365.
Lee, Y., Berry, C. M., & Gonzalez-Mul'e, E. (2019). The importance of being humble: A meta-analysis and incremental validity analysis of the relationship between honestyhumility and job performance. *Journal of Applied Psychology, 104*, 1535–1546. https://doi.org/10.1037/apl0000421
LePine, J. A., Erez, A., & Johnson, D. E. (2002). The nature and dimensionality of organizational citizenship behavior: A critical review and meta-analysis. *Journal of Applied Psychology, 87*(1), 52–65. Crossref. PubMed.
Lu, X. (2014). Ethical leadership and organizational citizenship behavior: The mediating roles of cognitive and affective trust. *Social Behavior and Personality: An International Journal, 42*(3), 379–390. https://doi.org/10.2224/sbp.2014.42.3.379
Lülfs, R., & Hahn, R. (2013). Corporate greening beyond formal programs, initiatives, and systems: A conceptual model for voluntary pro-environmental behavior of employees. *European Management Review, 10*(2), 83–98.
Lülfs, R., & Hahn, R. (2014). Corporate greening beyond formal programs, initiatives, and systems: A conceptual model for voluntary pro-environmental behavior of employees. *European Management Review, 10*(2), 83–98.
Maynes, T. D., & Podsakoff, P. M. (2014). An examination of the nature, antecedents, and consequences of an expanded set of employee voice behaviors. *The Journal of Applied Psychology, 99*(1), 87–112.
Mesmer-Magnus, J., Viswesvaran, C., & Wiernik, B. M. (2012). The role of commitment in bridging the gap between organizational sustainability and environmental sustainability. *Managing Human Resources for Environmental sustainability, 19*(4), 155–186.
Minbashian, A., & Luppino, D. (2014). Short-term and long-term within-person variability in performance: An integrative model. *Journal of Applied Psychology, 99*(5), 898–914. https://doi.org/10.1037/a0037402
Mitchell, T. R., & James, L. R. (2001). Building better theory: Time and the specification of when things happen. *Academy of Management Review, 26*(4), 530–547. https://doi.org/10.5465/AMR.2001.5393889
Moriano, J. A., Molero, F., Topa, G., & Mangin, J. P. L. (2014). The influence of trans-formational leadership and organizational identification on intrapreneurship. *International Entrepreneurship and Management Journal, 10*(1), 103e119. https://doi.org/10.1007/s11365-011-0196-x
Neessena, P. C. M., Jongb, J. P., Canielsa, M. C. J., & Vosca, B. (2020). Circular purchasing in Dutch and Belgian organizations: The role of intrapreneurship and organizational citizenship behavior towards the environment. *Journal of Cleaner Production, 280*(4), 124978. https://doi.org/10.1016/j.jclepro.2020.124978
Niehoff, B. P. (2005). *A theoretical model of the influence of organizational citizenship behaviors on organizational effectiveness* (pp. 385–397). New Research in Organizational Citizenship Behaviors.

Nielsen, T. M., Hrivnak, G. A., & Shaw, M. (2009). Organizational citizenship behavior and performance. *Small Group Research, 40*(5), 555–577. https://doi.org/10.1177/1046496409339630
Oikarinen, T., Hyypia, M., & Pihkala, T. (2007). Effects of HRM practices on employees within networks. *A paper prepared for 19th Nordic academy of management conference in Bergen*, Norway.
Ones, D. S., & Dilchert, S. (2009). Green behaviours of workers: A taxonomy for the green economy. *Annual Meeting of the Academy of Management, 3*(2), 321–343.
Ones, D. S., & Dilchert, S. (2012). Environmental sustainability at work: A call to action. *Industrial and Organizational Psychology, 5*(4), 444–466.
Organ, D. W. (1988). *Organizational citizenship behavior: The good soldier syndrome*. Lexington Books/D.C. Heath and Company.
Organ, D. W. (1990). The motivational basis of organizational citizenship behavior. *Research in Organizational Behavior, 12*(1), 43–72.
Organ, D. W. (1997). Organizational citizenship behavior: It's construct clean-up time. *Human Performance, OCB Measurement, 30510*, 85–97.
Organ, D. W. (2016). The roots of organizational citizenship behavior, in *Oxford handbooks online*, no. April 2018, 2016, pp. 1–20.
Organ, D. W., & Lingl, A. (1995). Personality satisfaction and organizational citizenship behavior. *The Journal of Social Psychology, 135*(3), 339–350.
Organ, D. W., Podsakoff, P. M., & MacKenzie, S. (2006). *Organizational citizenship behavior: Its nature, antecedents, and consequences*. Sage publications.
Paillé, P., & Boiral, O. (2013). Pro-environmental sustainability behavior at work: Construct validity and determinants. *Journal of Environmental sustainability Psychology, 3*(6), 118–128.
Paillé, P., & Mejía-Morelos, J. H. (2014). Antecedents of pro-environmental behaviours at work: The moderating influence of psychological contract breach. *Journal of Environmental Psychology, 38*, 124–131. https://doi.org/10.1016/j.jenvp.2014.01.004
Paillé, P., Boiral, O., & Chen, Y. (2013). Linking environmental management practices and organizational citizenship behaviour for the environment: A social exchange perspective. *The International Journal of Human Resource Management, 24*, 3552–3575. https://doi.org/10.1080/09585192.2013.777934
Paillé, P., Chen, Y., Boiral, O., & Jin, J. (2014). The impact of human resource management on environmental performance: An employee-level study. *Journal of Business Ethics, 121*(3), 451–466.
Penner, L. A., Midili, A. R., & Kegelmeyer, J. (1997). Beyond job attitudes: A personality and social psychology perspective on the causes of organizational citizenship behavior. *Human Performance, 10*, 111–131.
Pletzer, J. L., Oostrom, J. K., Bentvelzen, M., & de Vries, R. E. (2020). Comparing domain- and facet-level relations of the HEXACO personality model with workplace deviance: A meta-analysis. *Personality and Individual Differences, 152*, 109539. https://doi.org/10.1016/j.paid.2019.109539
Ployhart, R. E., & Vandenberg, R. J. (2010). Longitudinal research: The theory, design, and analysis of change. *Journal of Management, 36*(1), 94–120. https://doi.org/10.1177/0149206309352110
Podsakoff, P. M., & Mackenzie, S. B. (1994). Organisational citizenship behaviours and sales unit effectiveness. *Journal of Marketing Research, 31*, 351–363.28.
Podsakoff, P. M., Ahearne, M., & MacKenzie, S. B. (1997). Organizational citizenship behavior and the quantity and quality of work group performance. *Journal of Applied Psychology, 82*, 262–270.
Podsakoff, N. P., Whiting, S. W., Podsakoff, P. M., & Blume, B. D. (2009). Individual-and organizational-level consequences of organizational citizenship behaviors: A meta-analysis. *Journal of Applied Psychology, 94*(1), 122–141.
Prajogo, D., Tang, A. K. Y., & Lai, K. H. (2014). The diffusion of environmental sustainability management system and its effect on environmental sustainability management practices. *International Journal of Operations and Production Management, 34*(5), 565–585.

Priyankara, H. P. R., Luo, F., Saeed, A., Nubuor, S. A., & Jayasuriya, M. P. F. (2018). How does leader's support for environment promote organizational citizenship behaviour for environment? A multi-theory perspective. *Sustainability, 10*, 271. https://doi.org/10.3390/su10010271

Ramus, C. A., & Killmer, A. B. (2007). Corporate greening through prosocial extrarole behaviours – A conceptual framework for employee motivation. *Business Strategy and the Environment, 16*(8), 554–570.

Randall, M. L., Cropanzano, R., Bormann, C. A., & Birjulin, A. (1999). Organizational politics and organizational support as predictors of work attitudes, job performance, and organizational citizenship behavior. *Journal of Organizational Behavior, 20*(2), 159–174.

Rapp, A., Beitelspacher, L. S., Grewal, D., et al. (2013). Understanding social media effects across seller, retailer, and consumer interactions. *Journal of the Academy of Marketing Science, 41*, 547–566. https://doi.org/10.1007/s11747-013-0326-9

Renwick, D. W., Redman, T., & Maguire, S. (2013). Green human resource management: A review and research agenda. *International Journal of Management Reviews, 15*(1), 1–14.

Rhoades, L., & Eisenberger, R. (2002). Perceived organizational support: A review of the literature. *Journal of Applied Psychology, 87*(4), 698.

Rigtering, J. P. C., & Weitzel, U. (2013). Work context and employee behaviour as ante-cedents for intrapreneurship. *International Entrepreneurship and Management Journal, 9*(3), 337–360. https://doi.org/10.1007/s11365-013-0258-3

Robertson, J. L., & Barling, J. (2017). Toward a new measure of organizational environmental citizenship behavior. *Journal of Business Research, 75*, 57–66. https://doi.org/10.1016/j.jbusres.2017.02.007

Rotundo, M., & Sackett, P. R. (2002). The relative importance of task, citizenship, and counterproductive performance to global ratings of job performance: A policy-capturing approach. *Journal of Applied Psychology, 87*, 66–80.

Saunders, M. N. K., Lewis, P., & Thornhill, A. (2019). *Research methods for Businessed.* US, New York, Pearson.

Sawalha, N., & Kathawala, Y. (2019). Educator organizational citizenship behavior and job satisfaction moderation in the GCC expatriate-dominated market. *International Journal of Organizational Analysis, 27*(1), 19–35. https://doi.org/10.1108/ijoa-09-2017-1247

Sena, A., Cahyono, Y., & Purwanto, A. (2020). The influence of organizational culture, job satisfaction, and professional commitment on innovative behavior of flight instructors at the civil flight School in Indonesia. *Systematic Reviews in Pharmacy, 11*(9), 938–952. https://doi.org/10.31838/srp.2020.9.137

Shenkar, O. (2005). Organization behavior east and west. In K. Leung & D. Tsjovold (Eds.), *Handbook of Asian management research* (pp. 295–313). Kluwer.

Singer, J. D., & Willett, J. B. (2003). *Applied longitudinal data analysis: Modeling change and event occurrence*. Oxford university press.

Somech, A., & Drach-Zahavy, A. (2004). Exploring organizational citizenship behaviour from an organizational perspective: The relationship between organizational learning and organizational citizenship behavior. *Journal of Occupational and Organizational Psychology, 77*(3), 281–298.

Sturman, M. C. (2007). The past, present, and future of dynamic performance research. In J. J. Mortocchio (Ed.), *Research in personnel and human resources management* (Vol. 26, pp. 49–110). Emerald Group Publishing Limited.

Supriyanto, A. S., Ekowati, V. M., & Masyhuri, M. (2019). The relationship among spiritual intelligence, emotional intelligence, organizational citizenship behaviour, and employee performance. *Etikonomi, 2019*(18), 249–258.

Temminck, E., Mearns, K., & Fruhen, L. (2015). Motivating employees towards sustainable behaviour. *Business Strategy and the Environment., 24*, 402–212. https://doi.org/10.1002/bse.1827

Testa, F., Corsini, F., Gusmerotti, N. M., & Iraldo, F. (2020). Predictors of organizational citizenship behavior in relation to environmental and health & safety issues. *The International Journal of Human Resource Management, 31*(13), 1705–1738. https://doi.org/10.1080/09585192.2017.1423099

Triatmanto, B., Wahyuni, N., & Respati, H. (2019). Continual human resources empowerment through human capital and commitment for the organizational performance in hospitality industry. *Quality Access to Success, 20*(173), 84–91.

Turnley, W. H., & Fieldman, D. C. (2000). Reexamining the effects of psychological violations, unmet expectations, and job dissatisfaction as mediators. *Journal of Organizational Behaviour, 21*, 25–42.

Turnley, W. H., Bolino, M. C., Lester, S. W., & Bloodgood, J. M. (2003). The impact of psychological contract fulfillment on the performance of in-role and organizational 38 citizenship behaviors. *Journal of Management, 29*(2), 187–206. https://doi.org/10.1177/014920630302900204

Van Dyne, L., Graham, J. W., & Dienesch, R. M. (1994). Organizational citizenship behavior: Construct redefinition, measurement, and validation. *Academy of Management Journal, 37*(4), 765–802.

Van Dyne, L., Cummings, L. L., & McLean Parks, J. (1995). Extra-role behaviors: In pursuit of construct and definitional clarity (a bridge over muddied waters). *Research in Organizational Behavior, 17*, 215–285.

Vizano, N. A., Utami, W., Johanes, S., Herawati, A., Aima, H., Sutawijaya, A. H., Purwanto, A., Supono, J., Rahayu, P., Setiyani, A., & Widayati, C. C. (2020). Effect of compensation and organization commitment on tournover intention with work satisfaction as intervening variable in Indonesian industries. *Systematic Reviews in Pharmacy, 11*(9), 287–298. https://doi.org/10.31838/srp.2020.9.46

Wang, Y. L., Ellinger, A. D., & Wu, Y. C. J. (2013). Entrepreneurial opportunity recognition: An empirical study of R&D personnel. *Management Decision, 51*(1e2), 248–266. https://doi.org/10.1108/00251741311301803

Williams, L. J., & Anderson, S. E. (1991). Job satisfaction and organizational commitment as predictors of organizational citizenship and in-role behaviors. *Journal of Management, 17*(3), 601–617. https://doi.org/10.1177/014920639101700305

Witt, L. A., & Wilson, J. W. (1991). Moderating effect of job satisfaction on the relationship between equity and extra-role behaviors. *The Journal of Social Psychology, 131*(2), 247–252.

Xerri, M. J., & Brunetto, Y. (2013). Fostering innovative behaviour: The importance of employee commitment and organisational citizenship behaviour. *The International Journal of Human Resource Management, 24*(16), 3163e3177. https://doi.org/10.1080/09585192.2013.775033

Xie, J., Huang, Q., Wang, H., & Shen, M. (2019). Coping with negative workplace gossip: The joint roles of self-monitoring and impression management tactics. *Personality and Individual Differences, 151*, 109482. https://doi.org/10.1016/j.paid.2019.06.025

Yuriev, A., Boiral, O., Francoeur, V., & Paillé, P. (2018). Overcoming the barriers to pro-environmental behaviors in the workplace: A systematic review. *Journal of Cleaner Production, 182*, 379–394. https://doi.org/10.1016/j.jclepro.2018.02.041

Zbierowski, P. (2018). Social antecedents *of* entrepreneurial orientation–prosocial motivation and organizational citizenship behavior. *Journal of Positive Management, 9*(3), 83–100.

Zhang, J., Chen, Y., & Liu, J. (2016). Ethical leadership and OCBE: The influence of prosocial motivation and self-accountability. Paper presented at the *academy of management proceedings*, Briarcliff Manor, NY, 10510. https://doi.org/10.5465/ambpp.2016.15588abstract

José G. Vargas-Hernández is research professor at Instituto Tecnológico Mario Molina Unidad Zapopan, and earlier at University Center for Economic and Managerial Sciences, University of Guadalajara. Member of the National System of Researchers of Mexico. Professor Vargas-Hernández has a Ph.D. in Public Administration and a Ph.D. in Organizational Economics. He has been visiting scholar at Carleton University Canada, University of California Berkeley and Laurentian University, Canada. He holds a Ph.D. in Economics, Keele University; Ph.D. in Public Administration, Columbia University; studies in Organizational Behavior at Lancaster University and has a Master of Business Administration. He has Published 9 books and more than 300 papers in international journals and reviews (some translated to English, French, German, Portuguese,

Farsi, Chinese, etc.) and more than 300 essays in national journals and reviews. He has obtained several international awards and recognitions. He has also experience in consultancy. His main research is in organizational economics and strategic management. He teaches for several doctoral programs.

M. C. Omar C. Vargas-González serves as professor and head of the Department of Systems and Computing at the Technological National of Mexico Campus Ciudad Guzmán and professor of Telematics Engineering at the University Center of the South of the University of Guadalajara, with a master's degree in Computer Systems. He has received training in Entrepreneurship and Multidisciplinary Innovation at Arizona State University (2018) as well as a diploma in Ecosystems of Innovation and Entrepreneurship from the Harvard University School of Business. He currently conducts research in diverse areas such as Entrepreneurship, Economics, Statistics, Mathematics and Information and Computer Sciences, has collaborated in the publication of more than 15 articles in magazines and directs innovation and technological development projects.

Chapter 7
Urban Green Innovation

Public Interest, Territory Democratization, and Institutional Design

Vargas-Hernández and Simone Di Pietro

Abstract This chapter has the objective to analyze the urban green innovation based on the guarantee that public interests decentralize the infrastructure to democratize the territory and innovating the institutional design to address the complexity of the challenges in the city. The method employed is the critical analysis supported by review of the literature and consultation of experts in the field. It is concluded that the urban green innovation capacity planning has a critical role in urban innovation development in specific areas of economic growth, social inclusion and equality, environmental sustainability, health, education, business, etc. To achieve these aims, urban green innovation requires to guarantee the public interest, the democratization of the territory, and the new institutional design.

Keywords Urban green innovation · Public interest · Territory democratization · Institutional design

---

Vargas-Hernández (✉)
Department of Management, University Center for Economic and Managerial Sciences, University of Guadalajara, Zapopan, Jalisco, Mexico
e-mail: jose.vargas@zapopan.tecmm.edu.mx

S. Di Pietro
University Center for Economic and Managerial Sciences, Zapopan, Jalisco, Mexico

© The Author(s), under exclusive license to Springer Nature Switzerland AG 2023

S. Dhiman (ed.), *Sustainable Development and Environmental Stewardship*, https://doi.org/10.1007/978-3-031-28885-2_7

7.1 Introduction[1]

Cities are very different lively ecosystems brooding places of urban innovation, imagination, and creativity. Cities shape and are shaped by the vision of urban green innovation elements that anchor investments into environmental and sustainable development. Larger cities generate more innovations because they involve interactions between people of different socio-economic backgrounds (Arbesman et al., 2009; Granovetter, 1973). Large cities have more educated and transient people who share information and knowledge (Arbesman et al., 2009).

The objective of the urban green innovation is to improve the technologies and services of the labor market while restructuring old urban industrial areas creating new urban development by including economic, science, media, leisure, and living activities. Green innovative technologies can be complex as they change the pace of the city. Urban green innovation challenges sustainable and environmental development of the city across its scales and sectors. The multilevel conceptual analysis of urban green innovation takes into consideration the micro-, meso-, and macro-levels of the corresponding niche innovations, sociotechnical regimes and landscapes.

This chapter analyses the concept of urban green innovation from a social perspective of transformation of the infrastructure planning, based on territorial democratization that is capable of evolving institutional design in the interest of the public.

Defining urban green innovation is the first step in advocating investment in specific local spaces by encouraging external funding. The intention to introduce urban green innovation requires an enhanced level of corporate governance, environmental responsibility, and sustained implementation of the city's environmental strategies and policies to put urban green projects into practice by overcoming of environmental challenges and risks. Integrating environmental sustainability, economic growth, and social development issues into urban green innovation can be challenging. Other innovative urban green projects are alternative and renewable energy-saving buildings, neighborhoods, and public spaces.

There are different approaches that can be used, each one with different reasons and ends, for example, radical versus incremental, environmental performance. Urban green innovation essentially intents to respond to the environmental changes and new societal expectations and integrate sustainability issues into achievement of economic growth, social development, environmental sustainability, as well as community welfare and good.

Urban green innovation practices contribute to sustainable urban planning of infrastructure and urban green areas. These urban undertakings are innovation hubs for urban green areas having relevant impact on economy, technology, society, organization, etc. Urban green areas are the testing ground for innovation in several issues and disciplines. Urban green areas are hot spots for green innovation (Burch

[1] This chapter is the result of research completed, thanks to the support of European Union in the framework of the project "RE-CITY ITN REVIVING SHRINKING CITIES: innovative paths and perspectives towards livability for shrinking cities in Europe."

& De Luca, 1984; Krott, 1998). Innovation can be social, technical, socio-technical, ecological, socio-ecological, and environmental.

The urban innovation ecosystem includes communicating, implementing, and operating. The urban planning is interconnected and supported by a network of organizations, firms, local governments, educational centers, communities, etc., aimed at creating an inclusive economy based on innovation and encouragement by local business.

Urban innovation and entrepreneurship network systems in the local ecosystem play a crucial role in promoting and adapting to new situations (Granovetter, 1973). A green innovation ecosystem is formed with urban planning and development programs, existent labs and future Internet experimental facilities for users and citizens, policymakers, researchers, scientists, ICT business companies, etc. Urban cyberspace is crucial in technological innovation in management of digital cities to benefit people in green public spaces and developing civic networks in areas of economic regeneration and community building.

Urban green innovation may contribute significantly to urban sustainable development, providing an opportunity for research in a new direction for green environments (Cronon, 1991; Goudie, 1994; Forman, 1995).

There are three fundamental elements of urban green innovation: to guarantee public interest, to decentralize infrastructure to democratize the territory, and to innovate institutional design to address the complexity of the city.

7.2 To Guarantee Public Interest

Urban territorial sustainable planning is a means of public decision and expression of political will. It is aimed to regulate and organize land use and urban infrastructure and facilities through territorial democratization and safeguard of natural, environmental, and cultural resources. The main purpose of urban territorial sustainable planning is to improve the quality of life (Benabent, 2018).

Public interest is defined in the context of planning expertise as having equal access to common goods and services, such as the public spaces and urban green areas, etc. Public interest is the most commonly used term in democratic government and public administration. It is an amorphous concept typically not defined in specific terms that may change over time in accordance with the circumstances. Public interest is a commitment that has both short- and long-term implication of common interest.

Public interest is crucial in the analysis of green urban sustainable planning. Urban planning analysis is based more on quantifiable than qualitative factors. Xerez and Fonseca (2011) propose the adoption of a mixed system capable of integrating qualitative and quantitative research into the study of the social capital of the city and more specifically on the districts. The interviews and ethnographic observation would be accompanied by the production of cross-sectional data, with the aim of defining four axes: the network structure, trust and reciprocity in the resources of the neighborhood, and the community commitment.

However, normative planning is the tool in descriptive science, evident in the notion of public interest or common benefits of planning. An elaboration of the concept of public interest as a political decision must be the result of a debate on economic, sociopolitical, and ecological priorities of the society.

Defining the means and ends in planning is left to urban planners under the assumption that they know what is the best for the citizens (Banfield, 1973). Subjective motivation on participatory planning analysis may distort the objectivity of common good and realization of public interest. Urban planners rely more on facts that have scientific basis and on rules based on authority (Forester, 1987). Advocacy planning theory argues that determination of public interest of a society with many diverse interest groups are of contentious and controversial nature (Davidoff, 1973, 279).

Public interest is a common concern for the public good, the well-being, and welfare of the society and among the general public that have a stake in public affairs, management, and protection conducted by national, state, and local governments and agencies. Public interest is concerned with society as a single entity with the acceptance of sharing common ethical values. Public interest is related to the well-being and welfare of the general public where the society has a stake in opposition to the private interest related to the welfare of a person or firm. Public interest covers a wide range of issues, principles, and values of legitimate public concern in the best interest of the society. Thus, the matters and issues are of legitimate concern to the public.

Public interest is a fundamental criterion for stabilizing and determining the necessary and acceptable legitimation of political power and the best form of government in order to realize welfare and well-being. The best form of government is the one that is best suited to attain public interest for the entire society, nation, or group. Most of national and local governments are constitutionally obliged to act and serve in the public interest. The form of government best suited to public interest is democracy through mechanisms of representation and public debate. Political and government regimes where public interest is dependent on one or few people do not take into account the interest of the society, nation, or group. In a democracy, people have rights and obligations, among which is to contribute to the debates on what ought to be public interest in society.

Acting in the public interest has as components the objectives and outcomes as well as the process and procedures followed in the decision-making process, sometimes difficult to identify. Public interest definition can rest on the method to attain consensus. Clarification of the public interest requiring economic analysis of all viable alternatives to achieve the greatest aggregate utility becomes questionable. It is not always that the achievement of the greatest utility represents the public interest.

The analysis of the theoretical concepts of public interest offers to consider a nonsubjective and ethical concept to sustain the framework of green urban sustainable planning. For Montesquieu, the wise law-maker must formulate the laws based on the spirit of the people in correspondence to the public interest. There are different methods and processes to determine the public interest, either by the results or consequentialism, as utilitarianism and contractarianism, or by the procedure, as in the theory of communicative action. From the consequentialist perspective, public interest is not supported by public participation as it is essential in the theory of communicative action.

The concept of the public interest arises from utilitarianism, contractarianism, and the perspective of communicative action introduced by Habermas. All of them have in common to consider public interest as an achievement of a fair and just society. For utilitarianism, public interest means increasing well-being, although it does not explain how it should be distributed. On the other hand, contractarian proposes the principle of difference in distributive equity replacing the utilitarianism principle of efficiency in the public interest. To achieve the distributive justice in public interest is necessary to differentiate the treatment in the distribution of resources to guarantee equality between the members of the society.

The utilitarian perspective defines public interest by the consequences for those affected people. However, the concept of public interest defined by the sum of private and individual interest has some hurdles. Governments may regulate some practices for the public interest or good of society opposed to current practices. Also, not necessarily what is in the interest of the executive government is considered to be in the public interest. Public interest transcends individual interests when favoring an action that realizes public interest, although the members of the group do not agree. That is, what is public interest for the government is not for a large group of population, although this group can express what it considers is public interest.

In another perspective, Habermas replaces instrumental reason with communicative action-based reason, which considers that public interest emerges when the people accept the best argument through rational communication. Public interest communication multiplies information and discussion for a more informed citizenry geared toward democratic participation for the common good. Habermas' analysis makes imperative the distinction between lifeworld and system understood as economy—state, removing this one from democratic transformation. Also, the analysis limits participatory democracy to the lifeworld undergoing new threats from the system such as the colonization of information and communication technologies.

The libertarian-liberal perspectives of public interest, Marxism and postmodernism, reject the concept of public interest, allowing no place and leading to a critical questioning on the existence of public interest in urban planning. The analysis of philosophical theories on the concept reveals that it is a conflictive one, difficult-to-reach consensus, and its existence is questioned by the Marxism under the argument that the social classes have antagonist interests and postmodernism because individuals have different interests. The public interest is an ideological concept used to legitimize the social order.

The Marxist concept of public interest is equated with interests of the bourgeoisie which is protected by the capitalist state. The public interest in Marxism is only feasible in a classless state. Postmodernism rejects the existence of shared values of a plural society to represent the public interest. In postmodernism, the fragmentation of society does not necessarily reject the argument of public interest. The libertarian perspective argues that the concept of public interest interferes with individual liberties.

An issue that concerns the public interest does not necessarily means that the whole society can be affected or a large portion of it. Also, private interest may be of wider public interest under the assumption that damage to individual interest may damage the society as a whole. So, it is relevant to recognize personal interests through the public interests. People may have an agreement on the public interest in several critical issues despite their individual objectives. There is sometimes general consensus if people do not represent the public interest or the public good as in the case of a genocide. In this situation, the public interest acts not being favored by the majority.

An operative approach of public interest may result in shared solutions after the opinion of everyone is taken into account in transparent deliberations and after consensus is achieved. Decision-making regarding public interest should be close to the involved people and communities. On plural, cosmopolitan, and postmodern society, public and good interest should be defined on the basis of a consensus on actions and values after a social and political debate. However, it is rare if public interest cannot have consensus.

State, economics, business, society, and media as the main institutional forces of contemporary capitalism should be serving public interest in constructing a democratic society. Ecological integrity and sustainable society requires cooperative efforts among the institutions of the local and state governments, as well as economic, industry, business, and public interest groups to ensure public good of the concerned society. The public interest in green and sustainable urban planning cannot be defined in objective terms as it has unsatisfying results in terms of green innovation in urban green areas, aesthetic and social dullness, necessary urban infrastructure and services, etc.

Democracy is intricately intertwined within the state, society, and economy. In contemporary societies, the major institutional forces are the economy, state, society, and the media, which play a crucial role in democratic transformation and maintenance of social order by promoting democratization and serving the public. Public interest should be the expression of the social and political process, but it is not due to the dysfunctionalities of the democratic systems. Predominant and hegemonic interests of a corporate society will replace the public interest and good of the society.

Representatives of local authorities can be strengthened or weakened through recognition and institutional support of their ability to serve public interest. Official bodies of local governments and local civic organizations involved in formulation of public policy and legislative bills can put forth their agendas on issues of public interest. Also, they can plan and manage institutional transitions from weak governance to enabling national, state, and local governments to act in the public interest, although they cannot be expected to function only on the basis of public interest.

Identifying and determining when to act and serve in the public interest is an obligation of public officials to perform their functions and duties. Public officials must act and serve resolving the conflicts and incompatibilities derived from the flexibility of conduct standards.

Strikingly, the issue of public interest in socio-environmental affairs has been taken seriously by the local governments and communities. Stakeholder engagement in the public interest dialogue to clarify and prioritize the needs of the whole society should include local government policymakers, public interest social organizations and groups, industries and business, etc., all of them expecting environmental and social benefits. The public interest is specific, contingent, and situational regarding the needs, objectives, and values of a given time by society within the wider and shared framework.

Urban community organizations advocating local environmental and sustainable development are catalyzing urban resilience by making regulatory procedures and monitoring institutions in their deliberations of public interests. Urban community greening innovation is a community-based tool meant to strengthen resilience through social learning, adaptive management, and urban green innovation ecosystems. Innovativeness in urban green areas has developed from motivations to engage in environmental and sustainable development performance issues, risks, and challenges. This means producing and developing green innovation with environmental impact.

The concept of public interest applicable to green urban sustainable planning includes awareness of values and the consequences in the planning analysis. People may share the same public interest; however, this assumption may be questioned because they do not always share the same values and preferences. Public interest is used to justify the formulation, implementation, and evaluation of a green urban sustainable planning and other alternative planning proposals. The public interest as a concept is used to justify the green urban sustainable planning and the distribution of natural resources, although the concept is not always specific.

In situations in which local governments are delegated to perform as independent public and private agencies, they are faced with harshness at higher levels of uncertainty in order to acquire knowledge and to utilize their ability to the fullest extent. The exercise of public management in environments of high complexity and uncertainty requires that managers and public executives have the tools of analysis for the treatment of various public issues concerned with the public interest through the development of critical analytical skills and through training in applied heuristics.

Those who are users of public and social services from local government agencies and institutions do not always also have clarity about many interdependent and noncompetitive structures when there is only one choice for public interest. Then, it is assumed that local government agencies also have to make more choices than the citizens as consumers of goods and services themselves. It is also relevant in this order of structural relationships, the analysis and determination of satisfaction levels for the client-consumer-final citizen, as well as the determination of what will be most appropriate for the public interest and general social welfare.

The research consists of the evaluation of explanatory and comparative frames of reference, the proof of models centered on structures/cultures/organizations/spatial contexts, and the comparison of different instruments to achieve the verification of the same results. The reference model defining the urban green innovation ecosystem must describe the characteristics taking into consideration the different features of the cities to tailor the model of local urban green innovation accordingly. In addition, the new public management in local government organizations takes into account the organizational culture of the private sector, accountability for results rather than the traditional public sector, processes of accountability, and the vocabulary of efficiency and service rather than the justice of the public interest.

If the efficiency of the private market economy is questioned on its own, then this argument implies that private companies need a more critical and differentiated analytical approach to public interest before they are recommended as models of organizational efficiency to public organizations. These concerns are often equated to the public interest with the same interest of the current government; this is the public interest as per the current government. The democratic society is concerned about income disparities between citizens and their well-being, so it has to make policy decisions that often involve questionable negotiations.

There is a genuine concern for the application of the entrepreneurial model from the perspective of the new public management of democratic ideals, only because the public entrepreneur is capable of abandoning selfish behavior in favor of the public interest. It is worrying that the concept of public interest does not provide specifically enough guidance for ethical-centered behaviors. There is concern about the urgency of the behavior of public servants that to pursue the public interest may induce some of them to inject excessively their personal values into the decision-making processes, in such a way that they achieve their personal advantage instead of social benefit. The behavioral assumption on the public interest involved in the new management of the public entrepreneur is another area of concern.

An example of how the field of collaborative planning has evolved in recent decades is the northern European case. The schools of Versailles and that of Paris proposed a systematic way to qualify and quantify the spatial relations between the different elements of collaboration. Ravetz (1980) highlights the opposition generated to this type of urbanization processes generated in Great Britain in the 1970s. After the creation of numerous English cities around industrial activities, the new urban transformations required an approach that surpassed the modern architectural discourse to embrace the possibility of guaranteeing the inhabitants a space in which they could be identified, especially through conservation and restoration work.

Although the methods of quantitative spatial analysis supported by the new technological platforms certainly offer numerous advantages, the new challenge must consist in integrating the inhabitants into the decision-making processes, so as to allow the meeting of the parameters of efficiency with the quality of life, which is supported by identification mechanisms within the territory.

Negotiation in urban innovation processes among local authorities, city planners, land use conservatories, communities, neighborhoods, developers, and any stakeholder with interest can contribute to find innovative solutions concerning public interest for urban planning, design and development of public open spaces, urban green areas, and housing. The interest and expertise of each one of the stakeholders steer and enable to reframe of the problems and the developments of innovative solutions. Freedom and independence of stakeholder's participation to address politically sensitive problems are relevant factors in contributing to collaborative planning to explore innovative solutions based on the public interest. The expertise of stakeholders contributes to facilitate the access to broader and innovative research networks of public interest.

In turn, urban governance harnesses networks of urban green innovation ecosystems to sustain services to the city to locate resources and thus are challenging to urban resilience. Urban innovation drives urbanization influencing negatively the resilience of ecosystems, which needs to be reinvigorated. Innovation is supported by resilience and transformative capacity systems which are human dominated. Society innovative capacity in urban green spaces is based upon resilience and sustainability of the city. Urban green spaces are essential components of cities (Husqvarna Group, 2012) to promote ecosystems and human health (Tzoulas et al., 2007).

Integration of environmental sustainability into green area innovation is a key challenge. Green innovation is the result of interaction between innovation, sustainable and environmental development, and economic growth. The green areas innovation model can be refined into urban green innovation ecosystem supported by the theory of systems to determine city processes.

Finding urban green innovative solutions do not necessarily come out of confrontations based on ideological clash, although conflict resolution or third-party negotiation processes between the participants and stakeholders can be used in specific disputes concerning the public. The concept of user-driven open urban innovation ecosystems are sustained by the urban needs and interests of stakeholders, such as citizens, local governments, social organizations, and business. This concept of urban innovation ecosystems can bridge the gap between urban development priorities and technological research and experimentation based on public interest.

Popular mobilization on the street serves to produce political results in the street itself, because the institutions with their current configuration do not respond, they are co-opted, taken for other values and interests that are neither the interests of the population nor the public interest. The development of forms of participatory democracy of organized and well-informed citizens who participate in decision-making processes will enhance social policies based on the public interest of the society as a whole.

The public interest is an ethical concept that is the consubstantial essence of political discourse and a guide for political action attempting to justify the policies and actions of government on decision-making in the green urban sustainable planning and empirical evaluation processes. Public interest must be identified, demonstrable, and the implementation process and results must be evaluable. Public interest is a basic premise for a territorial democratic process of green urban sustainable planning. The public interest emanates from the action of the state because it can be demonstrated that it reflects that green urban sustainable planning. This is an instrument for the expression of territorial democratizing with the explicit objectives and consequences of decisions affecting all the individuals equally.

7.3 To Decentralize the Infrastructure to Democratize the Territory

The debate around the urban public space, the housing, and urban justice revolves around the object of struggles for the right to land, city infrastructure, and territorial order. Public space emerges as a situation of paradigmatic spatial transition, from a space that is a box or vehicle to a space that is an entity in itself. It is a collective presence rather than social movements in the city and in public spaces; it is a different type of claim to space.

The green urban resources and infrastructure for innovation, research, and innovation networks between governments, business, and higher education and research institutions are determinants of the city welfare. The position of the different actors in the innovation network facilitates the access to diverse information and knowledge flows and provides the potential to create and develop innovative opportunities (Burt, 2002, 2005). For example, green roofing is an innovative practice in design technologies and wastewater management. The use of innovative practices in green roof is in both promotional policies and construction as a tool in a broader plan to create green infrastructure for ecosystem services, although this may not completely mitigate the ecological footprint in urban ecosystems.

The dominant construction of territory, land, and urban space in the contemporaneity are the complementarity of capitalism and the large-scale colonialism brought about by geospatial concentrations and the creation of wide inequalities in cities. The spatial urban organization shows dysfunctionalities, such as the abandonment of public spaces, such that it hinders the agglomeration of people and the distancing of institutions from city centers making access difficult. Democratization has a very broad meaning.

The territorial unitary vision of urban spaces is problematic, in the sense that they are the reflection of the production of hegemonic imaginaries and fictions, from colonial fictions to nationalist fictions. Alongside authoritarian urban spaces, the spaces of the excluded coexist as a response, which gives rise to the struggle for public spaces that show many artists and many dimensions to accommodate urban social movements. The public space of the new social movements today is the space itself, the space itself is the value, it is the question of the political arena.

The extractivism of natural resources territorializes economic and political relations, giving rise to contradiction in the processes of economic globalization under the assumption that it is considered the deterritorialization of production, distribution, and consumption processes. The processes of deterritorialization are just one of the side of globalization as opposed to the processes of reterritorialization because there are certain elements inherent to process of production, distribution, and consumption in certain places.

The struggle for land, territory, and urban space is a struggle against the colonial and capitalist heritage of territorial space. Cities considered plurinational in geopolitical terms may have territorial autonomy that is not merely administrative or political, so they are not independent territories. This type of autonomy is the result of the recognition of the existence of other ways of administering the territory, cosmovisions, cultures, etc. The land and its natural resources today are within geopolitics of a territory much more complex than what existed before.

The struggles for land, territory, and urban space prosecuted by the new social movements form a front that is common to territorial fascism with its colonial domination and exploitation of the territory in the defense of ecological conceptions in the face of capitalist and colonialists pressure. Territorial fascism refers to the logic of the territory that crosses the cities and ends up fracturing them, giving rise to spaces within cities that have an abysmal line between civilized areas, private urbanizations that go against public spaces and wild areas, where the elites live. These tribal areas proliferate in cities that do not have the capacity to accommodate populations in an urbanistically reasonable, socially, and politically decent way.

The different forms of territorial fascism exist in spite of political and legal homogeneity, but these institutions are able to act in a totally different way irrespective of whether they consider the enemy territory or not in a civilized or savage way. The internal territory can be a very poor neighborhood or groups of terrorists. The internal territories are subject to forms of internal geopolitics that import the relations of international conflicts for the territory itself and that reproduces internal colonial relations.

A great innovation based on a new idea of territory that focuses on the neoliberal capitalist logic that validates the function of the territory as it is exploited is the consideration of the original groups that believes in the logic that no territory has commercial value. A great alteration occurred at the beginning of the twentieth century in the conception of the land, because previously there was a conception that there will be more human development of a territorial earth.

The social struggle around territorial land in urban centers in Latin America is part of the logic of geopolitical concentration of territory and land through colonial history and cultures. These situations that condition the current struggles over land and territory become relevant to the discourse of hegemonic and dominant practices. There are many international pressures for the exploitation of primary goods, foodstuffs, and speculation about minerals, and consequently, the pressure on land and territory is causing all political conquests to be undermined by the governments that have instituted them.

The counter-hegemonic movements fight for territory, land, and urban spaces. Movements that fight for a more equitable distribution of land are those of indigenous peoples, populations that today are largely considered indigenous or aboriginal. The territory is the very root of the cultural identity that is expressed in urban spaces where they try to recover the memory destroyed by capitalism and colonialism. The emergent political protagonism of colonialism considers that for certain social groups there is no dignity without territory. This is the case of indigenous peoples who try to claim respect for their culture and knowledge through their lands and territories.

The nature, the territory, and the use of the land and space have colonial, capitalist, and ecological dimensions. The ecological dimension marks the ecological limits of the other two dimensions. What is at the bottom of the cause is a change of civilization that forces to modify the habits of production, distribution, and consumption. The conceptual foundations of southern epistemologies, the ecology of knowledge, the sociology of absences and emergencies, and of intercultural translation well established as reference frameworks can help developed countries with a colonizing past, recognize more experiences and the knowledge to value the origins that come from the colonized territories.

In the traditional conception of urban territorial space, an attempt has been made to create the idea of an urban territorial space that would be outside the hegemonic space of subaltern colonialism that was constructed as such, rather than settler's colonialism, that is, colonialism of intense and direct occupation, within the very unequal colonial relationship. Colonialism created an arrogance that has incapacitated the colonizing countries to learn from experience and to teach the world because they despise all the innovations that may come from the colonized countries that are always considered inferior.

The pressures of the international institutions of neoliberalism argue that the comparative advantages of less-developed territories are mineral resources and that they should be exploited. For this reason, extraordinary initiatives have been taken to completely alter the development model based on extractivism to protect biodiversity that requires financing projects with tremendous political will. In the struggle for distribution of land as an agricultural resource for the political construction of a territory with its own cultural identity, it has become the reservoir of biodiversity precisely when the great orgy of natural resources occurs.

The concept of green bio-economy is rationalized based on territorial decision-making for smart city investments and capitalizes on business models (Belissent, 2010; McGeough & Newman, 2004). Smart city solutions must start with the city solutions and not smart solutions (Belissent, 2010). The term smart city was coined in the 1990s to suggest that urban planning and development was incorporating innovation, information, and communication technologies clubbed with globalization processes (Gibson et al., 1992).

Smart city is an initiative that promotes innovation and efficiency of urban planning and design of public services and facilitates access to government information and communication. To create, develop, and promote urban green innovation, values are needed in the relevant contribution of instruments for innovative communication. Innovation is a novel way of doing things embodied and implemented through recursive communication and alignment in urban social networks. Urban social networks underpin innovation of green areas.

The smart city model aims to foster the creation and development of knowledge, innovation, and creativity, increase the efficiency, and provides information access and cohesiveness for all the parties involved in sustainable urban development. The policy formulation and implementation of urban green innovation ecosystems must be supported by a sustainable urban planning involving structures and resources.

From another territorial planning perspective of green urban sustainable planning, smart-embedded devices are characteristic of smart cities to distinguish it from intelligent cities and used to create territorial innovation ecosystems by sharing and supporting cooperation between knowledge-intensive activities, institutions for learning and knowledge development, and web-based devices and applications to generate collective intelligence (Komninos, 2002, 2008a, b). Living lab (European Commission, 2010) is a user-driven open research and innovation approach to facilitate creative roles of users with the goal to involve communities of local urban users.

Open urban innovation and open business models are two concepts elaborated by Chesbrough (2003). The living labs concept is user-driven open urban innovation ecosystem which can be applied to smart cities embodying an open business model based on willingness and collaboration between the citizenry, local governments, social and nongovernmental organizations, and business enterprises to engage in innovative activities in a deterritorialized space.

The creation and transformation of the urban territory, according to industrial or commercial activities, have allowed the systematic planning model to flourish since the 1950s, continuing its predominance until today. However, the history of urban planning has evolved and continues to evolve in the direction of participatory planning in which the protagonists of the city, or everyday social actors, can express their influence in decision-making. Inevitably, urban planning represents a form of dominant hegemonic thought. The new challenge consists in reproducing systems of participatory urban planning that are characterized by a local ontology and epistemology.

The existence of successful local development processes is not possible without a strong identity that stimulates and spreads the potential of human initiatives (Arocena, 2002). The identity of a human group will be much stronger and capable of generating collective dynamics if the group has to overcome difficulties, if it has been able to transform threats into trump cards, and if it has won victories over adverse factors. On the contrary, history shows cases of collective decay when challenges disappear, when everything becomes too easy and too safe.

Finally, another important factor to consider is that territorial collaboration for urban green innovation systems is based on the concept of competitive advantage (Porter, 1990) and is driven by urban and regional development policies aiming to create the best conditions for sustainable development.

7.4 Innovating the Institutional Design

Innovative urban governments tend to be more democratic and accountable, supported by institutional and cultural frameworks that allow development of the population in general while providing local funds for initiatives to benefit low-income groups and communities (Satterthwaite, 2002).

Researchers have focused on distinctive theoretical approaches and practices of green area innovation by analyzing radical and incremental innovativeness from different perspectives such as energy based, material driven, pollution control, etc. Inovation can be incremental and radical changes in terms of ideas, practices, changing rules, and institutions, ways of organizing society. Both radical and incremental innovativeness in green areas have a relevant impact on addressing the challenges of the urban environmental and sustainable development. Therefore, urban planning must provide green areas innovation to contribute to sustainable development and economic growth.

The social capital that allows participatory planning certainly depends on the local context. The most educated urban communities normally have greater capacity to manage their territory through common initiatives, and the aspect of population size and density can be a positive factor in the effort to create neighborhood or city networks.

The uncertainty of urban development implies constant monitoring, learning, and adjustment in the objectives and intervention tools; the analysis methodology prioritizes the study of the interaction between variables of the urban system, proposing possible future scenarios and determining scenarios as objectives in the short and medium term. Concepts such as the common good are no longer taken for granted, but work is done by legitimizing and providing transparency to the process that determines the intervention criteria, while promoting the governance of the actors involved. Shared objectives and criteria are prioritized with a flexible implementation before plans and projects imposed on the territory. Inclusion now also means integrating the multiplicity of interests, capacities, and objectives of the actors involved in urban development.

The global, the totalizing, the statist, and the centralization bring criticism as they constitute the declining model. Alternative formulas emerge: "self-development," "endogenous development," "self-centered development," "integrated development," human-scale development, grassroots development, ecodevelopment, etc. It is in the local scene that the articulation between the singular and the universal, the place as a whole, is expressed at no other level (Arocena, 2002).The delimitation by municipalities corresponds only to a territorial subdivision that does not necessarily correspond to a local society. The local must refer to a local society, local initiatives, and local identity that stimulates and spreads the initiatives of local groups. Local actors are organized neighbors who try to improve the quality of life of a neighborhood or a specific area. To refer to local society is to talk about a human group that inhabits the same territory and shares identity traits and common problems.

The innovation ecosystem is a strategic concentration and interrelations of intensive knowledge-based activities provided by different institutions such as business incubators, technology parks, technology transfer centers, efficient energy centers (Barcelona Urban Studies, 2011; Duchesneau et al., 1979), etc. Urban planning and designing are challenged by innovative abilities and an innovative culture to capitalize on the creation of urban green innovation ecosystems to attract individuals and business (McGeough & Newman, 2004; Atkinson & Castro, 2008; OECD, 2003;

The Creative Class, 2011). The smart city can promote the public involvement framed at improving and exploiting the urban culture heritage.

The urban natural and green resources and infrastructures are basic elements in the innovation ecosystems that can be capitalized and may develop into new business models. Innovation ecosystems start from the formulation and implementation of urban green innovation policies supported by urban planning, resources, and structures. Urban and surrounding areas can evolve toward open, sustainable innovation ecosystems to boost research and experimentation of services driven by users in real-life environments.

Urban green area innovation is characterized as incremental versus radical, radical versus routine, revolutionary versus evolutionary, discontinuous versus incremental innovation, new versus extensions, pioneering versus modifying, original versus adapted, and basic versus improvements, by scholars and practitioners (Anderson & Tushman, 1990; Nord & Tucker, 1987; Baker & Sweeney, 1978; Stahl & Steger, 1977; Van de Ven, 1988; Zaltman et al., 1973). The context of the last 40 years is not in any way revolutionary, but perhaps counter-revolutionary. In a counter-revolutionary context, expectations are frustrated and democratic institutions do not respond.

Urban green innovation projects can capture an old and shrinking area of the city and turn it into restructured and revitalized creative metropolis. Attractive and cost-effective projects of green innovation should take into account the development of innovative building and green outdoor spaces. Redevelopment of vacant and abandoned urban spaces and buildings in shrinking cities for farming is a green innovation for revitalization. Adopting urban green innovation practices correlates with institutional design and policy changes that empower urban planning and development.

Some examples of urban green innovation projects are community gardens and farms, forestry projects, etc., which support biological diversity and spatial heterogeneity (Tidball & Krasny, 2007, 2009). They require investments, active participation, and involvement and decision-making by local residents. Community gardens in the form of green innovation exhibiting different types of management offer an opportunity for innovation in green space governance with innovative solutions to individuals, business, communities, etc. These different types of management are formed by local government, private organizations, health centers, schools, organized groups of gardeners, etc. (Lawson, 2005; Hou et al., 2009).

Urban green innovation can be steered toward sustainable knowledge, practices, institutions, and solutions. Citizenship and business demands for urban green innovation and green services quality are potential welfare of urban areas. An urban green innovation ecosystem engages user-driven innovation methods and requires the support of an open platform for heterogeneous technologies intended to be used for designing and implementing innovative and creative green spaces. However, innovative technologies require an innovative institutional design.

Operational innovative institutional mechanisms include differential land taxes and payments for environmental services to support sustainable urban greening activities that contribute to benefits such as carbon sequestration. Stressed urban green innovation systems require an institutional design shifting from the industrial economy and innovative bio-economy toward a more ecological economy to effectively support the generation and use of urban ecosystem services. Urban innovation processes can use industrialization aimed to reach sustainable economy. Urban research, experimentation, and innovation technology-oriented ecosystems may be aimed to contributing and developing potentially attractive environments to fulfill the needs of the citizenry.

For example, innovations in urban planning and policies on land use can integrate farming into multifunctional buildings. Urban farming in open rooftop generates several green-roof effects such as reducing the rooftop surface temperatures, reducing summer cooling load and heat losses, insulation against cold, etc. The spread of urban agriculture and gardens development and adaptation requires technological involvement, social organization techniques, and diffusion and extension of production techniques. Comprehensive urban planning of cities and food policies can include farming and agriculture as an urban green innovation.

The collaborative approach to urban green innovation ecosystems is based on sustainable partnerships among the stakeholders such as citizens, local government, leaders of the community, business firms, social organizations, etc. aimed to achieve resources and specific goals. The institutional design of the urban space is an object to attract considerable scientific–technological innovations interested in meeting the needs of people and supporting collaboration for the development of innovative solutions to sustainability issues. Smart city initiatives are characterized by public interventions rather than by new technology deployment in policies of innovation and social inclusion aimed at creating societal and institutional design conditions.

The public space is used to carry out demands in which the public was not in itself a vindication: it was the new work code, the rights of women, etc.. If the population goes through the noninstitutional space, it is because the institutions are not democratic or lack vitality and democratic force and therefore the understanding between institutions and institutional spaces is not achieved. Democratic institutional design does not fulfill their mission because they deviate from their functions. The people who have been expelled from the institutions are manifested in the streets depending on the capacity of democracy to respond. That is to say, the transition is manifested in a struggle for real democracy initiated by those who feel expelled and that is historically uncertain. What is claimed is an entry that implies a fundamental reform of the institutions.

Popular indigenous knowledge, rescued by the ecologies of knowledge, is knowledge that is often embedded in a practice that is born of struggle and only exists in the practical contexts. In the institutions of knowledge production, the theory and ecological knowledge as a practice opens spaces to multiplicity and diversity, insofar as it maintains that link to the social and moves away from privileging a certain type of knowledge, the knowledge that triumphed from the seventeenth-century scientific knowledge and the Eurocentric philosophical tradition.

The ecology of knowledge brings some hope in the postinstitutional design times because it is carried out in other instances than the traditional ones because the institutions no longer manage to accommodate the echoes of the new generations, which in some way makes new and diverse forms of action in the urban space, opening new political spaces.

The political struggle takes place in that space because the indignados believe that the institutional spaces were colonized by neoliberalism, neutralizing the right to political manifestation within the institutions, under the conceptualization of postinstitutionalist. The postinstitutional design is also translated into the occupation of spaces, and the logic is the same: It is a political response to a situation of frustration of expectations that were built in the last 40 years, obviously not accrediting institutions, nor the right to sustain them, because the right to private property and the right to public property are violated.

Social movements are engaging in cultural innovation challenging conceptual frameworks, that is, identity of the city and urban communities. The new forms of structural articulation between urban social movements and institutions, as well as lines of formalization between a micro- and a macropolicy, are essential components to improve urban green innovation processes. The acknowledgment of the role of social movements occurs in a neoliberal pressure boom for natural resources that cause the re-primarization of the economy, that is, a return to that idea, which is the curse from colonialism, that Latin America exports nature, exports commodities, exports natural resources, exports raw materials, and not industrial goods.

The occupation movement is more of a dimension of the postinstitutional design movement, which in this case is exploitation of private or public property. Private property belongs to an individual, public property is subject to rules of the state, so those who do not comply with the rules cannot occupy; these are the two dimensions of ownership. An opportunity for innovation of institutional design governance frameworks is a challenge that requires filling the knowledge gaps. Innovations in green urban systems at different scales and across sectors with the involvement of local society provide solutions to improve the quality of life of the urban communities.

A multidimensional measure of radical innovation is required to be applied across different institutional, community, and organizational settings with acceptable reliability and validity. Radical innovation of R & D projects is a multidimensional factor which can be measured using a construct of innovation radicalness described by the amount of technological uncertainty, technical expertise, business practices, and costs.

Local policies play an important role in creating the right institutional design setting to foster human capital for research and innovation capabilities to support the creation, establishment, and development of incubators for hi-tech start-ups connected to global-scaled innovation systems. Cities exhibit both weaknesses and strengths on innovation capabilities. Cities located in less-developed countries are more active in fostering innovation capabilities than cities in well-developed countries that are more active in hard domains. Governments are encouraged to increase investments in research and innovation to promote ICTs (van Winden et al., 2007).

7.5 Conclusions

This chapter has analyzed the interrelationships between the three fundamental principles of urban green innovation: guarantee the public interest, decentralize the infrastructure to democratize the territory, and innovating the institutional design to address the complexity of the challenges in the city.

Urban green innovation projects are interrelated to other environmentally friendly activities such as green infrastructure, energy efficiency, water quality, drinking water infrastructure, and waste water management. Urban green innovation projects are opportunities for producing goods, food, bioenergy, biomedicine, resource efficiency, farming technologies, new urban spaces, new forms of urban mobility, and transportation.

Local authorities of the city must promote the vision of the urban green innovation ecosystem as assemblies of policy making. Local authorities must provide support to enhance green innovation capacity and business intelligence through discussion, debates, and analysis of policies, research programs, and other forms to find solutions to meet the urban green spaces challenges.

The urban planning policies of local government have the potential to promote an urban green innovation ecosystem. Smart local government of a city has the capacity to generate service innovation and communication to deliver to local residents (González & Rossi, 2011). Cities more active in improving their capacity to sense and act through ICT systems are also less likely to differentiate soft domains initiatives related to innovation, human capital, and cultural heritage capabilities.

The innovation development on the use of technologies in ecosystems incorporates parameters like sustainable and environmental urban green innovation planning projects. Urban green innovation planning project is a strategy to stimulate economic growth. Research and development programs embody urban green innovation planning to meeting the needs and aspirations of citizens to provide them green public services. Urban innovation planning has a critical role in urban innovation development in specific areas of inclusion, environment, health, education, business, etc.

Development of innovative urban green planning is required to make use of some prospective studies and methodological tools focused on improving the urban green ecosystem in fields such as water, waste treatment, energy, etc. Urban green planning and design might integrate risk management in transition periods to incorporate innovative projects such as fighting against water to living with water (Rijke et al., 2008; Newman et al., 2011).

Fostering the city capacity for urban green innovation requires the implementation of human capital investment and improvement of quality of life initiatives supported by motivated local residents, innovative business, entrepreneurs and investors, talented persons, etc., able to start up new enterprises (Caragliu et al., 2009; Correia & Wünstel, 2011; Giffinger et al., 2007; Hollands, 2008; Rios, 2008; Toppeta, 2010).

Smart cities must comprise local urban green innovation ecosystems and the knowledge of innovation in bio-economy to face the challenge of securing high living standards. Future Internet technology arrangements in urban green environments involve large business and enterprises, micro, small, and medium enterprises (MSMEs), universities, research centers, etc. Future Internet facilities are used for developing and validating some service concepts and applications supported by the living standard approaches for smart cities.

Future Internet technologies engage users and citizens to enhance participation in the transformation process of individual and collective behaviors and social norms to discover and design urban green innovation projects. To this, learning alliance can operate in the context of research action referring to the risk management in urban development projects. Learning action alliances are used for urban green innovation in different sectors.

The model of urban green innovation ecosystems can evaluate its innovative capacity to identify complementarities and inconsistencies in urban planning and designs. All the opportunities should be explored to build on the innovative urban green capacity of the cities to develop and transform a multifunctional green infrastructure into a more urban green innovation ecosystem. Location of source of experimentation and innovation in green urban areas help to build capacities to face uncertainties and transitions in urban governance.

7.6 Cross-References

- Bio-Economy at the Crossroads of Sustainable Development
- Collaboration for Regional Sustainable Circular Economy Innovation
- Contributions of Urban Agro Ecological Agriculture to Ecosystem Services
- Ecopreneurship for Sustainable Development
- Environmental Intrapreneurship for Engaged Sustainability: Challenges and Pitfalls
- Environmental Stewardship
- From Environmental Awareness to Sustainable Practices
- Green Economic Growth based on Urban Ecology and Biodiversity
- Oceans and Impasses of "Sustainable Development"
- People, Planet, Profit
- Responsible Investing and Corporate Social Responsibility for Engaged Sustainability
- Responsible Investing and Environmental Economics
- Strategic Management Innovation of Urban Green Spaces for Sustainable Community Development
- Strategic Organizational Sustainability
- Sustainable Decision-Making: Moving Beyond People, Planets, and Profits
- Sustainable Living in the City
- Teaching Circular Economy: Overcoming the Challenge of Green-washing
- The Spirit of Sustainability
- The Sustainability Summit

- The Theology of Sustainability Practice
- Transformative Solutions for Sustainable Well-Being
- Urban Green Spaces as a Component of the Ecosystem

References

Anderson, P., & Tushman, M. L. (1990). Technological discontinuities and dominant designs: A cyclical model of technological change. *Administrative Science Quarterly, 35*, 604–633. https://doi.org/10.2307/2393511

Arbesman, S., Kleinberg, J. M., & Strogatz, S. H. (2009). Superlinear scaling for innovation in cities. *Physical Review E, 79*(016115), 1–5.

Arocena, J. (2002). *El Desarrollo Local: un desafío contemporáneo*. Taurus- Universidad Católica Segunda edición.

Atkinson, R., & Castro, D. (2008). *Digital quality of life* (pp. 137–145). The Information Technology and Innovation Foundation.

Baker, N. R., & Sweeney, D. J. (1978). Toward a conceptual framework of the process of organized technological innovation within the firm. *Research Policy, 7*, 150–174. https://doi.org/10.1016/0048-7333(78)90003-3

Banfield, E. C. (1973). Ends and means in planning. In A. Faludi (Ed.), *A reader in planning theory* (pp. 139–149). Pergamon. https://doi.org/10.1016/B978-0-08-017066-4.50014-X

Barcelona Urban Studies. (2011). *Barcelona Urban Studies*. http://geographyfieldwork.com/barcelona.htm. Accessed 3 Apr 2011.

Belissent, J. (2010). *Getting clever about smart cities: New opportunities require new business models*. Forrester for Ventor Strategy Professionals.

Benabent, F. M. (2018). *Public interest in political philosophy*. Córdoba Arenal Grupo Consultor S.L.

Burch, W. R., & De Luca, L. R. (1984). *Measuring the social impact of natural resource policies (216 p.)*. University of Mexico Press.

Burt, R. S. (2002). The social capital of structural holes. In M. F. Guillen, R. Collins, P. England, & M. Meyer (Eds.), *The new economic sociology: Developments in an emerging field* (pp. 148–189). Russell Sage.

Burt, R. S. (2005). *Brokerage and closure: An introduction to social capital*. Oxford University Press.

Caragliu, A., Bo, C., & Nijkamp, P. (2009). *Smart cities in Europe*. Research Memorandum.

Chesbrough, H. W. (2003). *Open innovation: The new imperative for creating and profiting from technology*. Harvard Business School Press.

Correia, L. M., & Wünstel, K. (2011). *Smart cities applications and requirements. White paper of the experts working group*. Net!Works European Technology Platform. http://www.scribd.com/doc/87944173/White-Paper-Smart-Cities-Applications. Accessed 22 Feb 2013.

Cronon, W. (1991). *Nature's metropolis: Chicago and the Great West*. WW Norton & Company.

Davidoff, P. (1973). Advocacy and pluralism in planning. In A. Faludi (Ed.), *A reader in planning theory* (pp. 277–296). Pergamon. https://doi.org/10.1016/B978-0-08-017066-4.50024-2

Duchesneau, T. D., Cohn, S. E., & Dutton, J. E. (1979). *A study of innovation in manufacturing: Determining process and methodological issues*. Univiversity of Maine, The Social Science Research Institute.

European Commission. (2010). *DG INFSO: Advancing and applying living lab methodologies*. European Commission.

Forester, J. (1987). Planning in the face of conflict: Negotiation and mediation strategies in local land use regulation. *Journal of the American Planning Association, 53*(3), 303–314. https://doi.org/10.1080/01944368708976450

Forman, R. (1995). *Land mosaics: The ecology of landscapes and regions*. Cambridge University Press. https://doi.org/10.1017/9781107050327
Gibson, D. V., Kozmetsky, G., & Smilor, R. W. (Eds.). (1992). *The Technopolis phenomenon: Smart cities, fast systems, global networks*. Rowman & Littlefield.
Giffinger, R., Fertner, C., Kramar, H., Kalasek, R., Pichler-Milanović, N., & Meijers, E. (2007). *Smart cities: Ranking of European medium-sized cities*. Centre of Regional Science (SRF), Vienna University of Technology.
González, J. A., & Rossi, A. (2011). *New trends for smart cities, open innovation mechanism in smart cities*. European Commission within the ICT Policy Support Programme. http://opencities.net/sites/opencities.net/files/contentfiles/repository/D2.2.21%20New%20trends%20for%20Smart%20Cities.pdf. Accessed 15 Jan 2013.
Goudie, A. (1994). *The human impact on the natural environment*. MIT Press.
Granovetter, M. (1973). The strength of weak ties. *American Journal of Sociology, 76*(6), 1360–1380. https://doi.org/10.1086/225469
Hollands, R. G. (2008). Will the real smart city please stand up? Intelligent, progressive or entrepreneurial? *City, 12*(3), 303–319. https://doi.org/10.1080/13604810802479126
Hou, J., Johnson, J. M., & Lawson, L. J. (2009). *Greening cities, growing communities: Learning from Seattle's urban community gardens*. University of Washington Press.
Husqvarna Group. (2012). *Husqvarna Global Garden Report 2012*. Retrieved March 31st, 2014 from http://husqvarnagroup.com/afw/files/press/husqvarna/Husqvarna_Global_Garden_Report_2012.pdf.
Komninos, N. (2002). *Intelligent cities: Innovation, knowledge systems and digital spaces*. Taylor & Francis.
Komninos, N. (2008a). *Intelligent cities and globalisation of innovation networks*. Routledge. https://doi.org/10.4324/9780203894491
Komninos, N. (2008b). *Intelligent cities and global innovation networks*. Routledge. https://doi.org/10.4324/9780203894491
Krott, M. (1998). Urban forestry: Management within the focus of people and trees. In M. Krott & K. Nilsson (Eds.), *Multiple-use of town forests in international comparison. Proceedings of the first European forum on urban forestry*, 5–7 May 1998, Wuppertal (pp. 9–19). IUFRO Working Group S.6.14.00. Wuppertal.
Lawson, L. J. (2005). *City bountiful: A century of community gardening in America*. University of California Press.
McGeough, U., & Newman, D. (2004). *Model for sustainable urban design with expanded sections on distributed energy resources*. Sustainable Energy Planning Office Gas Technology.
Newman, R., Ashley, R. M., Molyneux-Hodgson, S., & Cashman, A. (2011). Managing water as a socio-technical system: The shift from 'experts' to 'alliances'. In *Proceedings of the institution of civil engineers engineering sustainability issue* ES1.164 (pp. 95–102).
Nord, W. R., & Tucker, S. (1987). *Implementing routine and radical innovations- as substantive areas including employee attitudes*. Lexington Books.
OECD. (2003). *The e-government imperative: Main findings*. Policy Brief, Public Affairs Division, Public Affairs and Communications Directorate.
Porter, M. (1990). *The competitive advantage of nations*. Free Press. https://doi.org/10.1007/978-1-349-11336-1
Ravetz, A. (1980). *Remaking cities (375 pp.)*. Croom Helm.
Rijke, J. S., De Graaf, R. E., Van de Ven, F. H. M., Brown, R. R., & Biron, D. J. (2008). Comparative case studies towards mainstreaming water sensitive urban design in Australia and the Netherlands . In *Proceedings of the 11th international conference on urban drainage (ICUD)*, Edinburgh, Scotland, 31 August–5 September.
Rios, P. (2008). *Creating "the smart city"*. http://archive.udmercy.edu:8080/bitstream/handle/10429/393/2008_rios_smart.pdf?sequence=1. Accessed 01 Apr 2013.

Satterthwaite, D. (2002). Local funds and their potential to allow donor agencies to support community development and poverty reduction. *Environment and Urbanization, 14*(1), 179–188. https://doi.org/10.1177/095624780201400115

Stahl, M. J., & Steger, J. A. (1977). Innovation and productivity in R & D Tunng Management. Her past business experience associated individual and organizational variables. *R & D Management,* 7, 71–76, Feb 1977.

The Creative Class. (2011). *The Creative Class.* http://www.creativeclass.com/. Accessed 10 Mar 2011.

Tidball, K. G., & Krasny, M. E. (2007). From risk to resilience: What role for community greening and civic ecology in cities? In A. Wals (Ed.), *Social learning towards a more sustainable world* (pp. 149–164). Wageningen Academic Publishers.

Tidball, K. G., & Krasny, M. E. (2009). *From risk to resilience: What role for community greening and civic ecology in cities?* Wageningen Academic Publishers.

Toppeta, D. (2010). *The smart city vision: How innovation and ICT can build smart, "liveable", sustainable cities.* The Innovation Knowledge Foundation. *Think!Report,* 005/2010.

Tzoulas, K., Korpela, K., Venn, S., Yli-Pelkonen, V., Kazmierczak, A., Niemelä, J., et al. (2007). Promoting ecosystem and human health in urban areas using green infrastructure: A literature review. *Landscape and Urban Planning, 81,* 167–178. https://doi.org/10.1016/j.landurbplan.2007.02.001

Van de Ven, A. (1988). *Progress report on the Minnesota innovation research project.* University of Minnesota, Technical Report, Strategic Management Research Center.

van Winden, W., van den Berg, L., & Pol, P. (2007). European cities in the knowledge economy: Towards a typology. *Urban Studies, 44*(3), 525–549. https://doi.org/10.1080/00420980601131886

Xerez, R., & Fonseca, J. (2011). *Mixing methods in urban research: Exploring city and community social capital.* ISA RC 21 July 7–9, 2011, Amsterdam.

Zaltman, G., Duncan, R., & Holbeck, J. (1973). *Innovations and organizations* (p. 1973). Wiley.

Vargas-Hernández is research professor at Instituto Tecnológico Mario Molina Unidad Zapopan, and earlier at University Center for Economic and Managerial Sciences, University of Guadalajara. Member of the National System of Researchers of Mexico. Professor Vargas-Hernández has a Ph.D. in Public Administration and a Ph.D. in Organizational Economics. He has been visiting scholar at Carleton University Canada, University of California Berkeley and Laurentian University, Canada. He holds a Ph.D. in Economics, Keele University; Ph.D. in Public Administration, Columbia University; studies in Organizational Behavior at Lancaster University and has a Master of Business Administration. He has Published 9 books and more than 300 papers in international journals and reviews (some translated to English, French, German, Portuguese, Farsi, Chinese, etc.) and more than 300 essays in national journals and reviews. He has obtained several international awards and recognitions. He has also experience in consultancy. His main research is in organizational economics and strategic management. He teaches for several doctoral programs.

Simone Di Pietro participated in the RE-CITY ITN project financed by the European Union's Marie Curie HORIZON 2020 programme, working as a researcher at the University of Guadalajara and the Technische Universität Kaiserslautern and obtaining a doctorate in Space and Environmental studies. Previously he has three master's degrees in Socio-environmental Studies at FLACSO Ecuador, in Renewable Energies at the Universidad Politecnica de Cartagena and in Development Sciences and International Cooperation at the University of Rome La Sapienza. He worked as associate professor in the Pontificia Universidad Católica del Ecuador and as a departmental assistant in FLACSO Ecuador.

Chapter 8
Urban Green Spaces as a Component of an Ecosystem

José G. Vargas-Hernández, Karina Pallagst, and Justyna Zdunek-Wielgołaska

Abstract This chapter is aimed at analyzing a review of the empirical literature on some important features of urban green spaces such as the components, functions, services, community involvement, initiatives, and actions from an ecosystem perspective. The analysis begins from the assumption that urban green spaces are ecosystems of vital importance in enhancing the quality of life in an urban environment and supplying ecosystem services, such as biodiversity, climate regulation. Thus, the urban green space ecosystem is an important component of an ecosystem in any community development. Meeting the needs of users is related to the functions and services that urban green spaces provide to communities. Community involvement, engagement, and development require a methodology to ensure that the needs and aspirations of local users in the community are met. The methods employed in this analysis are a review of the empirical literature and documents, analysis of existing data on uses and users, interviews with authorities, and a more detailed examination of case-specific data. Also, as concluding remarks, some of the wider environmental, economic and social initiatives for local authorities and communities are suggested that might justify any funding for all the stakeholders that are represented

J. G. Vargas-Hernández (✉)
University Center for Economic and Managerial Sciences, University of Guadalajara, Guadalajara, Mexico

Núcleo Universitario Los Belenes, Zapopan, Jalisco, Mexico
e-mail: josevargas@cucea.udg.mx

K. Pallagst
IPS Department International Planning Systems, Faculty of Spatial and Environmental Planning, Technische Universität Kaiserslautern, Kaiserslautern, Germany
e-mail: karina.pallagst@ru.uni-kl.de; https://www.ru.uni-kl.de/ips/team/kpallagst/

J. Zdunek-Wielgołaska
Faculty of Architecture, University of Technology, Otwock, Poland

© The Author(s), under exclusive license to Springer Nature Switzerland AG 2023

S. Dhiman (ed.), *Sustainable Development and Environmental Stewardship*,
https://doi.org/10.1007/978-3-031-28885-2_8

and involved. Finally, the chapter proposes some of the opportunities, challenges, and further research.

Keywords Community involvement · Ecosystem · Functions · Services · Urban green spaces · Users

8.1 Introduction

The history of human life on earth is one of the living things surrounded by a natural environment that supplies water, fresh air, minerals, plants, vegetation, animals, and all the fruits of nature, etc., for enjoyment of everyday life. However, these natural and environmental resources do not last forever; most of them either become polluted and corrupted or are extinguished by consumption. Urban populations are facing ecologically vital threats from over-urbanization, such as water and air pollution, agricultural and forest lands with vegetation removal and groundwater overdraft. The ecological dimension of urban green spaces considers the objective and subjective components of a place providing a supportive habitat of biological diversity.

The renewal and increasing of urban green spaces considering the fast demographic growth and agglomeration should be accompanied by inhabitants' participation in environmental, social, cultural, and economic actions and objectives to promote the bio-economy in urban biodiversity and sustainable development. Inhabitants should be informed and motivated to participate in environmental, cultural, and educational activities and become active in the designing and planning of urban green spaces. Public support and political involvement of citizens for urban green space development requires various actions. Public urban green spaces are accessible to, and used by, all the citizens. Arrangements of public activities and actions on planned urban green spaces raise awareness among inhabitants of the city. For example, a public event can have the purpose of making users aware and educating them through experiencing and enjoying different activities organized by the urban green space.

In 2014, as many as 54% of the world's population were living in urban settings, and this is projected to reach 70% by 2050 (United Nations Department of Economic and Social Affairs – Population Division, 2014) By 2020, around 62% of world's population will live in urban areas, covering 2% of the world's land space and consuming 75% of natural resources. By 2030, two-thirds of the urban area that will exist will have to be built in sustainable urban environments.

In 1953, the seminal report on Park Life recognized urban green spaces as a vital component of urban environment and their role in social renewal (Comedia & Demos, 1995). The Urban Parks Programme was launched by the Heritage Lottery Fund and marked the change in attitude towards creating policy initiatives such as the Urban White Paper (Department of the Environment, 1996).

8.2 Components of Urban Green Spaces

An urban green space system is an important component in any community development, in housing, business, leisure areas, etc. (Baycan-Levent, 2002). Components of urban green areas are vegetation, water, accessibility, services of shelters, toilets, seating, playgrounds and sport areas, events and activities, environmental quality conditions and resources such as lighting, safety, litter bins, friendly staff, artistic features and artifacts such as sculptures, etc. The quality assessment of green spaces is measured by factors such as infrastructure, vegetation, accessibility, security, and equipment. Conditions that favor the use of urban green spaces are the distance in walking time (Herzele & Wiedeman, 2003), location and distribution, easy access, and proximity. Environmental enhancement makes better and more attractive urban green spaces by promoting inward investments, increasing the land value, and stimulating the economy of the community.

Urban green spaces are urban areas that were natural or semi-natural ecosystems that were converted into urban space by human influence (Bilgili & Gökyer, 2012). Urban green spaces are public and private open spaces in urban areas primarily covered by vegetation, which inspires active or passive recreational and sports activities or have an indirectly positive influence on the urban environment available for the users (Tuzin et al., 2002). Urban green spaces provide sustainable, diverse places where, according to the classic report Park Life, "people will find a sense of continuity, of relief from the pressure of urban living, places to be in touch with the natural cycle of the seasons and of wildlife and also places to meet and celebrate with others" (Comedia & Demos, 1995).

The mixed community green space is defined as the mix of overall community-level green spaces that significantly affect land surface temperature. However, there is inequitable distribution of heat and thermal discomfort (Huang et al., 2011). Inequitable distributions of community green space are related to socioeconomic status. There are no health impacts documented on the thermal comfort provided by green spaces. A mixed neighborhood green space is a mixed area of grass, trees, and vegetation.

Urban green spaces have different forms and types of open spaces, community parks and gardens, and landscaped areas. Some types of urban green spaces are formal green space, informal green space, natural green space, children's space, public participation, active sports space, recreation activities, and further land management policies. Urban green spaces exist in a variety of types, structures, and shapes. Urban green spaces include public parks, reserves, sports fields, streams, river banks and other riparian areas, greenways, walkways and trails, communal shared gardens, street trees and bushes, nature conservation areas, and less conventional spaces such as green walls, green alleyways, and cemeteries (Roy et al., 2012).

The broader notion of green space connotes turf grass-related residential, commercial and institutional surfaces and public facilities such as parks and playing fields. Turf grass is associated with the notion of green space that connotes turf-related surfaces as residential, commercial and institutional lawns and turf surfaces.

Urban green spaces connect the urban and the nature, while caring for the environment, social, and economic elements. Public forests and green roofs in public and community buildings and vacant and derelict land also provide ecosystem services. Productive land use ensures a long-term regeneration initiative to use green spaces properly for economic revenue by implementing sustainable urban initiatives such as drainage schemes. Green spaces include wilder, woodland-type, and untamed elements, with more areas for child development.

An urban green space can be considered a continuum without fences, hierarchies, and horizontally maintained at the same level of community-oriented service and a use-oriented approach. Green spaces are the spirit of the community. Urban green spaces are a focal point for communities (Greenspace, 2007). Urban green spaces contribute to building a sense of community among residents who are more likely to enjoy strong social ties. Green spaces promote interaction between people, developing social ties and community cohesion. Greenness in a neighborhood is one of the most important predictors of neighborhood satisfaction (Van Herzele & de Vries, 2011).

8.3 Ecosystems, Functions, and Services of Urban Green Spaces

Urban green spaces are ecosystems of vital importance in enhancing the quality of life in an urban environment. Urban green spaces supply ecosystem services such as biodiversity and climate regulation. Urban green spaces are essential for the quality of life, health, and well-being of citizens. Urban green spaces are critical for protecting wildlife, watersheds, meads, vegetation, providing air quality for a dense urban environment, and recreational activities. Cool islands in dense urban areas can be provided between green spaces. A dense green space is more effective at preventing nitrogen run-off, untreated human and industrial waste, toxic materials, and debris. Urban green spaces provide ecosystem services that can improve the conditions of the environment, pollution, and congestion of the large metropolitan area of Guadalajara.

Green spatial connectivity and density are associated with the cooling and pollution-mitigating capacity of the diversity of urban green space types and connected green path corridors. Green space density is described as the tree canopy cover (Feyisa et al., 2014), the relative percentage of vegetation (Ng et al., 2012). The cover patterns, densities, and balance of green spaces affect the urban heat island (Dobrovolný, 2013; Kong et al., 2014; Li et al., 2011). The density and size of the green space are highly interrelated and multi-scale, dependent on configuration. Greener spaces are cooler than non-green spaces and contribute to lower ambient temperatures (Srivanit & Hokao, 2013).

There is a significant association between the increased density and cooling effects of greenspaces (Dobrovolný, 2013; Feyisa et al., 2014; Hart & Sailor, 2009;

Ng et al., 2012; Perini & Magliocco, 2014; Vidrih & Medved, 2013; Weber et al., 2014; Zhang et al., 2013). The cooling range of green spaces into surroundings is influenced by the building density, arrangements, and heights (Li et al., 2012, 2013; Zoulia et al., 2009). Urban greening density is suggested to become optimal at ≥50% coverage (Ng et al., 2012). Urban greening initiatives are insufficient for achieving air quality and climate. Greening reduces heat stress and related illnesses (Bassil et al., 2010).

The reasons for visiting urban green spaces are mostly to enjoy a wide range of environmental elements such as flowers, trees, nature, fresh air, wildlife, watching cascades, educational opportunities, social activities, taking children to play, social interaction, to meet friends, picnics, meeting people, getting away from it all, passive walking and activities, shelter and sitting, etc. Shared parks and gardens may be set up to facilitate social links, collective participatory projects and collective cultural interventions, well-being recreational areas, and walkways connecting attractions and facilities.

Other reasons why people visit urban green spaces are to walk a dog, walk by the lake, river or creek, walk socially as part of a group, hanging out, passive enjoyment, sitting either on the grass or on seats, photography, messing about on swings, watching sport, reading, watching life go, smoking, sunbathing, an informal pursuit such as flying kites, fishing, etc. Walkable green spaces in urban areas are associated with a healthy environment and increasing green exercise. Walkable green spaces influence the longevity of urban senior citizens (Wolf, 2010).

Urban green spaces have beneficial physical, psychological, and health effects through physical activities and green exercises. Environmental determinants affect the use of green spaces, physical activities, and leisure. The amount of green spaces available to users in the living environment correlates with socioeconomic, demographic, and self-perceived health. Higher levels of greenness have been positively associated with lower stroke mortality. Perceived neighborhood greenness is positively associated with physical and mental health. Socioeconomic and cross-cultural variations may result from the unequal distribution of green spaces.

Another important reason to visit urban green spaces is the use of facilities such as cafes, restaurants, environmental centers, libraries, museums, sports such as football, tennis, etc., biking, skateboarding, cycling, and other forms of active enjoyment. Events are other important motive to visit urban green spaces, such as group music performance, concerts, Christmas carol concerts, orchestral performance, craft fairs, fun fairs, opera, circus, firework displays, bands playing, dancing, etc.

Making inhabitants aware of the existence of urban green spaces and using values contributing to urban inhabitants' lives in the form of a more balanced quality of life and lifestyle, encouraging physical and mental fitness, reduces tensions and conflicts, relieving the harshness of the urban environment, providing places for social and cultural interaction in informal contacts and more formal participation in social events, social inclusion, recreation, aesthetic pleasure, wildlife, and fostering community development.

Urban green spaces are natural meeting points for local inhabitants facilitating social inclusion and integration, community cohesion, social capital, civic society, supported by an increasing sense of identity and belonging (Konijnendijk et al., 2013a; Abraham et al., 2010). By providing a meeting place for social interaction and integration between community users, green spaces influence social capital. Inhabitants living near urban green spaces reduce health inequalities and have lower circulatory diseases (Mitchell & Popham, 2008). Inequitable distribution of green spaces correlates with the distribution of disadvantaged inhabitants.

Passive activities are the main reason why users visit urban green spaces such as passive or informal enjoyment of the environment, social activities and attending events, getting away from it all, walking activities including dog walking, active enjoyment including sport and specific activities. Surveys have shown that people are less stressed, communicate better and make sensible decisions when surrounded by green spaces.

Urban green spaces can be linked as wildlife corridors to facilitate the movement of fauna, preventing fragmentation and isolation of wildlife (Rouquette et al., 2013; Hale et al., 2012). Urban green spaces are home to many species, including those that are rare and threatened and the habitat for pollinators. More urban green spaces sustain more wildlife and biodiversity, providing a more favorable habitat, therefore requiring more protection from human interference (Cornelis & Hermy, 2004; Fuller et al., 2009; Schwartz et al., 2002; Baldock et al., 2015). Creation, protection, and development of urban green spaces are relevant elements of sustainable urban development.

Urban green spaces have an impact on human thermal comfort and air quality on human health (Cohen et al., 2012; Weber et al., 2014; Nowak et al., 2014). Comparisons of the impact of green space air quality and heat show that a greater predominance of trees mitigates urban heat islands, provides thermal comfort, and improves air quality. Green spaces reduce urban heat islands and air pollution, improving air quality in urban settings (Bowler et al., 2010). Community-level air quality is dependent on the tree population (Morani et al., 2011).

A community's green spaces are associated with reduced household pollution material exposure (Dadvand et al., 2012). Vegetation density in green spaces is associated with pollution mitigation (Yin et al., 2011; Dzierżanowski et al., 2011; Escobedo & Nowak, 2009; Nowak et al., 2013, 2014, Tallis et al., 2011; Tiwary et al., 2009; Tsiros et al., 2009). Bushes instead of trees may retain more pollution particles and reduce concentrations (Wania et al., 2012). A diversity of tree species including evergreen, conifer, and deciduous tree species has complementary air-pollution uptake patterns and provide maximum air-quality improvements (Manes et al., 2012). Trees and shrubs are more effective at removing pollutants than herbaceous perennials (Rowe, 2011).

A diversity of evergreen and conifer tree species provides complementary air-pollution mitigation. Coniferous trees are the best for capturing pollutant material (Tallis et al., 2011; Tiwary et al., 2009) and the evergreen, rather than deciduous trees in green spaces, provides more cooling and below comfort conditions in winter (Cohen et al., 2012; Zhang et al., 2013). Evergreen and deciduous trees remove more atmospheric (Alonso et al., 2011) than conifers.

The cooling capacity of green spaces is affected by multiple variables such as density, size, and shape associated with an increase in air quality. Urban green spaces reduce heat, ozone, and ultraviolet (UV) radiation and improve air quality (Roy et al., 2012; Konijnendijk et al., 2013b; Bowler et al., 2010). Absorbing pollutants improve air quality. Research based on modeling has weak evidence that capturing pollutants and particles by urban green spaces improves air quality (Konijnendijk et al., 2013a). Wong et al. (2013) reviewed the evidence for the relationship between green spaces, heat, and air quality considering variables such as green space type, climate, method, etc. Building orientation and heights affect cooling and air quality from green spaces.

Urban green spaces reduce the UHI effect by cooling the air on average by 1 °C and by providing shade. Cooling is influenced by plant type, green patch size and density, temperature, and wind (Armson et al., 2012; Cao et al., 2010; Feyisa et al., 2014; Fintikakis et al., 2011; Fröhlich & Matzarakis, 2013; Gaitani et al., 2011; Konijnendijk et al., 2013a; Lafortezza et al., 2009; Oliveira et al., 2011; Onishi et al., 2010; Vidrih & Medved, 2013). Many characteristics of green spaces affect the cooling capacity such as size, cover, shape, density, and spacing.

Green space scale is the area or size of green space including a single or multiple sites. The green area impact scale includes the site and the adjacent non-green areas. The percentage of covered greenspace (PLAND) equals the sum of the areas (m^2) of a specific land-cover class divided by a total landscape area, multiplied by 100 (Herold et al., 2003). There is a strong association between the size of green space and the cooling effects (Cao et al., 2010; Chen et al., 2014; Dobrovolný, 2013; Feyisa et al., 2014; Hart & Sailor, 2009; Li et al., 2012; Onishi et al., 2010; Susca et al., 2011; Weber et al., 2014). The size of the green space affects the urban cooling island because the cool air builds up and is emitted from the center (Vidrih & Medved, 2013) and it is stronger during the summer (Chen et al., 2014; Li et al., 2012; Onishi et al., 2010; Susca et al., 2011).

Increased community green space is related to lower surface and air temperatures and reduced air pollution (Bassil et al., 2010). Studies are consistent in finding low temperatures and reduced air temperature in urban green spaces (Bowler et al., 2010). Average temperatures are lower inside the urban green spaces, confirming their impact on urban heat (Yu & Hien, 2005). Mature trees remain relatively cool in an urban climate, in contrast to non-green impervious surfaces, by providing shade, thermal comfort, reduction of air temperature, and relief from the effects of heat islands (Meier & Scherer, 2012; Roy et al., 2012; Hwang et al., 2011; Lin et al., 2010; Lynn et al., 2009; Park et al., 2012; Shashua-Bar et al., 2012).

Temperature differences between green and non-green spaces are greater during the hot periods of the day (Hamada et al., 2013; Doick et al., 2014). The cooling effects are greater during the times of the hottest temperatures (Bowler et al., 2010; Cao et al., 2010; Cohen et al., 2012; Hwang et al., 2011; Meier & Scherer, 2012; Li et al., 2012; Oliveira et al., 2011; Park et al., 2012; Sung, 2013; Zhang et al., 2013).

Changes in surface temperatures from green space are related to urban heat islands, but are not an indicator of thermal comfort improvement and heat stress reduction. Higher land surface temperature is significantly associated with lower income communities with larger ethnic minorities and older adults (Huang et al., 2011). Increased green spaces increase energy flows, while decreasing land surface temperatures (Li et al., 2012, 2013; Zhou et al., 2011).

Air temperatures in warm humid climates are significantly cooler within the urban green spaces (Oliveira et al., 2011) compared with nongreen areas (Vidrih & Medved, 2013; Armson et al., 2012). Humidity tends to be higher in urban green areas than in inhabited zones. Densely inhabited areas without green spaces usually have an inadequate climate. Green spaces mitigate the effect of climate warming by providing shade. Replacing paved yards with urban green spaces reduces the heat island effect during the summer by moderating temperatures expected with climate change. Increased cover of community-level green spaces is associated with reduced air temperatures. The configuration and patch area of a community's green space have a relationship with personal exposure to air pollution at the household level, with cooler air temperatures and reduced urban heat island effects (Steeneveld et al., 2011; Li et al., 2012).

The main role and function of urban green spaces and gardens are to improve climate and reduce air pollution. The role and behavior of urban green spaces and gardens help in improving climate and reducing air pollution. Pollution in urban areas is dependent on the type of architecture and proximity to green spaces. An avenue with a green space is less polluted because dispersion is better, whereas narrow streets tend to be more polluted. Walkways with large green spaces are better protected from pollution (Eliasson, 2000). Planting more trees on street canyons may not be a good prescription where it may increase the concentration of pollutants (Escobedo & Nowak, 2009; McPherson et al., 2011).

The predominance of trees has the greatest cooling effects, provides thermal comfort and heat stress relief (Chen et al., 2014; Perini & Magliocco, 2014; Cohen et al., 2012; Zhang et al., 2013). Different scales and types of green spaces have diverse cooling effects on heat mitigation. A comparison of green space types and scales may cause the effects to overlap. Green space scales have differential scales (Cohen et al., 2012). Green spaces with trees provide greater cooling than spaces with grass (Chen et al., 2014). Higher concentrations of green spaces are associated with greater cooling (Rinner & Hussain, 2011). Connectivity between urban green spaces maximizes the cooling effects (Doick et al., 2014).

Other different types of green spaces are the green buildings that have a vegetated roof or wall serving for pollution, heat stress, and urban heat island mitigation. Green roofs and walls provide heat island and pollution mitigation services.

A green roof is a roof of a building covered with vegetation planted over a growing medium and a waterproof dispositive. Green roofs combined with insulation provide heat mitigation (Coutts et al., 2013) and good irrigation provides cooling (Zinzi & Agnoli, 2012). Green roofs affect air quality by removal of air pollution comparable with mitigation effects of urban forests (Speak et al., 2012; Baik et al., 2012). Green roofs provide cooling effects and reduce the heat island in the urban environment (Smith & Roebber, 2011; Susca et al., 2011). Green roofs and walls are an alternative in high-density urban areas for cooling and pollution mitigation.

Green roofs maximize air quality by plant selection such as creeping bent grass and red fescue that have a higher level of particle capture (Speak et al., 2012).

Green roofs do not have an effect on the street-level temperature, but decrease the cooling load of buildings (Perini & Magliocco, 2014). Green spaces with trees are more effective than grass surfacing and green roofs planted with grass at reducing temperatures and improving thermal comfort (Ng et al., 2012; Chen et al., 2009). Green roofs reduce storm water run-off (Mackey et al., 2012). Green roofs for heat mitigation cost more (Mackey et al., 2012; Coutts et al., 2013; Smith & Roebber, 2011; Zinzi & Agnoli, 2012). The impact of wind on pollution, mitigating the effects of urban green spaces, is complex, but green roofs located downwind of prevailing winds have significant mitigation effects (Baik et al., 2012; Speak et al., 2012). Wind increases the cooling and pollution-mitigating effects of green space.

Green walls have positive cooling effects (Speak et al., 2012; Baik et al., 2012) and mitigate urban heat island effects through the evapotranspiration of plants (Susca et al., 2011; Smith & Roebber, 2011). The cooling capacity of green walls increases with high temperatures (Hamada & Ohta, 2010; Koyama et al., 2013). Green walls with low wind speeds reduce air pollution in the street canyon (Amorim et al., 2013). Green walls are more effective than green roofs at mitigating in-canyon air pollution (Amorim et al., 2013; Buccolieri et al., 2011; Koyama et al., 2013).

Trees in an urban green infrastructure capture and sequester carbon, mitigating the negative effects of emissions. Carbon sequestration is the removal of the greenhouse gas carbon dioxide and its incorporation into plants. Any green space balances carbon, taking more than they return to the atmosphere (Nowak et al., 2013; Nowak & Crane, 2002). A forest in a green space maximizes carbon sequestration (Strohbach et al., 2012).

8.4 Methodological Considerations

Urban green spaces reflect the need for natural and landscaped areas within the cities. Cities have mixed land use such as residential areas, industrial areas, forest and agricultural areas, but most are man-made environments such as built-up areas and urban green areas, and water. Large cities have lost natural resources and invest more than medium cities that have more natural green areas (Tuzin et al., 2002). Urban green spaces have a critical value for planning and developing sustainable eco-cities. In cities with a higher rate of growth population density, urban green spaces tend to be reduced at the expense of the urbanization process. There are variations in areas given over to urban green spaces in cities, for example, Singapore has 47% and Sydney 46%.

Community involvement, engagement, and development require a methodology to ensure that local authorities meet the needs and aspirations of local users in the community. Some of the methods employed are the literature review, the survey of local authorities, structured interviews with authorities and a review of their documents, analysis of existing data on uses and users, and more detailed examination of a case study.

An analysis of the green space deals with the physical and quantitative, functional, ecological, environmental, economic, and quality aspects. Economic aspects are the expenses of development, costs of maintenance, financing, and budget sources. The quality of the urban green space experience needs to be studied from an interdisciplinary perspective, drawing on both natural and social sciences. Some of the physical quantitative indicators are the supply and distribution of natural and landscape resources of public green areas as a percentage of the city area, the square meter per capita, and structural and morphological characteristics. Quantitative evaluation of the relationship between urban population and urban green spaces takes into account functionality, the green space ratio, green space coverage, and green space area per capita (Xion-Jun, 2009). The quality aspects of urban green spaces are the suitability and quality of the site structure, design and provision, and the quality of the conditions.

Finding meaningful information on the uses and users of urban green spaces is hampered by inconsistencies with regard to information from local authorities. The use of model surveys to collect information from users of green spaces regarding the satisfaction of needs and aspirations needs to be developed through pilot studies and consultation by researchers and local authorities. Consultation on, and involvement in, environmental issues identifies the community's needs. Also, the results of research find evidence for the differentiation of the need to have a green space close to where they live, as opposed to where they work (Greenspace, 2007).

The user's perceptions of urban green areas matter to the community's image and decisions on uses. Perceptions of the image of urban green spaces affect uses and user aspirations and the value to the community of creating, designing, meeting the needs, and sustainably managing. The issues more related to designing are the variety, activities, spaces, sensory stimulation, vegetation, water, birds, animals, etc.

A sound basis for the collection and analysis of data is the means of finding out the priorities. A pool of data should be collected and analyzed to find out priorities in terms of type, quantity and quality, location, and accessibility. The amount of green space has been reduced by the trend toward more compact urban environment (Burton, 2003). The observed current trends on urban green spaces suggest increasing degradation, and without support, the process is not likely to be reversed. An expert study that is already available may help to compare and check the planning context and legislation.

8.5 Users of Urban Green Spaces

Meeting the needs of users is related to issues of awareness of needs, the nature of the facilities and their conditions, the opportunities for activities, events and playing, the provisions of conveniences such as toilets, shelters, seating, and refreshments. Users of urban green areas develop patterns of informal and passive activities, with peaks in the afternoons, weekends, and holidays on a daily basis. Involvement in urban green spaces leads to the creation of facilities to meet the needs of users who have more experience and demand higher quality. Facilities of urban green spaces must meet the environmental, socioeconomic, and psychological needs and attitudes of the user (Balram & Dragicevic, 2005). Meeting the users' needs at local environmental, social, and economic levels requires the development of local standards, such as the provision of urban green space per head.

In one piece of research, users of urban green spaces manifested psychological effects (Dunnett et al., 2002). A significant relationship was found between the use of green spaces and levels of stress (Grahn & Stigsdotter, 2003). Green spaces in the living environments also positively affect stress and quality of life. Urban green spaces reduce stresses for users and provide them with a pleasant positive distraction (Ulrich et al., 2010). The use of green spaces is associated with less stress. Viewing nature and urban green spaces ameliorates stress levels (Ulrich, 2002).

Natural green environments have restorative effects and pleasing stimuli promoting soft fascination (Forest Research, 2010). Also, users are happier and have greater well-being when they live in an urban area with large green spaces (White et al., 2013). The evidence between green spaces and physical activity is strong; although beneficial links have been reported between urban green spaces and emotional, psychological and mental health, and well-being, the evidence is weak. Large urban green spaces contribute to the physical and mental health and well-being of users.

There also appears to be seasonal patterns affected by the weather. Other reasons for using urban green spaces are for walking, including dog walking, passive and active enjoyment of the environment and sports, social encounters, and activities.

Accessibility to urban green spaces is related to ease of access by proximity with no physical barriers, transportation, open gates at an early hour, accessibility for disabled people, information on cues and path-finding features, maps, information at the entrance, path junctions, slopes, and cambers, and attendants for those with disabilities and visual impairment. Improving safety issues requires changes in the use of fencing, lighting, staff or rangers, removal of cars, restriction of cycling, roller-skating, and roller-blading, etc. Urban green areas are safer gathering places for children and young people, at least safer than being on the street.

User determinants such as gender and age affect accessibility and quality of urban green spaces and other environmental factors. Access to green spaces facilitates their use and increases the level of physical activities. Accessibility to green spaces has an impact on urban socioeconomic health inequalities. There are links between access to urban green spaces and social integration among older adults (Forest Research, 2010). The availability of green spaces is associated with increased survival of elderly people.

Unequal distribution of green spaces and reduced access to green environments are related to health inequalities and increases in pollution and intense heat (Alberti & Marzluff, 2004; Cohen et al., 2012; Girardet, 1996; Gregg et al., 2003; Grimm et al., 2008; Hough, 2004; Moore et al., 2003; Newman & Jennings, 2008). Deprivation levels are linked to access to green spaces. Distance from the green spaces is related to physical activity; thus, users living nearby report higher levels of physical activities, although there is no correlation with accessibility to green spaces. Proximity to green spaces is associated with self-reported health.

Increasing green spaces and optimizing spatial configuration mitigate urban heat (Choi et al., 2012; Rinner & Hussain, 2011). The ratio between the urban heat area and the urban cooling area increases with the distance from the urban green space (Choi et al., 2012). There is a negative correlation between the percentage cover of urban green spaces with a land surface temperature related to the distance where the closer it is, the stronger the cool island effects. Modifying variables that affect the relationship between green spaces and heat include density, distance, wind, temperature/season, the surrounding built-up environment, and precipitation.

Urban green space distribution inequities and neighborhood quality affect urban health inequalities. Inequalities in green space quality may affect urban health inequalities. There is evidence for the relationships among green space, heat, air pollution, and health (Lachowycz & Jones, 2011; Lee & Maheswaran, 2011). Heat and air pollution-related health inequalities are associated with green spaces. Urban green space distribution is related to health inequalities. There is evidence in the relationship between air pollution and heat mitigation from green space having positive effects on human health. Disparities and inequalities in distribution lead to pollution "hot spots" and green deserts (Escobedo & Nowak, 2009; Huang et al., 2011; Jesdale et al., 2013; Su et al., 2011).

All types of green space are associated with reductions in heat stress, urban heat islands, and air pollution. Green space density as the relative tree cover affects the relationship between green space and the mitigation of air pollution (Baik et al., 2012; Doick et al., 2014; Tsiros et al., 2009). Community green spaces are associated with lower exposure of air pollution at the household level (Dadvand et al., 2012). Greening has different impacts on heat and air pollution (Alonso et al., 2011; Nowak et al., 2014), on individual and household-level exposure to air pollution (Dadvand et al., 2012; Maher et al., 2013). Reductions of air pollution from green spaces are insignificant relative to urban-based emissions (Baró et al., 2014). Wind increases heat and air pollution, mitigating the effects of green spaces.

Unequal distribution of green spaces are related to health inequalities derived from heat and air pollution (Escobedo & Nowak, 2009; Huang et al., 2011; Jesdale et al., 2013; Su et al., 2011). The uneven distribution and quality of green spaces related to mitigation of heat and air pollution is associated with health inequalities. Green spaces have differential scales of health impacts associated with reduction in air pollution and heat (Bowler et al., 2010; Roy et al., 2012). A relationship has been identified between urban green spaces, air pollution, and health inequality (Su et al., 2011). Mitigation of pollution and heat from green spaces has a direct impact on health (Nowak et al., 2014).

Access to urban green spaces for the elderly, the disabled, children, women, and minority ethnic groups concerns issues such as ease of entrance, proximity, social inclusion, provision for the visually impaired, public transport, parking, moving safely, and surface design. Awareness and understanding of social inclusion in urban green areas is a recognition of the particular social and cultural needs and aspirations of users who are most likely to be excluded in society.

Some users of urban green areas are concerned about environmental quality issues such as litter, dog mess, graffiti and vandalism, lack of garbage cans; garbage and items such as condoms, food put out for birds that has been left lying around (Gregory & Baillie, 1998); smashed bottles and broken glass. Psychological issues related to the use of urban green areas prevent users from going alone because of the feeling of vulnerability, fears, safety concerns, laziness, loneliness, lack of confidence, inertia, etc.

Some negative economic impacts of urban green spaces may be a fall in profits and potential problems derived from gentrification of the area as a result of increasing the added value of land and increases in housing prices. Negative impacts identified with green spaces are the increased green density, which increases street canyon air pollution detrimental to health (Amorim et al., 2013; Morani et al., 2011). Other negative impacts of green spaces is the tree emissions of biogenic volatile organic compounds that increase the levels of ground-level ozone (Escobedo & Nowak, 2009; Roy et al., 2012). Green spaces with high BVOC-emitting tree species reduce ground-level ozone. Some negative impacts and trade-offs of green spaces are exposure to pollen and physical injuries.

Some of the personal issues that deter people from using urban green spaces are factors such as not having enough time, working unsocial hours, poor health and mobility, preferences for visiting other places, issues related to the location of urban green spaces, accessibility, user experience, and environmental quality. Other personal issues can deter users from going to urban green spaces, such as having their own park, changing circumstances, family and parental restrictions. To increase parental responsibility, training sports sessions for children and young people, encourage the active participation of parents. Users of urban green spaces are deterred by a lack of deficient facilities, the influence of undesirable people, safety issues and psychological concerns, dog mess, litter, graffiti, and vandalism.

The deterrent effects of other users are related to conflicts between children and young people, teenagers, with adults, drug users, undesirable characters, users drinking alcohol, verbal abuse, gay men, bikes and skateboards, gamblers, noisy people, and being crowded, etc. The study of the urban environment combines the sound levels, biodiversity, and green spaces. The results of this study confirm that the planning and designing of urban green spaces are enhanced by the ecological quality in issues such as noise levels of livable and sustainable communities (Williams et al., 2000; Girardet, 2004). The soundscapes of green urban spaces have been less well-studied.

The declining quality of urban green spaces contributing to a decline in the urban quality of life has been studied by Irvine et al. (2009). Dog mess is a critical concern in urban green spaces and requires special attention such as dog-free areas and areas for dogs, good positions of dog bins in suitable locations, dogs on the leash and controlled, dog toilets, proper use of fines, etc.

The most relevant emerging barriers are the resource issues rather than personal concerns regarding the lack of facilities, the lack of maintenance, including play opportunities for children; not enough to do, the negative influence of other green space users; dog mess and dogs not being on leash; physical safety and other psychological concerns such as fears concerning environmental quality, including litter, vandalism and graffiti, accessibility, poor public transport, distance, a lack of or poor facilities, a neglect of spaces and facilities, the conditions of play areas and play equipment, the lack of playing opportunities, inefficient staffing, poor conditions or a lack of toilets, seating, poor lighting, lack of provisions in spaces for children, the elderly, and for women.

Non-users of urban green spaces are people who have used them once in the last year or never. Infrequent users are those who have used these spaces only once in the last 6 months. Some of the reasons for non-use and infrequent use of urban green spaces are public drinking, vandalism and policies of care in the community, dog mess, perception of an unsafe environment, concern for personal safety and security, fear of violence, fear of bullying and racist attacks, dark passages, lack of lighting, poorly lit paths, emergency assistance and telephones, predominance of playing fields, lack of attractive activities and facilities, failure to provide activities and experiences demanded by users, the lack of character of many parks, unfamiliarity with landscapes and open space cultures, an uncomfortable feeling of "otherness" (DETR, 1996; MacFarlane et al., 2000; Thomas, 1999; McAllister, 2000).

8.6 Factors of Successful Community Involvement in Urban Green Spaces

Some relevant factors to improve the use of urban green areas are less dog mess, improved safety, better maintenance, better facilities, more events and activities, more staff and easier access to sites, a wider variety of things to do, dogs being on a leash, dog-free areas, more staff, provision of more seats, no motor vehicles, no cycling, no roller skating and roller blading, play areas, lower planting near paths, accessibility to an urban green space and being close to home, with parking facilities and good public transportation, an information center and information boards, displays boards, braille signs, maps signing posts with directions, etc.

Urban green spaces provide opportunities for all people to meet, regardless of their cultural, religious, ethnic origin, political ideology, sexuality, profession, etc. Urban green spaces are sites for community spirit, although different friends and users' groups have different levels of involvement, engagement, and development that range from the frankly adversarial to those that have already formed partnerships. Some community activities for children and young people could spring out around the urban green spaces, providing a free open gathering place, usually at weekends, during the summer time and holidays. Members of the friends' group can clean up the urban green space every weekend.

As a result of a consultative process, local authorities committed to partnerships must consider demonstrations of the willingness of community groups and residents to get involved in some of the initiatives and take responsibility for tasks at the urban green spaces. In fact, some participants are willing to volunteer and engage in the green space proposals, providing a unique experience.

Partnerships offer an opportunity for the coordination of environmental regeneration programs at a potentially low financial cost. For this purpose, a priority proposal is to establish a user community group to include representative volunteer local members to work on charity in partnership. Provision of other facilities that encourage the use of urban green areas are cafés, information centers, boards, toilets, sports areas, sporting events, dog litter bins, seating, parking, staffing including attendants, park wardens, and keepers, first aid facilities, boating, water features, and water plants.

Different models of partnerships between urban green spaces and communities require cultural change to move the emphasis onto community involvement and sense of ownership which results in caring, resourcing, involvement, creativity, and innovation. Some factors contributing to a successful involvement are the institutional culture of local authorities, community groups and users, resources and capabilities, sense of funding, investing, and ownership, voluntary commitment, and communication between stakeholders. The responsibility and ownership of urban green spaces should not be fragmented between different authorities and different structures to achieve more innovation, efficiency, and community involvement. Local authorities develop approaches to engaging and involving users through discussion groups, consultations, artistic events, sports activities, ethnic minority background activities, leisure programs, environmental and horticultural activities, community gardens, organic food growing projects, etc.

Community involvement and engagement in urban green spaces lead to enhanced quality of experiences and uses meeting the needs of users and long-term sustainability, giving access to additional funding and expertise. User groups usually set up priorities for urban green spaces where funding is available and consultation from local communities is required, tied as they are into yielding tangible results. Groups should be active to complement the capabilities of local authorities. One way to motivate is to provide grants for urban green space projects available to all groups. Urban green spaces are a catalyst for community projects because they revolve around the most relevant community issues, and the potential for environmental, social, and economic change is derived from adequate funding and managing of resources. Therefore, the prominence of urban green spaces in communities is in the promotion for funding in community regeneration initiatives.

The development of friends' and users' groups needs to be managed by requiring commitment from local authorities, but also from the community, moving from the concept of the local authority's duty to provide services because they are already funded by taxes. Well-managed urban green spaces have an impact on the urban fabric in benefiting urban environment and wildlife, promoting healthier lifestyles, increasing urban attractiveness and the urban value of land and infrastructure. Nature has beneficial effects on health and wellbeing and mood improvement (Hull & Michaels, 1995; Kaplan & Kaplan, 1989; Irvine & Warber, 2002), reducing stress (Ulrich, 1981), managing mental fatigue (Hartig et al., 1991) and opportunities for reflection (Kuo, 2001; Fuller et al., 2007).

Creative and innovative approaches to funding and resourcing of urban green spaces are required if designing the appropriate arrangement to make the best with the available resources (DETR, 1999). An innovative process is not exempted from conflicts. Conflicts arise amongst users and community organizations and groups who set up the trust. More innovative and creative local authorities are able to achieve more and better resources with less financial investment and spending. There are different methods of allocating, administering, and using the funding to be spent according to creative approaches aimed at enhancing the quality of life.

Access to additional funding is achieved by increasing the ownership of voluntary community involvement and group engagement with the local authorities. The multi-disciplinary and multi-agency team enables the sense of ownership by local groups of the community, increasing their capacity building, and partner agencies to take risks using the grant, focusing on long-term regeneration and renewal objectives.

Local authorities should ensure that the backgrounds, culture and environmental resources, new expertise, skills, and interests brought in are in harmony so that the potential to develop is self-fulfilled without leaving aside the commitment and voluntary efforts of traditional users. Urban green spaces are focus of community volunteer groups to achieve change, provide facilities and activities to local users, and involve and engage other users. Activities developed by community groups in urban green spaces are most essentially voluntary actions such as conservation and maintenance tasks, although volunteer maintenance is coordinated by rangers, and the feeling of ownership of upkeep is the responsibility of local authority.

Volunteers and trainees can be in charge of maintenance. Usually, voluntary community groups get involved in some of the routine operations and maintenance such as planting, grass cutting, cleaning, etc. Volunteering activities are more common in business groups conducted through staff initiatives. Active volunteers in the community need to be more motivated and negativity managed to achieve more active involvement, engagement and collaboration in a task-orientation approach with local authorities. A green space watch scheme run by volunteers can be set up in partnership with the police.

These arrangements can incur some risks, but can improve the facilities, infrastructure, maintenance, etc. However, after the initial investments, it is difficult to sustain the pace of change. Siting housing and business areas in green landscape environments is a means of promoting a green image, enhancing quality of life, and encouraging investment and economic activities. An eco-village can mix community development with providing good-quality accommodations for local residents.

Urban green space service delivery from environmental authorities may have a more holistic approach to policy and budget implementation. Resources available to local authorities and their efficient use make of better provision of quality service delivery. One of the main problems facing the urban green spaces is the capital and financial resources and budget decline in real terms by the spending per head of population for funding urban green space projects. Spending per head does not necessarily take into account the area of the green space. Comparison can be made with regard to the spending per head and per hectare of green space, despite the fact that there is no consistent methodology. Urban green space officers must have expertise in community involvement and engagement, with environmental training.

Community engagement and involvement occur with a change of institutional culture of the local government and changes in the users' culture. Sensory stimulation experiences, such as gardens to smell and touch, statues, warning sounds and colored items such as seats, litter bins, lights shining on the water at night stimulation through running water, scented plants, quacking ducks, calm flowery areas, etc., should be provided. Planting programs with volunteering users from a community not only inject color but also the sense of identity.

Diversity of vegetation and greenery is one of the most important elements of urban green spaces, color for aesthetic appreciation, grass, trees, flowers, natural trails, wild plants, an arboretum with labeled plants, plant names in Braille, and all varieties of plants such as tropical species. Animals within the urban green spaces are also an important element for children, such as birds, ducks, etc.

Determining the economic value of urban green spaces considers their natural resources. Some economic factors of urban green spaces include the production of wood, the supply of fruits, the economic value of the area, job creation, tourist attractions. Urban green spaces are of ecological value (Bilgili & Gökyer, 2012) (which has become a necessity together with aesthetic and recreational value). Evidence for the value of green ecological networks to wildlife is limited, although it has become an element of urban planning (Tzoulas et al., 2007). Ecological and environmental aspects are the biodiversity and ecological values, urban climate, and natural corridors.

This vision must be agreed and shared with all the users and stakeholders and local authorities. A vision can develop and protect the quality standards of using urban green spaces, in healthy and pleasant environments and improving new uses and ensuring sustainability with high ecological and environmental value for healthy living, offering well-designed and -maintained green space, meeting the demands of users, ensuring participative action and accessibility, stimulating socioeconomic development and quality of lifestyle in the community, and contributing to the spatial identity. The concept of economic development linked with the environment is one of the principles.

The spatial concept of urban green space incorporates green into the urban structure and is related to the concept of a green system, a network of corridors. A spatial concept for urban green space development describes and incorporates green issues, interconnects the existing urban spaces and the future desired network and their relationships with the hinterland of the city. Green spaces are related to and connected to green networks and green corridors, defining preservation, improvement and development areas, the neighboring countryside, the regional green network, pedestrian and cycling paths, etc.

The quality standard measures the amount of urban green spaces per inhabitant for each type based on providing appropriate sizes for different activities, security and protection, distance and accessibility based on the travel time and the willingness to walk. Regulations and standards ensure the quality standards of accessibility of users to urban green spaces. Guidelines and standards for the provision of quality services are set out. Some standards related to urban green spaces are recreation areas near to residential districts, larger recreation areas with multifunctional uses, protection for open spaces, nature protection, local climate, land use, and soil sealing (Stadt Leipzig, 2003). Combining various environmental factors results in rendering a standardized method of classifying urban green spaces virtually impossible.

Planning and design of urban green spaces consider the recreational and visual attraction, residential and business areas, leisure and tourism development (Dole, 1994). International tourism is attracted by the creativity, innovation and quality of urban green spaces, but also rolling programs of garden festivals. A scheme can include a festival space, market, fairs, etc., once the greens have been invigorated and their uses revitalized.

Urban green areas can provide countryside activities and educational activities to children outside school hours and to adults through training programs, workshops, and cultural events on urban regeneration initiatives ranging from horticulture, maintenance, school education visits on nature, art activities, lectures and training on environmental education, vocational qualifications in horticulture, animal husbandry, and a 4-week summer play scheme. Also, children are supported by their schools in some activities related to the environment, ecology, tree planting, etc., for example, providing an eLearning module to increase awareness and knowledge. Educational institutions can benefit from making use of urban green spaces for educational, sporting programs and community-based education activities for children, young people, and adults. Urban green spaces offer children the development of a social environment to improve cognitive and motor skills, and to achieve higher levels of creative play, socialization, more collaboration, and emotional resilience (Forest Research, 2010).

A partnership structure that enables a crosscutting integration of community group initiatives with officers of local authorities and the urban green space in a network to coordinate responsibilities, developing action plans and activities to improve biodiversity and improve the environment as a whole. The action plan describes the specific tasks for implementing and achieving each type and each issue, actions, timescale, potential funding sources, and partners. Local authorities of urban green spaces, acting as the eyes and ears of the friends and user's groups work as a partnership shaped by community orientation. Unintended consequences of urban green spaces are avoided with community-based decision-making (Jesdale et al., 2013; Su et al., 2011). Partnerships raise the quality of urban green space.

Partnerships among business, agencies, and communities with the local authorities make base-line funding available to achieve higher and better added value, far more than can be achieved by a local authority alone. Effective partnerships among local government, business, agencies, neighborhood organizations, and community groups can add financial and quality value to the green spaces. Identification of spatial, organizational, and financial problems with regard to the planning and management of urban green spaces, such as distribution, change of use, green corridors, and networks. Among the organizational problems are communication and cooperation problems. Financial problems are related to funding. Other important arrangements for increasing and making more efficient financial resources are among others, partnerships with grant-making foundations, private financial initiatives, community and business groups, targeted grant funding, and creative initiatives to increase revenue spending.

External funding and resources from externally funded capital programs amount to a small proportion of the budget required to maintain quality standards, although they are essentially crucial for capital works. Other forms of external funding are the so-called landfill tax credit scheme and private and business sponsorship that enable creation and operation of facilities and a wide range of financial private initiatives as a means of injecting private capital. Partnering to achieve external funding and expertise from community and business involvement is a formula for raising quality standards. An active sports program of events can attract funding so that it can be financially self-supporting.

Creative and innovative approaches for external funding from community and business groups are usually selective in their applications such as tackling deprivation. Local authorities have to change radically to find and make use of the best opportunities available for external funding through partnerships. Other relevant factors important for success are political support and networking support. Some factors contributing to external funding are the political will of the local authorities to match funding to urban green spaces by embracing an entrepreneurial culture and creativity of external funding officers to investigate sources and resources through partnership opportunities. Voluntary activities enable volunteers with creative, innovative and entrepreneurial capabilities and skills to contribute to urban renewal by pursuing personal development. Bringing the necessary external resources to the urban green spaces by managing change through the involvement and engagement of local residents requires professional input expertise to discuss and accept the evolving structure.

Private sponsorship should make more significant contributions in budgeting and enabling more facilities. The bulk of urban green space is transferred by contract to a private contractor, but accountability and quality monitoring roles are retained essentially through consultation mechanisms and to ensure public accountability and quality of service delivery.

Financial values result in increasing land prices, attracting more inward investments, economic growth and development, community economic spin-offs, etc. Urban green space is one of the main drives to attracting investments and multinational corporations that usually choose to build facilities taking into consideration the urban environment and landscape (Baycan-Levent & Nijkamp, 2009; Wuqiang et al., 2012).

Urban green spaces are the catalyst for social economic spin-offs in the community, such as sports activities, community centers, training programs, job creation, etc. Urban green space-based groups counting on the right individuals involved have the potential for spin-off effects in the community. Quality values are more intangible and may result in community strengthening and environmental quality. These programs and projects can be carried out in partnership with local business, industry, companies, and financial organizations in a continuing involvement with local schools, universities, research centers, museums, heritage organizations, local authorities, local community, neighborhoods and people, green and environmental societies and organizations, etc.

Partnerships among local authorities, funding agencies and institutions, community groups, and business can contribute time and resources to adding value and quality. Partnerships can be with voluntary sector support, voluntary sector led and managed, environmental/regeneration projects, and finally partnerships around a hub. Ground work trusts are locally based partnerships committed to national organizations as an area-wide player and as a network with local operators, although sometimes they have difficulties securing a long-term commitment and leave the community with aspirations to continue the project.

Trusts are an alternative for recreation and amenity facilities, environmental and wildlife, potential new business, and urban opportunities. Urban green spaces have the capacity to be attractive to local, national, and international leisure visitors, while playing a beneficial role for the brand of the city. Thus, they indirectly play a role in location business decisions. Research has found a positive correlation between urban green spaces and business location decisions (Woolley & Rose, undated for CABE), although there is little reliable evidence for the effect of green spaces on the decision to locate in a certain area and on economic growth and investments (Forest Research, 2010). Trusts and private finance initiatives are partners with communities in different situations, with different roles of partners, including appropriate safeguards. Trust partnerships provide assistance at the level of the friends' groups.

Value-added benefits that essentially result from community involvement and engagement are contradicted by the costs and problems derived from involving non-participative groups, because they require capacity building and development. The costs of urban green spaces for local authorities include all kinds of resources such as human capital, financial investments, material, knowledge, etc. Other costs are conflicting demands. Communities face the costs of responsibilities, the skills balance in services, the commitment of volunteers, etc.

Some costs associated with the involvement of local authorities in community development are a lack of long-term vision, the increase in workload without a complementary resource, a major demand on resources, greater expectations, motivating and maintaining momentum in capacity and supporting groups, over-reliance on volunteers and jealousy, identification of good leaders and representativeness in the community, lack of appropriate capabilities and skills, a hard learning process, maintaining volunteer commitment and responsibility, community development and maintenance, managing demands that conflict with and contradict constructive engagement, extending and delaying the process, job security, and successful community development. Volunteers receive training and are hired when there is funding, thus building capacities and promoting employment and ensuring commitment to the project.

8.7 Challenges and Opportunities

Some of the important challenges faced by urban green spaces are the structural and organizational changes in service delivery, re-engineering the staffing, redesigning and refurbishment, involving communities and agencies to promote the cause, providing guidance and sharing experience in good practice, avoiding duplication of responsibilities and stimulating active cooperation. Good practices are successful initiatives that are transferred and utilized in similar situations.

One of the most important challenges that mankind faces is sustainable urban green spaces where more than half of the world's population live and need to improve their lifestyle. Urban green spaces provide environmental, economic, social, political, cultural, and psychological services for the wellbeing and have an environmental sustainability impact on human activities, a role that cannot be ignored by policy makers. Use of urban green spaces improves wellbeing, reducing anxiety and depression. These impacts raise awareness of the rational use of natural resources. Environmental sustainability of any urban agenda must be considered when designing, planning, managing, and maintaining the distribution and qualitative improvements of urban green spaces as an integrated approach to providing a quality service and delivering it to users.

A variety of social and physical opportunities are available in urban green spaces, a particular recreational activity, getting away from the demands of daily life and relieving stress. Some social factors of urban green spaces to be included are the biodiversity of land uses, healthy and active lifestyles, inclusiveness and social justice (Thomas, 1999) and opportunities (Scottish Executive, 2001). Urban green spaces offer opportunities for aesthetic experiences with a positive psychosomatic effect that reduces attention deficits and other cognitive disorders, stress and mental fatigue, and blood pressure (Tzoulas et al., 2007).

Urban green spaces are opportunities for community dwellers to be surrounded by nature and biodiversity and gaining more enjoyment from spending time (Dallimer et al., 2012). Users are willing to pay for access to green spaces if they receive in return the opportunity to see the richness of nature in the various species (Dallimer et al., 2014). Urban green spaces provide opportunities for physical exercise and better air quality. Opportunities and potential for green urban spaces included the spatial, structural and morphological, functional, and ecological. The ecological aspects of urban green spaces comprise the maintenance of a healthy environment with water, air and soil, diversity of wildlife and resources, reduction of the impact of human activities, preservation of natural and cultural heritage, and promotion of private investments to conserve cultural heritage.

The spatial system of the urban green space requires conservation, restoration, maintenance, improvements, protection of existing and developments of new spatial forms (Kong et al., 2010), taking into consideration the impedance in the habitats posed by land use and landscape. A green infrastructure in a green network provides an overall value by offering opportunities for wildlife including various species of animals, birds, and insects. The urban green space infrastructure provides natural drainage, water interception, infiltration, storage, pollutant removal, surface flow and rainwater runoff reduction, and water quality. The storage of water in green areas is superior in quality to the runoff from streets, roads, and roofs (Hou et al., 2006; cited in Zhang et al., 2012).

Evaluation of the current conditions and economic perspectives is important for achieving results in relation to costs. Urban green spaces combine both monetary and nonmonetary valuations to assess the value. Green spaces are linked to residential and commercial property values. Cost-benefit analysis of urban green spaces is a critical factor in policy action (McPherson et al., 2011). The cost-benefit analysis of green spaces is a trade-off. A cost-benefit analysis of green spaces should take into consideration all the significant benefits such as energy consumption, reduction of pollutant materials, effects of heat stress island, reduction of greenhouse gas emissions, storm water runoff, mortality rates, etc. (Mackey et al., 2012).

8.8 Conclusion: Public Initiatives and Actions

Urban green spaces are wider initiatives of local authorities and communities with environmental, social, and economic objectives that can justify any funding for all the represented and involved stakeholders. The institutional structural framework of urban green spaces is a design concern of local authorities in response to providing services for the satisfaction of users' needs. Urban green spaces must be large enough to satisfy the urban users' needs and aspirations and distributed throughout the total urban area in such a way that better relationships with the environment can be sustained.

Therefore, promoting cooperation relationships through networks between urban green spaces and groups of friends is usually a political issue of high priority and commitment for local authorities. Some of the driving forces of urban green space initiatives behind community development are to improve the poor state of development, to generate employment, and return to the greenery as it was in the past.

Public initiatives and actions supported by local authorities aimed at inhabitants with regard to urban green spaces, parks and gardens in public spaces should demonstrate their attachment to sustainable development and the environment. Grass roots initiatives usually form community groups to work toward achieving better provision of services. Local and community initiatives in a green space develop as a result of inadequate provision for users' needs and aspirations by the local authorities or had not been developed because of a lack of resources.

Beyond this, there is also economic stimulation with the regeneration of the community. Urban green space stimulates social and economic regeneration of communities in a multi-agency area, beyond landscaping, which can be considered only a cosmetic change. The link between environmental regeneration and economic stimulation provides substantial infrastructure in housing, job creation, etc. Usually, policymakers underestimate the role of urban green spaces beyond the greenery or landscaping in urban regeneration to reinvigorate communities and neighborhoods, by improving lifestyles, making them more pleasant and attractive, increasing the land value, strengthening the community spirit and social networks, economic stimulation, etc. Urban green spaces play a relevant environmental, economic, social, and cultural role.

Green objectives must be integrated into spatial planning. Planning for distances between urban green spaces is required to provide climate cooling to communities and neighborhoods (Doick et al., 2014). Designing, planning, and sustainable management should address the resources for improving urban green areas focusing on meeting the users' needs in location, access and environmental quality, lighting, security and safety, playgrounds for children, the elderly, the disabled, and young people, dog-free areas, no vehicles, cycling or roller skating. Urban green spaces should be accessible, uniformly distributed, optimal in quality and quantity, large enough to accommodate the population's needs (Haq, 2011), sustainable, and livable. These features should be considered at the stages of designing, managing, maintaining, and protecting.

The designing of urban green spaces results in good quality and variety of activities in open spacious areas, trees and spaces, exit points, quiet areas, a good network of paths, meadows, water, formal areas, meeting places, monuments, hills, mazes, etc. Designing play areas with community engagement provides opportunities for skill development and major satisfaction of the final users. External experts in design and mediation are very welcome to the task.

Designing and planning of urban green spaces must move from traditional park railings and interconnected webs to planning land uses for multiple purposes, that is, recreational and conservation uses with other purposes such as wildlife corridors beside streams and roads, public gardens on top of buildings, reservoirs and waterparks; flood prevention with canoe courses, hides and ornithological habitats in conjunction with sewage farms. Conservation planning of urban green spaces should ensure natural flora, fauna, landforms, water, air, soil, etc., and protect them from other land uses. Urban green spaces comprise habitats supporting a wide range of species, some of them with a conservation concern (Park & Lee, 2000; Mörtberg & Wallentinus, 2000).

Clean outdoor green spaces should provide facilities for pedestrian and cycle routes to promote well-being and health and encourage physical activities such as walking, jogging, trim trails, running, and cycling. Walking in green spaces, self-guided or led by guides for health walks on prescription, is a service that can be provided. Other offers are to promote healthy living by providing safe routes to school or business, facilitating journeys among home, the school, and the community. A green space surrounding schools lowers the levels of traffic-related pollution (Dadvand et al., 2015). Also, it is recommended to promote healthy lifestyles by growing vegetables and fruits in community urban green spaces.

Creating a sense of shared ownership for all users and stakeholders is crucial for communities using urban green spaces and may result in innovativeness, creativity, resourcing, funding, and care. Maintenance of facilities such as cafés involving community engagement recognizes the value of the services provided. The focus is on innovative and creative approaches to delivering urban green space services in the local communities. Innovation lies in applying principles with a more holistic approach to the urban green space as part of a wider network to meet the needs and aspirations of urban users. A green network improves environmental quality and safety by providing green routes and adjacent buildings, which can be business units, chapels, youth and children's facilities and diversification of activities to take advantage of further funding opportunities. There is no correlation between innovation practices and levels of spending.

This approach sets the framework for the greater potential of a range of creativity and innovation with the sense of empowerment and ownership streaming from the partnership collaboration between the local authorities and the users. To investigate creativity and innovation in creating alternative models and developing funding partnerships may increase budgeting and the use and spending of resources. On a plot scheme, green spaces can be adopted by groups or individuals, residents' and tenants' associations in agreements with local authorities. Plots of any size ranging across a large area of green space, taking over parts of a street by the participants of the program adopting a plot, for example, who have the responsibility for clearing the site and maintenance. Despite residents maintaining the green area, the local authorities have the responsibility for the bulk of planting and landscaping and the community may use it for public events.

Urban green space renewal action plans require political attention and have a wide range of environmental, economic, and social regeneration, an increase in resources and investment and educational benefits, and contributes to improving the urban landscape and its use for recreation and enjoyment. Planned activities in urban green spaces for development operations may include plant trails, biological corridors, botanical gardens, shared gardens, play areas for recreational use.

Urban spaces that could potentially receive biodiversity should be inventoried, including open spaces, abandoned wastelands, etc. Regarding water as an element of urban green spaces, fountains and waterfalls with sound, water for children's play, ponds with wildlife, rivers, streams, boating lakes, etc., should be included. A skate park, wheelchair activities, a graffiti wall, outdoor chess and draughts, and a community fishing space could also be provided.

Sport facilities in urban green spaces require changing facilities in good condition, so that users can dress correctly, play on football pitches, access to bowling greens, tennis, basketball courts, facilities for organized clubs and for casual teams, and the option to join in activities without being a member. Programs involving the elderly, such as playing bowls, children and young people, such as football, and women, etc., contribute to better social cohesion.

Spaces for musical performances provide opportunities for better cultural awareness in the community, such as active event programs concerts, theater, fairs, bandstands, etc.

An inventory should be made of resources and maintenance and building facilities for which the urban green space is responsible, finding new ways of reducing the costs of maintenance without a loss of green assets, such as using woodland spaces for recreation and protection from climatic conditions, also involving the extensive agricultural and agro-ecological use of urban green spaces and afforestation of derelict land green spaces.

8.9 Research Gaps

There is a research gap in the empirical literature, including a lack of data on the optimal size, characteristics, distribution, and the influence on health effects of green spaces (Bowler et al., 2010). Research into inequalities in urban greener environments is necessary to improve health equity. However, there is little research on the effects of green areas on carbon capture, although study on green spaces and pollution looks at the link between carbon capture in green spaces' capacity for pollution and the absorption of particles. Research into the impact of urban green spaces on air pollution has been limited, showing moderate evidence that mitigates sulfur oxides, nitric oxides, carbon oxides, and particulate matter (Konijnendijk et al., 2013a; Yin et al., 2011). The urban green infrastructure contributes to carbon capture by building up carbon reserves in the soil (Forest Research, 2010).

Green barriers are useful in protection from traffic emissions, but require further research to clarify the effects of green street canyon geometries, wind speeds, velocity of air pollutants velocity, types of vegetation, etc. Furthermore, a research gap has been reported by Bowler et al. (2010) on the cooling effect on nongreen areas adjacent to urban green spaces. The impact of small green urban areas on heat have also been less frequently explored (Oliveira et al., 2011; Bowler et al., 2010).A final gap in the research on urban green spaces is the reduction of a habitat to one independent variable, such as levels of vegetation (Kuo et al., 1998a, b), overlooking the structural complexity of biodiversity patterns interacting with social and psychological benefits and bypassing the intangible benefits associated with socioeconomic factors (Pickett et al., 2001; Hope et al., 2003; Martin et al., 2004; Kinzig et al., 2005).

8.10 Cross-References

- Collaboration for Regional Sustainable Circular Economy Innovation
- Environmental Stewardship. Achieving Purpose and Community Cohesion Through
- Just Conservation. In Defense of Environmentalism
- Responsible Investing and Environmental Economics. Green Finance and the Transition to a Green Economy
- Sustainable Living in the City. The Case of the Urban Ecovillage
- Smart Cities. New Urbanism and New Agrarianism as a Path to Sustainability
- Strategic Management Innovation of Urban Green Spaces for Sustainable Community Development
- The Theology of Sustainability Practice. How Cities Create Hope

References

Abraham, A., Sommerhalder, K., & Abel, T. (2010). Landscape and wellbeing: A scoping study on the health-promoting impact of outdoor environments. *International Journal of Public Health, 55*(1), 59–69. https://doi.org/10.1007/s00038-009-0069-z

Alberti, M., & Marzluff, J. M. (2004). Ecological resilience in urban ecosystems: Linking urban patterns to human and ecological functions. *Urban Ecosystems, 7*(3), 241–265. https://doi.org/10.1023/B:UECO.0000044038.90173.c6

Alonso, R., Vivanco, M. G., González-Fernández, I., Bermejo, V., Palomino, I., Garrido, J. L., et al. (2011). Modelling the influence of peri-urban trees in the air quality of Madrid region (Spain). *Environmental Pollution, 159*(8–9), 2138–2147.

Amorim, J. H., Rodrigues, V., Tavares, R., Valente, J., & Borrego, C. (2013). CFD modelling of the aerodynamic effect of trees on urban air pollution dispersion. *Science of the Total Environment, 461*, 541–551. https://doi.org/10.1016/j.scitotenv.2013.05.031

Armson, D., Stringer, P., & Ennos, A. R. (2012). The effect of tree shade and grass on surface and globe temperatures in an urban area. *Urban Forestry & Urban Greening, 11*(3), 245–255. https://doi.org/10.1016/j.ufug.2012.05.002

Baik, J. J., Kwak, K. H., Park, S. B., & Ryu, Y. H. (2012). Effects of building roof greening on air quality in street canyons. *Atmospheric Environment, 61*, 48–55. https://doi.org/10.1016/j.atmosenv.2012.06.076

Baldock, K. C. R., et al. (2015). Where is the UK's pollinator biodiversity? The importance of urban areas for flower-visiting insects. *Proceedings of the Royal Society B, 282*, 20142849. https://doi.org/10.1098/rspb.2014.2849

Balram, S., & Dragicevic, S. (2005). Attitude towards urban green spaces; integrated questionnaire survey and collaborative GIS techniques to improve attitude measurement. *Elsevier: Landscape and Urban Planning, 71*(2–4), 147–162.

Baró, F., Chaparro, L., Gómez-Baggethun, E., Langemeyer, J., Nowak, D. J., & Terradas, J. (2014). Contribution of ecosystem services to air quality and climate change mitigation policies: The case of urban forests in Barcelona, Spain. *Ambio, 43*(4), 466–479.

Bassil, K. L., Cole, D. C., Moineddin, R., Lou, W., Craig, A. M., Schwartz, B., & Rea, E. (2010). The relationship between temperature and ambulance response calls for heat-related illness in Toronto, Ontario, 2005. *Journal of Epidemiology and Community Health, 65*, jech-2009.

Baycan-Levent, T. (2002). Demographic transition and urbanization dynamics in Turkey. In H. S. Geyer (Ed.), *International textbook of urban systems: Studies of urbanization and migration in advanced and developing countries* (pp. 329–361). Edward Elgar Publishing.

Baycan-Levent, T., & Nijkamp, P. (2009). Planning and management of urban green spaces in Europe: Comparative analysis. *Journal of Urban Planning and Development, 135*, 1. https://doi.org/10.1061/(ASCE)0733-9488(2009)135:1(1)

Bilgili, B. C., & Gökyer, E. (2012). Urban green space system planning. In D. M. Ozyavuz (Ed.), *Landscape planning*. InTech Publisher.

Bowler, D., Buyung-Ali, L., Knight, T., & Pullin, A. (2010). Urban greening to cool towns and cities: A systematic review of the empirical evidence. *Landscape and Urban Planning, 97*(2010), 147–155. https://doi.org/10.1016/j.landurbplan.2010.05.006

Buccolieri, R., Salim, S. M., Leo, L. S., Di Sabatino, S., Chan, A., Ielpo, P., et al. (2011). Analysis of local scale tree–atmosphere interaction on pollutant concentration in idealized street canyons and application to a real urban junction. *Atmospheric Environment, 45*(9), 1702–1713.

Burton, E. (2003). Housing for an urban renaissance: Implications for social equity. *Housing Studies, 18*(4), 537–562. https://doi.org/10.1080/02673030304249

Cao, X., Onishi, A., Chen, J., & Imura, H. (2010). Quantifying the cool Island intensity of urban parks using ASTER and IKONOS data. *Landscape and Urban Planning, 96*(4), 224–231. https://doi.org/10.1016/j.landurbplan.2010.03.008

Chen, H., Ooka, R., Huang, H., & Tsuchiya, T. (2009). Study on mitigation measures for outdoor thermal environment on present urban blocks in Tokyo using coupled simulation. *Building and Environment, 44*, 2290–2299. https://doi.org/10.1016/j.buildenv.2009.03.012

Chen, A., Yao, X. A., Sun, R., & Chen, L. (2014). Effect of urban green patterns on surface urban cool islands and its seasonal variations. *Urban Forestry & Urban Greening, 13, 646.*

Choi, H. A., Lee, W. K., & Byun, W. H. (2012). Determining the effect of green spaces on urban heat distribution using satellite imagery. *Asian Journal of Atmospheric Environment, 6*(2), 127–135. https://doi.org/10.5572/ajae.2012.6.2.127

Cohen, P., Potchter, O., & Matzarakis, A. (2012). Daily and seasonal climatic conditions of green urban open spaces in the Mediterranean climate and their impact on human comfort. *Building and Environment, 51*, 285–295. https://doi.org/10.1016/j.buildenv.2011.11.020

Comedia and Demos. (1995). *Park life: Urban parks and social renewal*. Comedia.

Cornelis, J., & Hermy, M. (2004). *Landscape and Urban Planning, 69*, 385–401. https://doi.org/10.1016/j.landurbplan.2003.10.038

Coutts, A. M., Daly, E., Beringer, J., & Tapper, N. J. (2013). Assessing practical measures to reduce urban heat: Green and cool roofs. *Building and Environment, 70*, 266. https://doi.org/10.1016/j.buildenv.2013.08.021

Dadvand, P., De Nazelle, A., Triguero-Mas, M., Schembari, A., Cirach, M., Amoly, E., Figueras, F., Basagana, X., Ostro, B., & Nieuwenhuijsen, M. (2012). Surrounding greenness and exposure to air pollution during pregnancy: An analysis of personal monitoring data. *Epidemiology, 120*(9), 1286–1290.

Dadvand, P., et al. (2015). The association between greenness and traffic-related air pollution at schools. *Science of The Total Environment, 523*, 59–63.

Dallimer, M., et al. (2012). Biodiversity and the feel-good factor: Understanding associations between self-reported human well-being and species richness. *Bioscience, 62*, 47–55. https://doi.org/10.1525/bio.2012.62.1.9

Dallimer, M., et al. (2014). Quantifying preferences for the natural world using monetary and non-monetary assessments of value. *Conservation Biology, 28*, 404–413. https://doi.org/10.1111/cobi.12215

Department of the Environment. (1996). *People, parks and cities – A guide to current good practice in urban parks*. HMSO.

DETR. (1996). *English house conditions survey*. DETR London.

DETR. (1999). *By design, urban design in the planning system: Towards better practice*. Available from www.odpm.gov.uk

Dobrovolný, P. (2013). The surface urban heat Island in the city of Brno (Czech Republic) derived from land surface temperatures and selected reasons for its spatial variability. *Theoretical and Applied Climatology, 112*(1–2), 89–98. https://doi.org/10.1007/s00704-012-0717-8

Doick, K. J., Peace, A., & Hutchings, T. R. (2014). The role of one large greenspace in mitigating London's nocturnal urban heat island. *The Science of the Total Environment, 493*, 662–671. https://doi.org/10.1016/j.scitotenv.2014.06.048

Dole, J. (1994). Greenscape 5: Green cities, architects' journal. In G. Haughton & C. Hunterm (Eds.), *Sustainable cities* (pp. 61–69). JKP.

Dunnett, N., Swanwick, C., & Woolley, H. (2002). *Improving urban parks, play areas and green spaces*. Department for Transport, Local Government and the Regions. [pdf] Available at: http://www.ocs.polito.it/biblioteca/verde/improving_full.pdf. Accessed Apr 2013.

Dzierżanowski, K., Popek, R., Gawrońska, H., Sabo, A., & Gawroński, S. W. (2011). Deposition of particulate matter of different size fractions on leaf surfaces and in waxes of urban forest species. *International Journal of Phytoremediation, 13*(10), 1037–1046. https://doi.org/10.1080/15226514.2011.552929

Eliasson, I. (2000). The use of climate knowledge in urban planning. *Landscape and Urban Planning, 48*, 31–44. https://doi.org/10.1016/S0169-2046(00)00034-7

Escobedo, F. J., & Nowak, D. J. (2009). Spatial heterogeneity and air pollution removal by an urban forest. *Landscape and Urban Planning, 90*(3–4), 102–110. https://doi.org/10.1016/j.landurbplan.2008.10.021

Feyisa, G. L., Dons, K., & Meilby, H. (2014). Efficiency of parks in mitigating urban heat Island effect: An example from Addis Ababa. *Landscape and Urban Planning, 123*, 87–95. https://doi.org/10.1016/j.landurbplan.2013.12.008

Fintikakis, N., Gaitani, N., Santamouris, M., Assimakopoulos, M., Assimakopoulos, D. N., Fintikaki, M., et al. (2011). Bioclimatic design of open public spaces in the historic Centre of Tirana, Albania. *Sustainable Cities and Society, 1*(1), 54–62.

Forest Research. (2010). *Benefits of green infrastructure*. Report to Defra and CLG. [pdf] Available at: http://www.forestry.gov.uk/pdf/urgp_benefits_of_green_infrastructure_main_report.pdf/$file/urgp_benefits_of_green_infrastructure_main_report.pdf. Accessed Apr 2013.

Fröhlich, D., & Matzarakis, A. (2013). Modeling of changes in thermal bioclimate: Examples based on urban spaces in Freiburg, Germany. *Theoretical and Applied Climatology, 111*(3–4), 547–558. https://doi.org/10.1007/s00704-012-0678-y

Fuller, R. A., Irvine, K. N., Devine-Wright, P., Warren, P. H., & Gaston, K. J. (2007). Psychological benefits of greenspace increase with biodiversity. *Biology Letters, 3*(4), 390–394. https://doi.org/10.1098/rsbl.2007.0149

Fuller, R. A., et al. (2009). *Diversity and Distributions, 15*, 328–337.

Gaitani, N., Spanou, A., Saliari, M., Synnefa, A., Vassilakopoulou, K., Papadopoulou, K., et al. (2011). Improving the microclimate in urban areas: A case study in the Centre of Athens. *Building Services Engineering Research and Technology, 32*(1), 53–71.

Girardet, H. (1996). *The Gaia atlas of cities: New directions for sustainable urban living*. Gaia. UN-HABITAT.

Girardet, H. (2004). *Cities people planet: Liveable cities for a sustainable world*. Wiley.

Grahn, P., & Stigsdotter, U. A. (2003). Landscape planning and stress. *Urban Forestry & Urban Greening, 2*(1), 1–18. https://doi.org/10.1078/1618-8667-00019

GreenSpace. (2007). *The park life report. The first ever public satisfaction survey green spaces: The benefits for London bibliography of Britain's parks and green spaces*. [pdf] Available at: http://www.greenspace.org.uk/downloads/

Gregg, J. W., Jones, C. G., & Dawson, T. E. (2003). Urbanization effects on tree growth in the vicinity of New York City. *Nature, 424*(6945), 183–187. https://doi.org/10.1038/nature01728

Gregory, R. D., & Baillie, S. R. (1998). Large-scale habitat use of some declining British birds. *Journal of Applied Ecology, 35*(5), 785–799. https://doi.org/10.1046/j.1365-2664.1998.355349.x

Grimm, N. B., Faeth, S. H., Golubiewski, N. E., Redman, C. L., Wu, J., Bai, X., et al. (2008). Global change and the ecology of cities. *Science, 319*(5864), 756–760. https://doi.org/10.1126/science.1150195

Hale, J. D., et al. (2012). *PLoS ONE, 7*, e33300. https://doi.org/10.1371/journal.pone.0033300

Hamada, S., & Ohta, T. (2010). Seasonal variations in the cooling effect of urban green areas on surrounding urban areas. *Urban Forestry & Urban Greening, 9*(1), 15–24. https://doi.org/10.1016/j.ufug.2009.10.002

Hamada, S., Tanaka, T., & Ohta, T. (2013). Impacts of land use and topography on the cooling effect of green areas on surrounding urban areas. *Urban Forestry & Urban Greening, 12*(4), 426–434. https://doi.org/10.1016/j.ufug.2013.06.008

Haq, S. M. A. (2011). Urban green spaces and an integrative approach to sustainable environment. *Journal of Environmental Protection, 2*(5), 601–608. https://doi.org/10.4236/jep.2011.25069

Hart, M. A., & Sailor, D. J. (2009). Quantifying the influence of land-use and surface characteristics on spatial variability in the urban heat Island. *Theoretical and Applied Climatology, 95*(3–4), 397–406. https://doi.org/10.1007/s00704-008-0017-5

Hartig, T., Mang, M., & Evans, G. W. (1991). Restorative effects of natural environment experiences. *Environment and Behavior, 23*(1), 3–26. https://doi.org/10.1177/0013916591231001

Herold, M., Liu, X., & Clarke, K. C. (2003). Spatial metrics and image texture for mapping urban land use. *Photogrammetric Engineering & Remote Sensing, 69*(9), 991–1001. https://doi.org/10.14358/PERS.69.9.991

Herzele, V., & Wiedeman, T. (2003). A monitoring tool for the provision for accessible and attractive green spaces. *Elsevier Sciences: Landscape and Urban Planning, 63*(2), 109–126. https://doi.org/10.1016/S0169-2046(02)00192-5

Hope, D., et al. (2003). Socioeconomics drive urban plant diversity. *Proceedings of the National Academy of Sciences, 100*(15), 8788–8792. https://doi.org/10.1073/pnas.1537557100

Hou, L. Z., Ding, Y. Y., Feng, S. Y., Zhang, S. H., Chen, J. G., & Liao, R. H. (2006). Comparison of water quality of rainwater runoff from different underlying surface in Beijing city, China. *Water and Wastewater, 22*(23), 35–38. (in Chinese).

Hough, M. (2004). *Cities and natural process*. Routledge.

Huang, G., Zhou, W., & Cadenasso, M. L. (2011). Is everyone hot in the city? Spatial pattern of land surface temperatures, land cover and neighborhood socioeconomic characteristics in Baltimore, MD. *Journal of Environmental Management, 92*, 1753–1759. https://doi.org/10.1016/j.jenvman.2011.02.006

Hull, R. B., & Michaels, S. E. (1995). Nature-based recreation, mood change and stress restoration. *Leisure Sciences, 17*(1), 1–14.

Hwang, R. L., Lin, T. P., & Matzarakis, A. (2011). Seasonal effects of urban street shading on long-term outdoor thermal comfort. *Building and Environment, 46*(4), 863–870. https://doi.org/10.1016/j.buildenv.2010.10.017

Irvine, K. N., & Warber, S. L. (2002). Greening healthcare: Practicing as if the natural environment really mattered. *Alternative Therapies in Health and Medicine, 8*(5), 76–83.

Irvine, K. N., Devine-Wright, P., Payne, S. R., Fuller, R. A., Painter, B., & Gaston, K. J. (2009). Green space, soundscape and urban sustainability: An interdisciplinary, empirical study. *Local Environment, 14*(2), 155–172. https://doi.org/10.1080/13549830802522061

Jesdale, B. M., Morello-Frosch, R., & Cushing, L. (2013). The racial/ethnic distribution of heat risk-related land cover in relation to residential segregation. *Environmental Health Perspectives, 121*, 811–817. https://doi.org/10.1289/ehp.1205919

Kaplan, R., & Kaplan, S. (1989). *The experience of nature: A psychological perspective*. Cambridge University Press. (Republished by Ann Arbor, MI: Ulrich's, 1995).

Kinzig, A., et al. (2005). The effects of human socioeconomic status and cultural characteristics on urban patterns of biodiversity. *Ecology and Society, 10*(1), 23. https://doi.org/10.5751/ES-01264-100123

Kong, F., Yin, H., Nakagoshi, N., & Zong, Y. (2010). Urban green space network development for biodiversity conservation: Identification based on graph theory and gravity modeling. *Landscape and Urban Planning, 95*(1–2), 16–27. https://doi.org/10.1016/j.landurbplan.2009.11.001

Kong, F., Yin, H., James, P., Hutyra, L. R., & He, H. S. (2014). Effects of spatial pattern of greenspace on urban cooling in a large metropolitan area of eastern China. *Landscape and Urban Planning, 128*, 35–47. https://doi.org/10.1016/j.landurbplan.2014.04.018

Konijnendijk, C. C., Annerstedt, M., Nielsen, A. B., & Maruthaveeran, S. (2013a). Benefits of urban parks: A systematic review. *IFPRA World, 2012*(6), 10–12.

Konijnendijk, C. C., Annerstedt, M., Nielsen, A. B., & Maruthaveeran, S. (2013b). *Benefits of urban parks: A systematic review*. A report for IPFRA, Canada.

Koyama, T., Yoshinga, M., Hayashi, H., Maeda, K., & Yamauchi, A. (2013). Identification of key plant traits contributing to the cooling effects of green façades using free-standing walls. *Building and Environment, 66*, 96–103. https://doi.org/10.1016/j.buildenv.2013.04.020

Kuo, F. E. (2001). Coping with poverty: Impacts of environment and attention in the inner city. *Environment and Behavior, 33*(1), 5–34. https://doi.org/10.1177/00139160121972846

Kuo, F. E., Bocacia, M., & Sullivan, W. C. (1998a). Transforming inner-city neighbourhoods: Trees, sense of safety, and preference. *Environment and Behavior, 30*(1), 28–59. https://doi.org/10.1177/0013916598301002

Kuo, F. E., et al. (1998b). Fertile ground for community: Inner-city neighbourhood common spaces. *American Journal of Community Psychology, 26*(6), 823–851. https://doi.org/10.1023/A:1022294028903

Lachowycz, K., & Jones, A. P. (2011). Green space and obesity: A systematic review of the evidence. *Obesity Reviews, 12*(5), e183–e189. https://doi.org/10.1111/j.1467-789X.2010.00827.x

Lafortezza, R., Carrus, G., Sanesi, G., & Davies, C. (2009). Benefits and well-being perceived by people visiting green spaces in periods of heat stress. *Urban Forestry & Urban Greening, 8*(2), 97–108. https://doi.org/10.1016/j.ufug.2009.02.003

Lee, A. C. K., & Maheswaran, R. (2011). The health benefits of urban green spaces: A review of the evidence. *Journal of Public Health, 33*(2), 212–222. https://doi.org/10.1093/pubmed/fdq068

Leipzig, S. (2003). *Umweltqualitatsziele und -standards fur die Stadt Leipzig*. Amt fur Umweltschutz. http://www.leipzig.de/imperia/md/content/36_amt_fuer_umweltschutz/umweltziele.pdf. Accessed July 2008

Li, J., Song, C., Cao, L., Zhu, F., Meng, X., & Wu, J. (2011). Impacts of landscape structure on surface urban heat islands: A case study of Shanghai, China. *Remote Sensing of Environment, 115*(12), 3249–3263.

Li, X., Zhou, W., Ouyang, Z., Xu, W., & Zheng, H. (2012). Spatial pattern of greenspace affects land surface temperature: Evidence from the heavily urbanized Beijing metropolitan area. *China. Landscape Ecology, 27*(6), 887–898. https://doi.org/10.1007/s10980-012-9731-6

Li, X., Zhou, W., & Ouyang, Z. (2013). Relationship between land surface temperature and spatial pattern of greenspace: What are the effects of spatial resolution? *Landscape and Urban Planning, 114*, 1–8.

Lin, T.-P., Matzarakis, A., & Hwang, R.-L. (2010). Shading effect on long-term outdoor thermal comfort. *Building and Environment, 45*(1), 213–221.

Lynn, B. H., Carlson, T. N., Rosenzweig, C., Goldberg, R., Druyan, L., Cox, J., et al. (2009). A modification to the NOAH LSM to simulate heat mitigation strategies in the New York City metropolitan area. *Journal of Applied Meteorology and Climatology, 48*(2), 199–216.

MacFarlane, R., Fuller, D., & Jeffries, M. (2000). Outsiders in the urban landscape? An analysis of ethnic minority landscape projects. In J. F. Benson & M. H. Roe (Eds.), *Urban lifestyles: Spaces, places, people*. Balkema.

Mackey, C. W., Lee, X., & Smith, R. B. (2012). Remotely sensing the cooling effects of city scale efforts to reduce urban heat island. *Building and Environment, 49*, 348–358. https://doi.org/10.1016/j.buildenv.2011.08.004

Maher, B. A., Ahmed, I. A., Davison, B., Karloukovski, V., & Clarke, R. (2013). Impact of roadside tree lines on indoor concentrations of traffic-derived particulate matter. *Environmental Science & Technology, 47*(23), 13737–13744. https://doi.org/10.1021/es404363m

Manes, F., Incerti, G., Salvatori, E., Vitale, M., Ricotta, C., & Costanza, R. (2012). Urban ecosystem services: Tree diversity and stability of tropospheric ozone removal. *Ecological Applications, 22*(1), 349–360. https://doi.org/10.1890/11-0561.1

Martin, C. A., Warren, P. S., & Kinzig, A. P. (2004). Neighborhood socioeconomic status is a useful predictor of perennial landscape vegetation in residential neighborhoods and embedded small parks of Phoenix, AZ. *Landscape and Urban Planning, 69*(4), 355–368. https://doi.org/10.1016/j.landurbplan.2003.10.034

McAllister, S. (2000). Institutionalised racism in the landscape: The exclusion of ethnic minorities from landscape processes. Unpublished manuscript, University of Sheffield, Department of Landscape.

McPherson, E. G., Simpson, J. R., Xiao, Q., & Wu, C. (2011). Million trees Los Angeles canopy cover and benefit assessment. *Landscape and Urban Planning, 99*(1), 40–50. https://doi.org/10.1016/j.landurbplan.2010.08.011

Meier, F., & Scherer, D. (2012). Spatial and temporal variability of urban tree canopy temperature during summer 2010 in Berlin, Germany. *Theoretical and Applied Climatology, 110*(3), 373–384. https://doi.org/10.1007/s00704-012-0631-0

Mitchell, R., & Popham, F. (2008). Effect of exposure to natural environment on health inequalities: An observational population study. *Lancet, 372*, 1655–1660. Retrieved from: http://www.thelancet.com/journals/lancet/article/PIIS0140-6736(08)61689-X/fulltext. https://doi.org/10.1016/S0140-6736(08)61689-X

Moore, M., Gould, P., & Keary, B. S. (2003). Global urbanization and impact on health. *International Journal of Hygiene and Environmental Health, 206*(4), 269–278. https://doi.org/10.1078/1438-4639-00223

Morani, A., Nowak, D. J., Hirabayashi, S., & Calfapietra, C. (2011). How to select the best tree planting locations to enhance air pollution removal in the MillionTreesNYC initiative. *Environmental Pollution, 159*(5), 1040–1047. https://doi.org/10.1016/j.envpol.2010.11.022

Mörtberg, U., & Wallentinus, H.-G. (2000). Red-listed forest bird species in an urban environment – Assessment of green space corridors. *Landscape and Urban Planning, 50*(4), 215–226. https://doi.org/10.1016/S0169-2046(00)00090-6

Newman, P., & Jennings, I. (2008). *Cities as sustainable ecosystems: Principles and practices.* Island Press.

Ng, E., Chen, L., Wang, Y., & Yuan, C. (2012). A study on the cooling effects of greening in a high-density city: An experience from Hong Kong. *Building and Environment, 47*, 256–271.

Nowak, D. J., & Crane, D. E. (2002). Carbon storage and sequestration by urban trees in the USA. *Environmental Pollution, 116*, 381–389. https://doi.org/10.1016/S0269-7491(01)00214-7

Nowak, D. J., et al. (2013). Carbon storage and sequestration by trees in urban and community areas of the United States. *Environmental Pollution, 178*, 229–236. https://doi.org/10.1016/j.envpol.2013.03.019

Nowak, D. J., Hirabayashi, S., Bodine, A., & Greenfield, E. (2014). Tree and forest effects on air quality and human health in the United States. *Environmental Pollution, 193*, 119–129. https://doi.org/10.1016/j.envpol.2014.05.02

Oliveira, S., Andrade, H., & Vaz, T. (2011). The cooling effect of green spaces as a contribution to the mitigation of urban heat: A case study in Lisbon. *Building and Environment, 46*(11), 2186–2194. https://doi.org/10.1016/j.buildenv.2011.04.034

Onishi, A., Cao, X., Ito, T., Shi, F., & Imura, H. (2010). Evaluating the potential for urban heat-Island mitigation by greening parking lots. *Urban Forestry & Urban Greening, 9*(4), 323–332. https://doi.org/10.1016/j.ufug.2010.06.002

Park, C. R., & Lee, W. S. (2000). Relationship between species composition and area in breeding birds of urban woods in Seoul, Korea. *Landscape and Urban Planning, 51*(1), 29–36. https://doi.org/10.1016/S0169-2046(00)00094-3

Park, M., Hagishima, A., Tanimoto, J., & Narita, K. I. (2012). Effect of urban vegetation on outdoor thermal environment: Field measurement at a scale model site. *Building and Environment, 56*, 38–46. https://doi.org/10.1016/j.buildenv.2012.02.015

Perini, K., & Magliocco, A. (2014). Effects of vegetation, urban density, building height, and atmospheric conditions on local temperatures and thermal comfort. *Urban Forestry & Urban Greening, 13*(3), 495–506. https://doi.org/10.1016/j.ufug.2014.03.003

Pickett, S. T. A., et al. (2001). Urban ecological systems: Linking terrestrial, ecological, physical, and socioeconomic components of metropolitan areas. *Annual Review of Ecology and Systematics, 32*, 127–157. https://doi.org/10.1146/annurev.ecolsys.32.081501.114012

Rinner, C., & Hussain, M. (2011). Toronto's urban heat island – Exploring the relationship between land use and surface temperature. *Remote Sensing, 3*(6), 1251–1265. https://doi.org/10.3390/rs3061251

Rouquette, J. R., et al. (2013). *Diversity and Distributions, 19*, 1429–1439. https://doi.org/10.1111/ddi.12120

Rowe, D. B. (2011). Green roofs as a means of pollution abatement. *Environmental Pollution, 159*(8), 2100–2110. https://doi.org/10.1016/j.envpol.2010.10.029

Roy, S., Byrne, J., & Pickering, C. (2012). A systematic quantitative review of urban tree benefits, costs, and assessment methods across cities in different climatic zones. *Urban Forestry and Urban Greening, 11*(4), 351–363. https://doi.org/10.1016/j.ufug.2012.06.006

Schwartz, M. W., et al. (2002). Conservation's disenfranchised urban poor. *Bioscience, 52*, 601–606. https://doi.org/10.1641/0006-3568(2002)052[0601:CSDUP]2.0.CO;2

Scottish Executive. (2001). *Rethinking open space, I stationery office*. Kit Campbell Associates.

Shashua-Bar, L., Tsiros, I. X., & Hoffman, M. (2012). Passive cooling design options to ameliorate thermal comfort in urban streets of a Mediterranean climate (Athens) under hot summer conditions. *Building and Environment, 57*, 110–119. https://doi.org/10.1016/j.buildenv.2012.04.019

Smith, K. R., & Roebber, P. J. (2011). Green roof mitigation potential for a proxy future climate scenario in Chicago, Illinois. *Journal of Applied Meteorology and Climatology, 50*(3), 507–522. https://doi.org/10.1175/2010JAMC2337.1

Speak, A. F., Rothwell, J. J., Lindley, S. J., & Smith, C. L. (2012). Urban particulate pollution reduction by four species of green roof vegetation in a UK city. *Atmospheric Environment, 61*, 283–293. https://doi.org/10.1016/j.atmosenv.2012.07.043

Srivanit, M., & Hokao, K. (2013). Evaluating the cooling effects of greening for improving the outdoor thermal environment at an institutional campus in the summer. *Building and Environment, 66*, 158–172. https://doi.org/10.1016/j.buildenv.2013.04.012

Steeneveld, G. J., Koopmans, S., Heusinkveld, B. G., Van Hove, L. W. A., & Holtslag, A. A. M. (2011). Quantifying urban heat island effects and human comfort for cities of variable size and urban morphology in the Netherlands. *Journal of Geophysical Research Atmospheres (1984–2012), 116*(D20), 2156.

Strohbach, M. W., et al. (2012). The carbon footprint of urban green space – A life cycle approach. *Landscape and Urban Planning, 104*, 220–229. https://doi.org/10.1016/j.landurbplan.2011.10.013

Su, J. G., Jerrett, M., de Nazelle, A., & Wolch, J. (2011). Does exposure to air pollution in urban parks have socioeconomic, racial or ethnic gradients? *Environmental Research, 111*(3), 319–328. https://doi.org/10.1016/j.envres.2011.01.002

Sung, C. Y. (2013). Mitigating surface urban heat Island by a tree protection policy: A case study of the woodland, Texas, USA. *Urban Forestry & Urban Greening, 12*(4), 474–480. https://doi.org/10.1016/j.ufug.2013.05.009

Susca, T., Gaffin, S. R., & Dell'Osso, G. R. (2011). Positive effects of vegetation: Urban heat Island and green roofs. *Environmental Pollution, 159*(8), 2119–2126. https://doi.org/10.1016/j.envpol.2011.03.007

Tallis, M., Taylor, G., Sinnett, D., & Freer-Smith, P. (2011). Estimating the removal of atmospheric particulate pollution by the urban tree canopy of London, under current and future environments. *Landscape and Urban Planning, 103*(2), 129–138. https://doi.org/10.1016/j.landurbplan.2011.07.003

Thomas, H. (1999). Urban renaissance and social justice. *Town and Country Planning, 68*(11), 332–333.

Tiwary, A., Sinnett, D., Peachey, C., Chalabi, Z., Vardoulakis, S., Fletcher, T., et al. (2009). An integrated tool to assess the role of new planting in PM10 capture and the human health benefits: A case study in London. *Environmental Pollution, 157*(10), 2645–2653.

Tsiros, I. X., Dimopoulos, I. F., Chronopoulos, K. I., & Chronopoulos, G. (2009). Estimating airborne pollutant concentrations in vegetated urban sites using statistical models with microclimate and urban geometry parameters as predictor variables: A case study in the city of Athens, Greece. *Journal of Environmental Science and Health, Part A, 44*(14), 1496–1502. CrossRef

Tuzin, B., Leeuwen, E., Rodenburg, C., & Peter, N. (2002). Paper presented at the 38th international planning congress on "the pulsar effect" planning with peaks, Glifada, Athens, 21–26 September 2002.

Tzoulas, K., Korpela, K., Venn, S., Yli-Pelkonen, V., Kazmierczak, A., Niemela, J., & James, P. (2007). Promoting ecosystem and human health in urban areas using green infrastructure: A literature review. *Landscape and Urban Planning, 81*(2007), 167–178. https://doi.org/10.1016/j.landurbplan.2007.02.001

Ulrich, R. S. (1981). Natural versus urban scenes: Some psychophysiological effects. *Environment and Behavior, 13*(5), 523–556. https://doi.org/10.1177/0013916581135001

Ulrich, R. S. (2002). *Health benefits of gardens in hospitals*. International Exhibition Floriade. 2002. Retrieved from: http://www.planterra.com/research/research_3.php

Ulrich, R., Quan, X., & Zimring, C. (2010). The role of the physical environment in the hospital of the 21st century: A once-in-a-lifetime opportunity. Report prepared for TriPoint Hospital Center. Retrieved from: tinyurl.com/healthdesignstudy as cited in http://www.cleveland.com/healthfit/index.ssf/2010/09/blueprint_for_healing_%2D%2D_hospi.html

United Nations (Department of Economic and Social Affairs – Population Division). (2014). World Urbanization Prospects: The 2014 revision, highlights (ST/ESA/SER.A/352).

Van Herzele, A., & de Vries, S. (2011). Linking green space to health: A comparative study of two urban neighbourhoods in Ghent. *Belgium. Population and Environment, 34*(2), 171–193. https://doi.org/10.1007/s11111-011-0153-1

Vidrih, B., & Medved, S. (2013). Multiparametric model of urban park cooling island. *Urban Forestry & Urban Greening, 12*(2), 220–229. https://doi.org/10.1016/j.ufug.2013.01.002

Wania, A., Bruse, M., Blond, N., & Weber, C. (2012). Analysing the influence of different street vegetation on traffic-induced particle dispersion using microscale simulations. *Journal of Environmental Management, 94*(1), 91–101. https://doi.org/10.1016/j.jenvman.2011.06.036

Weber, N., Haase, D., & Franck, U. (2014). Zooming into temperature conditions in the city of Leipzig: How do urban built and green structures influence earth surface temperatures in the city? *The Science of the Total Environment, 496*, 289–298. https://doi.org/10.1016/j.scitotenv.2014.06.144

White, M., Alcock, I., Wheeler, B., & Depledge, M. (2013). *Would you be happier living in a greener urban area? A fixed effects analysis of panel data*. European Centre for Environment and Human Health. [online] Available at: http://www.ecehh.org/publication/would-you-be-happier-living-greenerurban-area. Accessed Apr 2013.

Williams, K., Burton, E., & Jenks, M. (2000). *Achieving sustainable urban form*. E&FN Spon.

Wolf, K. L. (2010). Active living – A literature review. In *Green cities: Good health*. College of the Environment, University of Washington. Retrieved from: http://depts.washington.edu/hhwb/Thm_ActiveLiving.html

Wong, G., Greenhalgh, T., Westhorp, G., Buckingham, J., & Pawson, R. (2013). RAMESES publication standards: Meta-narrative reviews. *BMC Medicine, 11*(1), 20. https://doi.org/10.1186/1741-7015-11-20

Woolley, H., & Rose, S. (undated). *The value of public space. How high quality parks and public spaces create economic, social and environmental value*. CABEspace. [pdf] Available at: http://webarchive.nationalarchives.gov.uk/20110118095356/ http://www.cabe.org.uk/files/the-value-of-publicspace.pdf. Accessed Apr 2013.

Wuqiang, L., Song, S., & Wei, L. (2012). Urban spatial patterns based on the urban green space system: A strategic plan for Wuhan City, P. R. China Shi Song. www.intechopen.com

Xion-Jun, W. (2009). Analysis of problems in urban green space system planning in China. *Journal of Forestry Research, 20*(1), 79–82. https://doi.org/10.1007/s11676-009-0014-2

Yin, S., Shen, Z., Zhou, P., Zou, X., Che, S., & Wang, W. (2011). Quantifying air pollution attenuation within urban parks: An experimental approach in Shanghai, China. *Environmental Pollution, 159*(8), 2155–2163. https://doi.org/10.1016/j.envpol.2011.03.009

Yu, C., & Hien, W. (2005). Thermal benefits of city parks. *Energy and Buildings, 38*(2006), 105–120.

Zhang, B., Xie, G., Zhang, C., & Zhang, J. (2012). The economic benefits of rainwater runoff reduction by urban green spaces: A case study in Beijing, China. *Journal of Environmental Management, 100*(2012)), 65–71. https://doi.org/10.1016/j.jenvman.2012.01.015

Zhang, Z., Lv, Y., & Pan, H. (2013). Cooling and humidifying effect of plant communities in subtropical urban parks. *Urban Forestry & Urban Greening, 12*(3), 323–329. https://doi.org/10.1016/j.ufug.2013.03.010

Zhou, W., Huang, G., & Cadenasso, M. L. (2011). Does spatial configuration matter? Understanding the effects of land cover pattern on land surface temperature in urban landscapes. *Landscape and Urban Planning, 102*(1), 54–63. https://doi.org/10.1016/j.landurbplan.2011.03.009

Zinzi, M., & Agnoli, S. (2012). Cool and green roofs. An energy and comfort comparison between passive cooling and mitigation urban heat Island techniques for residential buildings in the Mediterranean region. *Energy and Buildings, 55*, 66–76. https://doi.org/10.1016/j.enbuild.2011.09.024

Zoulia, I., Santamouris, M., & Dimoudi, A. (2009). Monitoring the effect of urban green areas on the heat island in Athens. *Environmental Monitoring and Assessment, 156*(1–4), 275–292. https://doi.org/10.1007/s10661-008-0483-3

José G. Vargas-Hernández is research professor at Instituto Tecnológico Mario Molina Unidad Zapopan, and earlier at University Center for Economic and Managerial Sciences, University of Guadalajara. Member of the National System of Researchers of Mexico. Professor Vargas-Hernández has a Ph.D. in Public Administration and a Ph.D. in Organizational Economics. He has been visiting scholar at Carleton University Canada, University of California Berkeley and Laurentian University, Canada. He holds a Ph.D. in Economics, Keele University; Ph.D. in Public Administration, Columbia University; studies in Organizational Behavior at Lancaster University and has a Master of Business Administration. He has Published 9 books and more than 300 papers in international journals and reviews (some translated to English, French, German, Portuguese, Farsi, Chinese, etc.) and more than 300 essays in national journals and reviews. He has obtained several international awards and recognitions. He has also experience in consultancy. His main research is in organizational economics and strategic management. He teaches for several doctoral programs.

Karina Pallagst is professor for International Planning Systems at Kaiserslautern University's faculty of Spatial and Environmental Planning. Previously she worked at UC Berkeley's center for Global Metropolitan Studies (GMS) and the Institute of Urban and Regional Development (IURD) as the program director of the Shrinking Cities International Research Group. Prior to this appointment, she was a senior research specialist at the Dresden-based Institute of Ecological and Regional Development (IOER). Prof. Pallagst holds a Ph.D. from Kaiserslautern University and a postdoctoral degree from Dresden Technical University. Her research focuses on international comparative urban development, shrinking cities, urban growth, planning cultures, planning theory and border studies. She serves on numerous think tanks, working groups and committees regarding spatial planning and international urban development. She is a co-founder of the Centre for Border Studies of the University of the Greater Region.

Justyna Zdunek-Wielgołaska, Ph.D. Eng. Arch., is associate professor at the Urban and Spatial Planning Department at Faculty of Architecture, Warsaw University of Technology. Beneficiary of Fulbright scholarship (visiting scholar at University of California, Berkeley, USA in the period 2004-2005); also, a beneficiary of the Institut Für Stadt- und Regionalplanung Technische Universität scholarship, Berlin, Germany, and a laureate of the Start for Young Academics programme of the Foundation for Polish Science (2007). She was a leader and member of multiple research projects. Justyna Zdunek-Wielgołaska is also the author of surveys on the teaching quality in the Faculty of Architecture at the Warsaw University of Technology. She was also lecturer at Warsaw University and at the School of the Architecture, University of Detroit Mercy, USA. He conducts research in the field of urban planning on green infrastructure, urban sprawl, suburbs and protection of urban and architectural heritage, in particular concerning Warsaw from the interwar period. Author of many articles, participant of international conferences.

Chapter 9
Mushroom Packages: An Ecovative Approach in Packaging Industry

Younsung Kim and Daniel R. Ruedy

Abstract The challenge of global sustainability presents enormous opportunities for product innovation and creative response. Concerns about the rising volume of packaging wastes, particularly, request the redefinition and redesign of conventional products in the packaging industry. Ecovative Design, being founded by a young mechanical engineering student in 2007, develops an array of environmentally friendly materials that perform like plastics but are made from mushrooms. The mushroom packaging is renewable and biodegradable and can be made with crop waste brought from local farms. Drawing on the case of Ecovative Design, this chapter aims to describe a creative destruction approach to redress sustainability challenges and to explore pivotal factors that fostered the company's ability to decarbonize and dematerialize in support of sustainability. We find that a company's technical competency with systems-thinking skills and stakeholder engagement capacity can lead to the realization of ecological modernization principle.

Keywords Sustainable packaging · Technical competence · Systems thinking · Ecovative Design · Creative destruction

9.1 Introduction

Sustainability in the packaging industry has never been more important. A $400 billion USD industry globally, packaging includes the manufacture and transport of paperboard and plastics (Ernst & Young, 2013). Most conventionally designed packaging is of low cost and spends little more than a year in a linear route from producer to consumer life cycle (Hopewell et al., 2009; Niero et al., 2017). In the United States, more than 35 million tons of packaging paperboard and 160 million tons of packaging plastic are produced annually to contain, cradle, and cushion goods as they make their way through the supply chain (Geyer et al., 2017; Richtel, 2016). As these packaging materials are discarded and enter the waste stream, they

Y. Kim (✉) · D. R. Ruedy
Department of Environmental Science and Policy, George Mason University, Fairfax, VA, USA
e-mail: ykih@gmu.edu; druedy@gmu.edu

© The Author(s), under exclusive license to Springer Nature Switzerland AG 2023

S. Dhiman (ed.), *Sustainable Development and Environmental Stewardship*,
https://doi.org/10.1007/978-3-031-28885-2_9

are disposed of in landfills, recycled, or incinerated for waste-to-energy (WTE) recovery.

Being highly visible to consumers, the packaging industry receives intense scrutiny worldwide, along the entire supply chain, from raw materials to end-of-life processes, for reducing its societal and environmental impacts (Hillier et al., 2017). The increasing concerns about the rising volume of packaging wastes are also aligned with the unproductive use of resources and materials. Both packaging producers and their primary customers—original equipment manufacturers (OEM), e-commerce and brick-and-mortar retailers, and wholesalers—face pressures to reduce materials in packaging so as to mitigate packaging's price fluctuations linked to global commodities. As such, packaging producers are forced to seek out and manage stable supply chain sources due to the instability of conventional packaging feedstock such as wood, pulp, and petrochemicals, which would lead to reduced environmental impacts (Ernst & Young, 2013).

Enhancing resource efficiency has become a conventional approach when a firm attempts to address sustainability challenges for the packaging industry (Ernst & Young, 2013; Hillier et al., 2017). Noticing emerging market potentials of bio-based alternatives, some other firms have considered a more innovative approach such as developing biomaterials that may have a potential for replacing conventional, petroleum-based materials (Haneef et al., 2017; Hillier et al., 2017). Once succeeded, sustainability-promoting innovations would help firms reap greater environmental benefits and create sustainable value for shareholders (Hart, 2005). However, a few firms only consider the innovative approach positively, and prior literature has rarely discussed what factors would explain why a firm could undertake innovations for sustainability. Drawing on the case of Ecovative Design, a biomaterials company headquartered in New York, USA, this study explores key factors prompting the firm to innovate their package products. Founded in 2007, Ecovative Design develops an array of environmentally friendly packaging materials by growing fibers on waste like cotton seed, wood fiber, and buckwheat hulls.

In the following sections, we first introduce the Schumpeterian creative response framework as an eco-innovation, technology-based approach to contemporary social and ecological crises, which can result in ecological modernization (York & Rosa, 2003; Janicke, 2008; Mason, 2011). Creative response framework has been used to explain the causes and the consequences of innovation in economics and in the economy. It deviates from the continuous improvement-based approach to adopting new ideas and systems within the assumption of static economic changes (Schumpeter, 1947). In the subsequent section, we examine the unsustainability of packaging products. In doing so, we underpin the current waste treatment practices of packaging materials and describe the challenges for recycling packaging products. We then explain the product innovation procedures taken by Ecovative and identify key factors of its successful practices by taking an inductive reasoning approach. Our findings indicate that technical competence combined with systems-thinking skills would prompt a firm to undertake innovative sustainability changes in its products and processes. In addition, a firm's ability to engage with governments to garner financial support and market recognition would help a firm to pursue sustainability-driven innovations for the packaging industry.

9.2 Creative Destruction for Sustainability

Global sustainability challenges present prodigious opportunities for creative responses from market disrupters. Current society–nature interactions are not sustainable, in that they negatively affect both vital ecological systems and human welfare and induce irreversible, long-term damages. Achieving global sustainability then involves a sheer transition to economically revolutionary processes and routines that would be necessary to remain viable without overwhelming the society and environment. It particularly implies a structural realignment of our dominant economic development paths away from energy- and material-intensive processes relying on fossil fuels toward ecological modernization in which societies modify their institutions in order to internalize environmental impacts and ecologically transform material process and consumption processes (Mol, 1995; Hajer, 1995; Mason, 2011).

Creative destruction was coined by Schumpeter (1947), as he noticed creative responses that brought economic developments with dynamic changes in economy. According to traditional economists, economic changes are explained by indicating specific conditioning or causal factors within a given historical development. For instance, classical economic theory predicted that an increase in population would result in a fall in per capita income, as population growth may have no other effect than that predicated by classical theory. However, this is not necessarily true in actual instances. Population growth may rather derive new developments with increasing income per capita. In other cases, a protective duty may have no other effect than to increase the price of the protected commodity and, in consequence, its output. Unlike the classical economists' prediction, however, it may also induce a complete reorganization of the protected industry which eventually results in an increase in output so great as to reduce the price below its initial level. In observation of such historical economic developments and societal changes, Schumpeter reported the neglected area of economic change, creative and energizing *reactions* to changes in "condition" that can cause novel achievements being outside of the range of existing practices (Schumpeter, 1947).

As such, creative response is distinct from adaptive response. Adaptive response is the process by which an economy adapts itself to a change in its data by simply adding new resources or modestly modifying the currently existing practices. Creative response is something *beyond* the range of expanding existing practices or applying the ordinary rules. In general, the three fundamental characteristics would dictate creative response: (1) It can only be fully appreciated ex post, never being understood ex ante; (2) it shapes the whole course of subsequent events and their long-run outcomes; and (3) in terms of frequency, intensity, and success, it is influenced by the quality of the personnel available at a social and sector level and by decisions, actions, and patterns of behaviors of individuals or of groups. In this light, entrepreneurship aligned with creative response is the pivotal mechanism of inducing long-term sustained economic change in a capitalist society (Schumpeter, 1947).

Creative response framework is a suitable theoretical approach to putting an emphasis on innovative responses to environmental and sustainability challenges in a society. Given the emergence and prevalence of wicked environmental problems (Lackey, 2007) for the late twentieth and twenty-first centuries, different organizational responses from defiance to proactiveness have been observed (Kim & Darnall, 2016). Scholars in the field of sustainability management have utilized a wide array of theoretical approaches advocated in the mainstream management sciences (Starik & Kanashiro, 2013), and creative response framework has been uniquely poised to explore new ideas, experiments, or innovative sustainability practices such as biomimicry, leapfrog technology, sustainable technology, and closed loops (Braungart & McDonough, 2002; Hart, 2005; Larson, 2000).

The social actor's energizing effects that induce new developments and exhaustive disruptions can be attributed to a different set of characteristics. Typically, sustainability responses involving modest changes for pollution involve rationalization in terms of the impact on the industry structure. That means firms' green initiatives would be flourished as long as they can yield relatively short-term benefits and appeal stakeholders that concern pollution prevention and product stewardship without restructuring the existing industry. However, the biggest leaps in performance may not be driven from firms' modest sustainability responses. Rather, firms pursuing emerging technologies, new markets, new partners, new customers not served before, and entirely new stakeholders would practice "beyond greening" initiatives that entail discontinuity (Hart, 2005). For instance, the automobile industry looks nothing like it did 10–15 years ago, as companies aiming at designing sustainability mobility like Tesla pushed the industry divested from carbon-intensive, gasoline-based engine vehicles to electric ones (Jay & Gerand, 2015). What makes re-invention and creative destruction possible will depend on a whole new set of skills and capabilities that have the potential of being inherently clean and sustainable (Hart, 2005).

Some fundamental elements for sustainability-oriented innovation involve technical competency and stakeholder engagement for market innovation (Larson, 2000; Hart, 2005; Jay & Gerand, 2015; Kim & Darnall, 2016). Not all firms would equip with such capabilities, because the process of navigating a systems-based sustainability solution would not be appealing to most firms in the mainstream market. Some visionary companies can seize the opportunity to drive redefinition and redesign of their industries toward sustainability, and innovative upstarts can unseat established firms. The value of creative disruptive process would be reaped over the long term, as it would be able to exceed competitive gains from incremental changes in sustainability. Large incumbent firms would be less likely to shift their underlying portfolio away from what used to be its core competencies and move toward the entirely new skill sets to play a completely new, different game (Prahalad & Hart, 2002; de Soto, 2000; Christensen, 1997).

9.3 The Need for Ecological Modernization in the Packaging Industry

Market-Oriented Sustainability Opportunity

Unsustainability of Plastics An estimated 9150 million tons of plastics have been produced since large-scale production began around 1950, of which 30% remains in use, 9% has been recycled, and 60% is in landfills or the natural environment (Geyer et al., 2017). It is also projected that by 2050, the world will produce 28,600 million metric tons of plastic polymer resins based on observed trends (Geyer et al., 2017). The recent analyses of the global plastic packaging industry place production at 78 million tons annually at a total value of $260 billion USD (World Economic Forum et al., 2016), and it is expected to reach $490 billion by 2026 (BusinessWire, 2018).

Plastic packaging does not biodegrade by design. Although exposure to sunlight is known to weaken and fragment plastic, the environmental impact of the resulting millimeter-sized fragments has not been well understood (Tudryn et al., 2018; Geyer et al., 2017). Plastic manufacturing is also dependent on nonrenewable petroleum resources for its feedstock, which indicates plastic prices fluctuating closely with the cost of petroleum (Tudryn et al., 2018). Energy consumption to sustain rising plastic demand extends past production. Transportation of plastic resins from manufacturers to downstream wholesalers also comes at a cost. The US Environmental Protection Agency (USEPA) estimates on average that plastic shipments travel 497 miles, consume 0.49 million British thermal units (BTUs) of energy, and emit 0.04 metric tons of CO_2 per short ton of plastic packaging waste (USEPA, 2016a).

Approximately 50% of all plastics produced are manufactured into single-use items and packaging products among them (Hopewell et al., 2009). However, plastic recycling is increasing worldwide at 0.7% annually (Geyer et al., 2017), and the recycling practice is widely accepted to reduce lifecycle net CO_2 emissions by as much as 27% over plastic manufactured from virgin raw material feedstock (Hopewell et al., 2009). Recycling rates vary among plastic types. Polyethylene terephthalate (PET), which comprises plastic water bottles and some packaging components, is frequently recycled at a recovery rate of 19.5% by weight (USEPA, 2015). However, polystyrene and its variants—expanded polystyrene (EPS) and extruded polystyrene (XPS, commonly known as Styrofoam ™)—are the least recycled plastics, faring at just 0.9% recovery by weight. Polystyrene's high bulk (EPS is 95% air by weight) makes its transport costly and logistically difficult, and polystyrene's polymer structure makes it an unattractive candidate economically for primary and secondary recycling. When plastic recycling facilities are unavailable or the recovered plastic is of poor quality, waste-to-energy (WTE) incineration may be a preferred disposition option over landfilling. Yet when adjusted for the utility emissions WTE avoids, there remain net anthropogenic CO_2 emissions of 1.27 and 1.25 $MtCO_2$/ton for polystyrene and mixed plastics, respectively (USEPA, 2015).

Irrespective of treatment options, all plastic waste management practices (e.g., WTE incineration, recycling [including plastic fraction collection and transport to facility], landfill, etc.) have negative impacts on the environment and human health. However, it is evident that no perfect waste management solution exists, and finding the most optimal way is challenging (Rigamonti et al., 2014). In addition, on a global scale, the prospects of improving sustainability of plastics, ultimately in a hope to create a circular economy, seem to be bleak in large part due to increasing demand for consumer goods, change in consumers' lifestyle with rapid urbanization, and rising consumerism in emerging economies (De los Rios & Charnley, 2017). Through a full life cycle assessment, only 5% of the value of recovered plastics is retained, costing the global economy between $80 and $120 billion USD in losses annually (World Economic Forum et al., 2016).

In response to the rigidity of the mature plastics market and its well-established supply and processing systems, plastics companies in the 1980s began blending plant starch-based polymers into environmentally friendly, theoretically biodegradable polyethylene blends (Iles & Martin, 2013). When biodegradability of these bioplastics fell short of standards, they were instead remarketed as "renewably sourced," since they did replace a fraction of the petroleum-based feedstock of conventional plastics with plant-based material.

Bioplastic production is typically specialized to a single commodity feedstock, such as corn or sugarcane, and the production of which may carry its own environmental costs. Diversion of these commodities to industrial uses such as bioplastic manufacture also creates competition with stocks going to food supply, instigating complex socioeconomic policy problems. For example, North America and South America are already using 37% and 27% of domestic sugar crops for ethanol production, respectively (Golden et al., 2015). Furthermore, though their manufacture is very similar to conventional plastics, significant infrastructure investment is required where plastics manufactured from petroleum-based feedstock are already well established. The wide adoption of bioplastics has failed to materialize, and prices of bioplastics remain higher than petrochemical-derived plastics (PricewaterhouseCoopers, 2010), with a total production capacity of about four million metric tons per year (Geyer et al., 2017). Given these limitations, best-case projections predict bioplastics will at peak satisfy no more than 20% of global plastics demand (AT Kearnery, 2012).

Unsustainability of Paperboard Paperboard containers travel an average of 675 miles from manufacturer to primary customer, consuming 0.67 million BTUs of energy and emitting 0.05 mT of CO_2 per short ton shipped (USEPA, 2015). Though recycling is preferred to divert paperboard from landfill, paperboard that is contaminated or otherwise not suitable for recycling drives the retention rate down from an ideal 1:1 ratio (i.e., 100% of recovered paperboard is recycled) to an actual retention rate of 93.5% (USEPA, 2016b). Though a paperboard product may be labeled as 100% post-consumer content, the increment not retained means long-term paperboard recycling operations will yield diminishing returns over time.

Compared to plastics, paperboard's recycling record is better at 89.5% in 2014 and, as testament to the market forces at play, up significantly from just 55% in 1993 (USEPA, 2014). However, the majority of paperboard recycled is by retailers, as fewer than 50% of American consumers are estimated to utilize curbside paperboard recycling programs (Feiner, 2017). This fraction of cardboard waste thus remains difficult to divert from landfill to recovery.

Regulatory Demands for the Sustainable Packaging Industry

Aside from market forces, some government policy and regulatory actions have pushed packaging producers, OEMs, retailers, and consumers to adopt more sustainable business behaviors and models.

In the European Union (EU), the Registration, Evaluation, Authorisation, and Restriction of Chemicals (REACH) Regulation passed in 2006 charges private industry to determine and register the physical, environmental, and toxicological properties of substances used in quantities greater than 1 ton annually in the manufacture of goods. Suppliers and manufacturers are then required to communicate this information to downstream users, empowering OEMs and consumers to account for sustainability in their purchase decisions. As a driver for minimization, diversion, and recovery of paperboard packaging, the 1999 EU Landfill Directive called for a 50% reduction by 2009, in landfilled biodegradable waste from 1995 levels, and a 65% reduction by 2016 (PricewaterhouseCoopers, 2010).

In the United States, federal regulations pertaining to the packaging industry fall under the specialized Federal Food, Drug, and Cosmetic Act but are only relevant to packaging material that comes into contact with food. Other statutes include the Toxic Substances Control Act (TSCA), the Emergency Planning and Community Right-to-Know Act (EPCRA), the Clean Air Act (CAA), the Clean Water Act (CWA), and the Resource Conservation and Recovery Act (RCRA). Though these latter statutory authorities provide for reporting requirements and adherence to some emission limits at various points in the supply chain, none expressly regulate environmental or sustainability standards for packaging. An exception is the Federal Trade Commission's (FTC's) Guides for the Use of Environmental Marketing Claims which prohibit packaging producers from making deceptive claims about compostability, degradability, recyclability, and recycled content (16 C.F.R. 260).

In terms of recycling promotion policies, half of the states have data on curbside consumer recycling programs, totaling 4371 and serving a population 87.9 million, roughly a quarter of whom live in California (Van Haaren et al., 2010). Nationwide, average per capita municipal solid waste generation is estimated at 1.28 tons annually, of which just over a quarter is recycled (Van Haaren et al., 2010). With the void of direct federal regulation, state and local governments have taken initiatives to implement their own policies; both Washington, DC, and New York have implemented bans on EPS. Acknowledging the increasing landfill volume that packaging waste occupies and in a significant step toward meeting an ambitious 80% waste reduction goal by 2032, the Washington, DC, Department of Energy and Environment banned the use of EPS in packaging effective January 1, 2016, for food containers and January 1, 2017, for all packaging containers (DOEE, 2014). Pursuant to Local Law 142 of 2013, New York City's Department of Sanitation recently determined food-service foam cannot be effectively recycled and recommended a city-wide ban starting November 12, 2017, with enforcement beginning on May 14, 2018 (DSNY, 2017).

Eco-Innovation Opportunity Cumulated

In 2016, e-commerce giant Amazon delivered in excess of a billion packages (Green, 2017), evidencing the tremendous success of e-commerce, but an ominous milestone from a sustainability perspective. As a whole, the e-commerce industry is valued at $350 billion and doubled in size between 2011 and 2016 (Richtel, 2016). Leading online grocery markets are projected to double by 2020 (WEF, 2016). Amazon for years has epitomized wasteful shipping and cardboard use—justified or not—based on its practices of packaging goods within cardboard shipping containers bearing its distinctive logo. Indeed, the successes of Amazon and e-commerce in general seem to suggest a boom in cardboard production and associated waste generation; e-commerce now accounts for 10% of all US retail (Howland, 2017). But reports from the industry group Fibre Box Association (FBA) indicate that cardboard quantities shipped by US companies have actually decreased modestly since 1995 (Dove, 2017).

However, a net decrease in cardboard shipped does not mean waste has been reduced or eliminated. When intrinsic costs of packaging materials are low and shipping fees are fixed, retailers and e-commerce sellers have little incentives to adopt sustainable business practices or materials into their operations and distribution. The rise of e-commerce and home delivery direct to consumers means more cardboard and plastic packaging—as well as more secondary shipping containers—can make their way directly to front stoops and business-receiving departments. The result is wasteful, producing environmentally costly overpackaging and growing waste streams.

Both consumers and shippers appear to have taken notice of this trend. Amazon's customer container feedback program has received more than 33 million responses—including comments, complaints, and photographs—from customers (Richtel, 2016). In an effort to embrace an emerging positive marketing opportunity, Amazon introduced a "frustration-free packaging" option that its customers can select with their other shipping instructions (Gopaldas, 2015). The option instructs Amazon's fulfillment services to forgo secondary shipping packaging to rely instead on carefully designed, minimalist OEM packaging and is credited with eliminating 83 million unnecessary cardboard boxes from shipments in 2016 alone (Pierce, 2017). Furthermore, Amazon has begun utilizing machine learning in the algorithms to determine packaging material and method (Green, 2017). Informed by both cost data and customer comments, the algorithms seek to continuously optimize packaging configuration for reduced packaging and shipping costs and wastes at the consumer end (Green, 2017).

Likewise, both UPS and FedEx have implemented billing systems based on dimensional weight—a measure of package volume relative to its actual weight (UPS, 2017). Such measures are widely believed to incentivize OEMs and e-commerce retailers to optimize their packaging solutions to lightest and smallest possible.

While diversion of paperboard and plastic packaging waste to recycling or WTE incineration is certainly preferred over landfill disposal, many nuances influence the sustainability of this producer-to-consumer business model. Despite the good news of increased cardboard and plastic recycling or WTE rates, no matter how efficient or what material retention percent is reached, recycling at best delays waste disposal rather than eliminates it, and there remains not insignificant CO_2 emissions associated with conventional recycling and WTE incineration (Geyer et al., 2017; USEPA, 2016a). Bioplastics then represents an important shift from traditional plastics manufacture, but it and other renewably sourced packaging—incrementally greener than conventional petrochemical plastics or paperboard sourced from virgin wood pulp—seem to constitute only adaptive responses to changing economic conditions (Schumpeter, 1947).

Significant advances in sustainable packaging can be only made through source reduction, and ideally elimination. For example, USEPA estimates that source reduction of polystyrene from both current feedstock mixes and virgin inputs could reduce greenhouse gas (GHG) emissions by 2.5 $MtCO_2$ per short ton of source reduced (USEPA, 2015). For paperboard, the reduction is more dramatic at 5.59 and 8.1 $MtCO_2$ for mixed recycled and virgin feedstock, respectively (USEPA, 2015).

When upstream packaging suppliers are distanced from end users, sustainability in upstream manufacturing may not have much incentive to improve (Foerstl et al., 2015). However, as stakeholder interest in sustainability increases, as it has for Amazon, holdout firms place themselves at competitive risk by not adopting sustainable practices and may further fail to engage untapped customer bases with nascent interests in sustainably sourced materials and products (Foerstl et al., 2015). Facing pressure from expanding regulation and interest from primary and secondary customers for sustainable alternatives, the $400 billion USD packaging industry is therefore primed for disruption (Ernst & Young, 2013). When assigned a monetary value, these regulatory- and altruistic-driven interests comprise a portion of a staggering $10 trillion USD in projected cumulative eco-innovation investment worldwide by 2020 (Boons et al., 2013).

9.4 Defining and Developing Sustainable Package

Sustainable Packaging Considering Business Performance and Environmental Concerns

Sustainable development and sustainability have become the focus of mainstream management studies and practices. However, the concept of sustainability has not yet been understood very clearly by corporate managers or the general public. This is in part because the concept of sustainability has been adapted to address very different challenges, ranging from the planning of sustainable cities to sustainable livelihoods, sustainable agriculture to sustainable fishing, and the efforts to develop common corporate standards in the UN Global Compact and in the World Business Council for Sustainable Development. Despite this creative ambiguity and openness to interpretation, sustainable development has evolved a core set of guiding principles and values, based on the Brundtland Commission's standard definition to meet the needs, now and in the future, for human, economic, and social development within the restraints of the life support systems of the planet (Kates et al., 2005).

As much as sustainability or sustainable development is vague in its definition, there is no clear understanding about what constitutes "sustainable packaging." A widely agreed-to and accepted understanding would be critical in the societal pursuit for promoting sustainability in the packaging domain and leading to associated business development (James et al., 2005).

As one of the semantic efforts to define sustainable packaging, Verghese and Lewis (2005) undertook a stakeholder survey in partnership with the Sustainable Packaging Alliance (SPA). The SPA's sustainable packaging definition took into consideration the role packaging plays in our social and economic systems. It also accounted for the need to meet environmental goals and reduce harm to humans and ecosystems. As such, the SPA's sustainable packaging definition includes four levels, which are society, packaging material, packaging system, and packaging component. It also identifies four different principles: *effective*, *efficient*, *cyclic*, and *safe*. The *effective* principle means that products should be packaged as they would be delivered from producers to consumers. The *efficient* principle seeks to maximize material and energy efficiency in every step of packaging, storage, transport, and handling. The *cyclic* principle is aligned with a closed-loop system, increasing recycling, reuse, and ease of disassembly and assembly. Lastly, the *safe* principle aims to minimize the human and ecological risks from packaging components, being subject to the precautionary principle (The precautionary principle is defined by Principle 15 of the Rio Declaration. Under the precautionary principle, if there are threats of serious or irreversible damage, lack of full scientific certainty shall not be used as a reason for postponing cost-effective measures to prevent environmental degradation (United Nations 1992)) (Pielke, 2002; Raffensperger & Tickner, 1999). Table 9.1 summarizes the four-level definition of sustainable packaging. Packaging is assumed to be *sustainable* and support sustainable development if the four principles are met. While this is an earlier attempt for conceptualization, the concept

Table 9.1 SPA's sustainable packaging definition

Principle	Description	Levels at which the principle is applied
Effective	It adds real value to society by effectively containing and protecting products as they move through the supply chain and by supporting informed and responsible consumption	Society
Efficient	Packaging systems are designed to use materials and energy as efficiently as possible throughout the product life cycle. This should include material and energy efficiency in interactions with associated support systems such as storage, transport, and handling	Packaging system
Cyclic	Packaging materials are cycled continuously through natural or (industrial) technical systems, minimizing material degradation and/or the use of upgrading additives	Packaging material
Safe	Packaging components do not pose any risks to human health or ecosystems. When in doubt, the precautionary principle applies	Packaging component

Source: Adapted from James et al. (2005)

well represents the multifaceted dimensions of packaging in consideration of the elements of sustainability, economy, society, and the environment (Verghese & Lewis, 2005).

As another well-recognized definition, sustainable packaging is also defined by the Sustainable Packaging Coalition (SPC), a project of GreenBlue which is dedicated to the sustainable use of materials in society. The SPC's sustainable packaging concept highlights a closed-loop system and promotes the five principles of (1) responsible sourcing, (2) optimization for efficiency, (3) effective recovery, (4) non-toxic, and (5) low impact. Further outlining the five principles, eight criteria were developed to define sustainable packaging (Table 9.2).

The eight criteria presented blend broad sustainability and industrial ecology objectives with business considerations and strategies that address the environmental concerns related to the life cycle of packaging. These criteria relate to the activities of the packaging value chain and define the areas in which we actively seek to encourage transformation, innovation, and optimization.

Indicators for Sustainable Packaging Development

Sustainable packaging development would not be straightforward even with a clear definition. Translating the definition into more specific targets or performance indictors would be useful to implement sustainable development principles in product packaging development. The proposed key performance indicators for the SPA's four sustainability packaging principles underscore two focal points, to reduce product waste and to improve functionality. Table 9.3 provides 21 indicators that can assist in the process of reaching a state of sustainability in packaging.

Table 9.2 SPC's definition of sustainable packaging

Criterion	Characteristic
1	Is beneficial, safe, and healthy for individuals and communities throughout its life cycle
2	Meets market criteria for performance and cost
3	Is sourced, manufactured, transported, and recycled using renewable energy
4	Optimizes the use of renewable or recycled source materials
5	Is manufactured using clean production technologies and best practices
6	Is made from materials healthy throughout the life cycle
7	Is physically designed to optimize materials and energy
8	Is effectively recovered and utilized in biological and/or industrial closed-loop cycles

Source: SPC (2011)

Table 9.3 SPA's sustainable packaging indicators

Sustainable packaging principle	Sustainable packaging indicator
Effective	Reduces product waste
	Improves functionality
	Prevents overpackaging
	Reduces business costs
	Achieves satisfactory return on investment (ROI)
Efficient	Improves product/packaging ratio
	Improves efficiency of logistics
	Improves energy efficiency (embodied energy)
	Improves materials efficiency (total amount of material used)
	Improves water efficiency (embodied water)
	Increases recycled content
	Reduces waste to landfill
Cyclic	Returnable
	Reusable (alternative purpose)
	Recyclable (technically recyclable and system exists for collection and reprocessing)
	Biodegradable
Safe (clean)	Reduces airborne emissions
	Reduces waterborne emissions
	Reduces greenhouse gas emissions
	Reduces toxicity
	Reduces litter impacts

Source: Lewis et al. (2007)

Several sustainability packaging assessment tools have been used to evaluate and compare packaging with other options. The Packaging Impact Quick Evaluation Tool (PIQET), developed by SPA, is a tool for rapid environmental impact assessment of packaging systems. This web-based software tool uses environmental indicators, based on the LCA methodology. PIQET functions as a credible, business-ready tool for multi-criteria packaging environmental decision-making and guides materials selection, packaging redesign or packaging performance, and evaluation of environmental requirements (Verghese & Lewis, 2005).

One of the applications of the sustainability packaging assessment tool can be found in innovating novel food packaging systems. The tool helped scholars identify optimum sustainable packaging design for food that should balance potential reductions in food loss, the ratio of the environmental impact of the food to the impact of the packaging, the handling of food waste, and the handling of packaging waste (Wikström & Williams, 2010). In addition, the evaluation tool surprisingly recognized reusable plastic containers for a variety of fresh produce (Levi et al., 2011; Menesatti et al., 2012; Singh et al., 2006) as a more sustainable option when compared to commonly used corrugated paper boxes (Park et al., 2014).

Inputs and guidance from professionals in different disciplines— such as designers, engineers, technologists, marketers, and environmental managers—would inform a multidisciplinary, collaborative, and holistic approach and benefit the development process of product packaging systems.

The role of packaging technologists in the industry seems to be particularly important for sustainable packaging, as they can provide comprehensive and credible information to others within and external to the organization. This information ranges from packaging material characteristics, to packaging functionality in distribution and use, to processability in manufacturing and filling, and finally to environmental impact (SPC, 2011). A significant degree of understanding and skills is needed to appropriately collect and analyze such information and to be able to present the findings to internal company decision makers (SPC, 2011).

Sustainable packaging assessment tools, a wide array of stakeholders, and technologists' information can thus be instrumental in transforming packaging system for sustainability. A closed-loop flow of packaging materials can be economically robust and provide benefits throughout its life cycle, constituting a sustainable packaging system.

In the following section, we discuss the sustainable packaging development process using a case study of Ecovative Design. The company develops an array of environmentally friendly materials that perform like plastics but are made from mushrooms. The mushroom packaging is renewable and biodegradable and can be made with crop waste brought from local farms, showing the four principles of sustainability packaging: effective, efficient, cyclic, and safe.

9.5 Case Study: Ecovative Design

Company History

Ecovative is a biotech company with the mission of designing the future of sustainable materials using Mycelium Biofabrication Platform. It took its root at Rensselaer Polytechnic Institute (RPI) in Troy, New York, created by Eben Bayer, now CEO, and Gavin McIntyre, now Chief Scientist (Zeller & Zocher, 2012). In 2006, the two were classmates in Inventor's Studio, an undergraduate class instructed by RPI Professor Burt Swersey. After a semester of unsuccessful ideas, Bayer pitched the concept of a mushroom-based biopolymer as a replacement for traditional plastic insulation to Swersey. Spurred by the entrepreneurial spirit of their academic backgrounds in product design and engineering, and sharing concern for environmental responsibility, Bayer later approached McIntyre with his idea for a novel mushroom-based material to replace conventional plastics. Though the two had job offers, Swersey provided guidance and funds for the pair to pursue the idea (VentureWell, 2014). Instead of following traditional career paths after graduation, the two then founded Ecovative in 2007 (Knapp, 2015).

In 2008, Bayer and McIntyre travelled to Amsterdam, Netherlands, to compete in the National Postcode Lottery Green Challenge (NPCLGC) where they won the top prize of $750,000 USD for a technology to mitigate climate change (VentureWell, 2014). After winning the NPCLGC, Bayer and McIntyre transitioned their technology to the packaging industry as a cost-competitive replacement for polypropylene and polystyrene-based foams.

In the years since, Bayer, McIntyre, and their team have evaluated countless fungi species for desired ecological traits and tested their growth in various substrate media, from woodchips Bayer observed at his Vermont farm growing up to agricultural wastes of cotton burr and corn stalks.

During the stage of product development, Ecovative has experimented with shaping the material during growth by applying a mold, akin to the additive process of three-dimensional printing, or through post-growth subtractive processes, such as machining or cutting, to produce or refine bulk growth to a specified shape (Bayer 2011). To augment volume and strength, agricultural waste or wood salvage may be incorporated to create a composite material (Tudryn et al., 2018).

As of 2018, Ecovative has filed for 15 patents and obtained 9 patents with the biofabrication technologies and processes to develop high-performance, sustainable materials, and products. The company's MycoFLEX platform is licensed to other manufacturers making packaging like Dell and Ikea. The platform has been also used for different consumer products that have sustainability challenges, including biofabricated leather that uses the network of mycelial fibers to create the look and texture of a hide from a cow and 3D-print artificial hearts and other body parts in the regenerative medicine industry (Peters, 2018). The company employs nearly 50 employees, operating 2 warehouses, Eco-HQ in Troy, New York, at 32,000 square feet, and Eco-East in Green Island, New York, at 20,000 square feet, and generates over $one million USD in revenue and one million pounds of manufactured materials annually (Ecovative Design, 2018).

Ecovative's Product Innovation

The Fungi kingdom represents tremendous diversity at an estimated 1.5 million species (Hawksworth, 2001). Among its other superlatives, the kingdom boasts the largest living organism on earth: a specimen of *Armillaria ostoyae* discovered in the Blue Mountains of Oregon is believed to be 2384 acres in size and between 2400 and 8650 years old (Casselman, 2007). Abundant fungi serve important ecological roles in the natural environment and are a source of food and medicines such as antibiotics for humans. Fungi have been growing mycelia, their unassuming chitinous root structures, for millennia.

What distinguishes Ecovative's materials is not the mycelia themselves but its leaders' visionary of harnessing fungi's natural, biological growth for niche applications ripe for disruption. As Bayer said to a TED audience in 2010, "the things that these organisms do are far more technologically advanced than anything we can dream of doing today with nanotechnology, with silicon technology, with anything we've gotten, and then by harnessing these innate properties of these organisms, these things that you would be used to seeing in your daily lives, like grass, we can do some incredible things for our planet and some really incredible things for the people living on this planet" (Bayer, 2011).

In 2015, Forbes Magazine named Bayer to its eponymous "30 Under 30 List in Manufacturing" (Knapp, 2015). When asked about his vision for industrial materials grown from mushrooms, Bayer recounted observing mushrooms' characteristic clumping of wood chips he would shovel for fuel while growing up on a Vermont farm (Schiffman, 2013). Yet even as its materials enter the automotive, furniture, and construction markets, Ecovative maintains its original vision as an expanded polystyrene (EPS) replacement, selling its specialized Mushroom® packaging in stock shapes including corner protectors and standard molded shippers. Ecovative also offers manufacture of custom packaging molds above a volume threshold (Ecovative Design, 2018).

Under Bayer and McIntyre's leadership, Ecovative leveraged and received numerous government grants and contracts, including US Department of Agriculture's (USDA) Wood Innovation Grant for "Scaled Demonstration of Biological Resin System to Expand the Non-structural Engineered Wood Market," seven awards from US EPA Small Business Innovation Research (SBIR) program, and in June 2017, $9.1 million by Defense Advanced Research Projects Agency's Engineered Living Materials (ELM) program (USDA, 2017; USEPA, 2017; Ecovative Design, 2018). Beyond funding research and development, government assistance provided Ecovative with a test bed to refine and optimize substrate blends to produce material of desired physical specifications, valuable third-party cost and performance data, and exposure to potential stakeholders and niche customers (USDA, 2013; Holt et al., 2012). Among them is the National Oceanic and Atmospheric Administration's (NOAA) specialized requirement for single-use readily biodegradable launch vehicles for its tsunami buoys (USDA, 2013) which are impractical to recover from the ocean once deployed from ships.

Furniture Ecovative has recently explored processes to layer and press its materials into densities and rigidities suitable for furniture and wood flooring (Knapp, 2015) and, in 2016, launched Ecovative Interiors with a limited product line. Offered directly to consumers, MycoBoard is currently used in chair backs by furniture designer Gunlocke and is also available in finished interior acoustic tiles (Ecovative Design, 2018). Preliminary evaluations of Ecovative's prototype construction materials— structural insulating panels among them—found them comparable or superior to EPS (USEPA, 2012).

Automotive Bayer and McIntyre have also identified the high plastic demand of the automotive industry as a potential market and designed and tested sound-absorbing acoustic panels for automotive bodies and interiors (Pelletier et al., 2013). Results are favorable. In evaluating blends of various agricultural waste substrate, Pelletier et al. (2013) found that even Ecovative's worst-performing composite, utilizing waste cotton bur fiber as a substrate, demonstrated acoustic absorption up to 75% at 1000 Hz. Incentivized by fuel efficiency performance standards, automakers are increasingly turning to lightweight plastic components as alternatives to steel in manufacturing. Plastics contributed to just 6% of vehicle weight in 1970. In 2010, that share was up to 16% and is projected to increase to 18% by 2020 (AT Kearney, 2012). Once vehicles reach the end of life, recent EU legislation requires the recycling of 60% of plastic components (AT Kearney, 2012). Ecovative's solution is not only recyclable but also fully biodegradable, enabling automakers to meet and exceed increasingly stringent standards.

Construction As one of the Ecovative's first incarnations of a viable construction material, Greensulate ™ was marketed as a biodegradable alternative to conventional EPS-based insulating panels (USEPA, 2009). As Ecovative scaled up its manufacturing and increased its visibility, it gained the attention of architectural designers and engineers. In 2014, architect David Benjamin installed a 40-foot-tall "Hy-Fi" pavilion at the Museum of Modern Art in New York, constructed from mycelium and corn stalk waste (Peters, 2014). The design team was able to consider local sources of agricultural waste and, once the temporary installation was ready for removal, arranged for the bricks to be recycled in the Queens Community Gardens.

Though the "Hy-Fi" pavilion was perhaps more an installation art than a building prototype, it did demonstrate proof of concept and drew attention to construction applications for Ecovative's Mushroom® materials. Perhaps most importantly, Ecovative demonstrated the Hy-Fi's construction feedstock could be sourced from local agricultural operations. By working with communities to refine its supply chains in favor of local sources, Ecovative and its partners could tap into nascent consumer preferences for and pioneer a locally sourced construction materials market, similar to those that popularized now ubiquitous locally grown produce and farm-to-table restaurants.

9.6 Discussion

Key Driving Forces of Ecovative's Success

Unsustainable practices in the packaging industry are not considered new due to petroleum-based plastic's dominance, but their social and environmental costs have not been fully internalized (Ernst & Young, 2013; Hillier et al., 2017). However, the rising volume of packaging plastics along with increasing consumption of paperboard packaging explains why diverse stakeholder groups including consumers have raised serious concerns over the unsustainability of packaging materials through the supply chain (Geyer et al., 2017; Richtel, 2016). Some firms in the packaging industry have thus acted to lower social and environmental impacts of packaging, although most responses lead modest changes within the existing industrial practices. Common are projects related to resource efficiency improvement, recycling, or waste-to-energy recovery. Seen in this light, Ecovative's response involving the entrepreneurial procedures of fabricated biomaterials from mycelia is creative and disruptive. It can shape the contour of packaging industry, creating situations from which there is no bridge to those situations that might have emerged in its absence. Ecovative's creative response leaves us with a question of what drives the firm's creative response. This inductive case study pivots the firm's unique capabilities, technical competence for systems thinking, and the ability of engaging the stakeholder matrix. Each element is investigated in detail below.

Technical Competence for Systems Thinking Ecovative has harnessed the power of mushroom to create natural, biodegradable packaging and materials for potential use in multiple industries. Rather than produce, manage, and ideally minimize its own waste stream, Ecovative has focused on the development of technology which is capable of drawing on the waste streams from agricultural processes (Holt et al., 2012). The company's technological innovation is radical and unique in that it can horizontally integrate agricultural organizations, itself, and primary and secondary customers in a novel, innovative way (Larson, 2000).

Ecovative's ability to disrupt existing practices employed by incumbents originates from the founding team's profound understanding about natural ecosystem and organisms, competitive imagination, and entrepreneurial leadership. Bayer touts the opportunities of organisms' innate properties and biological processes, many of which cannot be duplicated by technology: "with biology you can tell these organisms to do extra things; to make a compound while you're growing. That's our long-range vision for biofabrication" (Knapp, 2015). He also held entrepreneurial leadership, through which he successfully organized a core group of people willing to take risks of failures and to take advantage of a sustainability-driven innovation opportunity that is not quite regarded as appealing.

Because packaging is a low-value good, it is generally not profitable to transport to long distances to primary or secondary customers. To an extent, this characteristic has protected packaging producers in developed countries from lower prices of overseas competitors (Ernst & Young, 2013) and insulated established, geographic regions in the United States from innovators like Ecovative, were it to follow a conventional supply chain model.

Ecovative's packaging solution instead upends the traditional manufacturing packaging supply models and connects various nodes into a novel, cross-sector supply chain. Its packaging's primary raw material, agricultural waste, is readily available nationwide. Ecovative's packaging products do not require the capital and infrastructure—not to mention energy-intensive manufacturing and associated waste streams—of traditional cardboard or even bioplastics. Should they be so emboldened, retailers can eliminate the need for a packaging manufacturer, transporters, as well as associated emissions and waste at each node, by producing their own packaging materials in-house and supplanting distant suppliers of paperboard pulp or plastic pellets with a local provider of agricultural waste feedstock (Verghese & Lewis, 2007). With little infrastructure investment or utility costs, an OEM could conceivably source its agricultural waste locally within tens of miles (rather than the hundreds of miles characteristic of both paperboard and plastic packaging) and produce its own customized packaging in-house under license from Ecovative. Bayer recognizes supply chain economics are just as important as product performance: "on the raw materials side, we've been able to drive the raw material costing down to be at price parity with the styrene and polyethylene that's used in packaging so we're just finally getting to realize that dream of a triple-threat value proposition" (VentureWell, 2014).

Engaging the Stakeholder Matrix Though Ecovative's applications of fungi mycelium for materials are radical, the fundamental problem of unsustainability in packaging it sought to address was not. Independent of their lifecycle environmental performance or adoption by industry, high post-consumer content materials and bioplastics likely lowered the acceptance threshold of biomaterials for Ecovative's future customers. With exception of stakeholders already entrenched in the industry, Ecovative's stakeholders included government and industry incumbents, which is very unique to new entrants to the industry (Hall & Martin, 2005; USEPA, 2009). Bayer is very familiar with the importance of engaging the stakeholder network: "We unite a lot of stakeholders. Commercially, we work with really big companies either on direct customer access or on future projects, and we also work with the government. We've sort of connected to as many stakeholders in our space as we can" (VentureWell, 2014).

Ecovative's interactions with government entities crossed both primary (as bona fide or potential customers of Ecovative's bio-based labeled products) and secondary stakeholder (as influencers of regulatory action) domains. Driven by Executive Order 13693, Planning for Federal Sustainability in the Next Decade, the U.S. Department of Agriculture's (USDA) BioPreferred® program sets minimum standards for percent bio-based content among more than 97 product categories (e.g., cleaners, carpet, paint, etc.) and mandates their purchase by federal agencies and contractors (USDA, 2015b; Golden et al., 2015). With federal consumption of goods and services valued at $445 billion USD annually (Golden et al., 2015), through USDA's BioPreferred® certification, Ecovative enjoys a competitive advantage to a somewhat captive primary stakeholder customer (USDA, 2015a, b). Ecovative's choice to participate in USDA's Agricultural Research Service (ARS) and EPA's Small Business Innovative Research (SBIR) programs engaged these agencies as secondary stakeholders and valuable co-producers, providing capital, lending technical support, and helping to establish objective third-party performance data to assess commercial viability.

Being equally important to collaborating with governmental agencies, paramount to navigating growth is carefully selecting supply chain partners and understanding the requirements of primary and secondary customers (Golden et al., 2015). Ecovative's unique supply chain links together diverse stakeholders in novel relationships. Ecovative entered into an existing packaging industry market by large, established firms. Among them was Sealed Air, a $7 billion company and purveyor of packaging staples like Bubble Wrap®, and a customer base that includes distribution giants US Foods and Kroger Corporation (Sealed Air, 2012). An unlikely ally, Sealed Air, has become a manufacturing partner to Ecovative in operations of its Eco-East facility.

9.7 Conclusion

The unsustainability of packaging industry has been under high scrutiny from various stakeholders (Ernst & Young, 2013). Being highly visible to consumer, the packaging industry is thus compelled to decarbonize and dematerialize the industry across the entire supply chain, from raw materials to end-of-life processes for reducing its societal and environmental impacts (Hillier et al., 2017). Identifying the need to modify their production practices and processes, some firms in the packaging industry have attempted to seize a market opportunity for environmentally friendly packages and bio-based materials that can replace petroleum-based ones.

More than a decade ago, Ecovative Design, a biomaterials company growing packaging products with mycelium, envisioned the need of responsible product designs in packaging. Rather than incrementally changing the status quo of conventional packaging products, the company has brought an entirely new material solution to market by harnessing the natural mycelia root structures of fungi. Ecovative is then able to produce materials comparable in cost and performance to EPS and engineered cardboard securely bound by pressure physically and enzymatically by a completely natural, chitinous biopolymer (Haneef et al., 2017). The company enters the automotive, furniture, and construction markets, illustrating boundless product application options that have not been yet investigated or claimed (Ecovative Design, 2018). Companies partnering with Ecovative for sustainable packaging alternatives include large retailers like Dell and Ikea. Innovative firms seeking bio-based materials (e.g., biofabricated leather) and growing meat without livestock are collaborating with Ecovative to find out the optimal applications of the Ecovative's innovative MycoFLEX platform. The medical start-ups focusing on regenerative medicines also plan to use the Ecovative's approach in an attempt to develop 3D-print artificial hearts and other body parts (Peters, 2018).

Relying on the inductive case study approach, this study finds that Ecovative's creative innovation would be attributed to technical competence for systems thinking and the firm's capacity of engaging the stakeholder matrix. The two elements facilitate getting closed-loop thinking be transformed into the technological process of growing trays of mycelia in controlled temperature, humidity, airflow, and other factors. Those factors would entirely determine the shape and density of mycelia. Utilizing the Ecovative's approach for biofabrication technologies presents an immense market prospect, because biofabrication has been emerged as a twenty-first-century manufacturing paradigm, but its development has been in its infancy (Mironov et al., 2009). A huge innovation opportunity for sustainability thus relates to sustained firm value that is not achievable solely through continuous, incremental improvement (Hart, 2005).

In sum, the Ecovative's approach pivots on the significance of entrepreneurial firms' innovative initiatives based on their eco-innovation capabilities, technical competence for systems thinking, and ability of engaging with diverse stakeholders. The approach has proven enough to disrupt the ruts of existing practices in packaging industry. Extending this approach and encouraging firms' creative response for discontinuity, society can advance its sustainability and radically lower social and environmental impacts from today's old, existing unsustainable practices.

9.8 Exercises in Practice

- Identify a sustainability challenge in an industry that requires a systems-thinking approach to innovate the entire industry's unsustainable practices. Discover which kind of technical competence is needed to stymie the traditional production patterns across the unsustainable supply chain.
- Explore available opportunities to engage with stakeholders that may provide essential financial support and technical advice during the process of ecological modernization and creative disruption for sustainability.

9.9 Key Lesson for Engaged Sustainability

This case study illustrates the potential of technical competence along with a systems-thinking skill could upend the highly mature industrial supply chain. Engaging with a diverse set of stakeholders would also be another element to break through the unsustainable pattern of industry practices.

9.10 Reflection Questions

- How can new market entrants seize the opportunity to innovate a highly mature industry that could be rigid in bringing new innovative changes and ecological modernization?
- How can an innovative firm balance direct stakeholder engagement while protecting its intellectual property?
- What other industries could follow Ecovative's licensing model to get their sustainability ideas to market?
- What challenges may be presented if new market entrants target an incumbent solution and its associated supply chain?
- How can regulators not only incentivize firms with technical competence but also promote the scalability of the sustainability solutions to enhance industry-wide transformation for sustainability?
- Are there any examples that are synonymous with Ecovative in other industries? If then, what are the features that would make Ecovative distinct from those illustrations?

9.11 Cross-References

- Collaboration for Regional Sustainable Circular Economy Innovation
- Ecopreneurship for Sustainable Development: The Bricolage Solution
- Environmental Entrepreneurship for Engaged Sustainability: Challenges and Pitfalls

References

Bayer, E. (2011). *Eben Bayer: Drinking trees [Video file]*. Retrieved from https://www.youtube.com/watch?v=VTsH8qgIb80

Boons, F., Montalvo, C., Quist, J., & Wagner, M. (2013). Sustainable innovation, business models and economic performance: An overview. *Journal of Cleaner Production, 45*, 1–8.

Braungart, M., & McDonough, W. (2002). *Cradle to cradle. Remaking the way we make things*. North Point Press.

BusinessWire. (2018). *The global market for plastic packaging (2017–2026) to grow at 5.4% CAGR-environmental concern is hampering growth*. https://www.businesswire.com/news/home/20180828005731/en/Global-Market-Plastic-Packaging-2017-2026-Grow-5.4

Casselman, A. (2007, October 4). Strange but true: The largest organism on earth is a fungus. *Scientific American*. https://www.scientificamerican.com/article/strange-but-true-largest-organism-is-fungus/

Christensen, C. (1997). *The innovator's dilemma*. Harvard Business School Press.

De los Rios, I. C., & Charnley, F. J. S. (2017). Skills and capabilities for a sustainable and circular economy: The changing role of design. *Journal of Cleaner Production, 160*, 109–122.

De Soto, H. (2000). *The mystery of capital*. Basic Books.

Department of Energy & Environment (DOEE). (2014). *Sustainable DC omnibus amendment act of 2014*. https://doee.dc.gov/foodserviceware

Department of Sanitation, New York City (DSNY). (2017). *Determination on the recyclability of food-service foam pursuant to local law 142 of 2013*. https://www1.nyc.gov/assets/dsny/docs/2017-05-12FoamDetermination_FINAL.pdf

Dove, L. (2017). *Has online shopping changed how much cardboard we use?* https://science.howstuffworks.com/environmental/green-science/online-shopping-cardboard-consumption-industry-amazon.htm

Ecovative Design. (2018). *How it works*. https://ecovativedesign.com/how-it-works

Ernst & Young. (2013). *Unwrapping the packaging industry: Seven factors for success*. http://www.ey.com/Publication/vwLUAssets/Unwrapping_the_packaging_industry_%E2%80%93_seven_factors_for_success/$FILE/EY_Unwrapping_the_packaging_industry_-_seven_success_factors.pdf

Feiner, L. (2017, August 12). Why online shopping may not save the cardboard box. *Boston Globe*. https://www.bostonglobe.com/business/2017/08/11/why-online-shopping-may-not-save-cardboard-box/s2euhXKXpRhDWoOzmu2CNO/story.html

Foerstl, K., Azadegan, A., Leppelt, T. L., & Hartman, E. (2015). Drivers of supplier sustainability: Moving beyond compliance to commitment. *Journal of Supply Chain Management, 51*(1), 67–92.

Geyer, R., Jambeck, J. R., & Law, K. L. (2017). Production, use, and fate of all plastics ever made. *Science Advances, 3*(7), 1–5.

Golden, J. S., Handfield, R. B., Daystar, J., & McConnell, T. E. (2015). An economic impact analysis of the US biobased products industry: A report to the congress of The United States of America. *Industrial Biotechnology, 11*(4), 201–209.

Gopaldas, A. (2015). Creating firm, customer, and societal value: Toward a theory of positive marketing. *Journal of Business Research, 68*, 2446–2451.

Green, D. (2017). Amazon fixed the most annoying thing about receiving online orders. *Business Insider*. http://www.businessinsider.com/amazon-fixes-packaging-to-be-more-efficient-2017-12

Hajer, M. A. (1995). *The politics of environmental discourse: Ecological modernization and the policy process*. Oxford University Press. ISBN 0-19-82769-8.

Hall, J. K., & Martin, M. J. C. (2005). Disruptive technologies, stakeholders, and the innovation value-added chain: A framework for evaluating radical technology development. *Research & Development Management, 35*(3), 273–284.

Haneef, M., Cesuracciu, L., Cahale, C., Bayer, I. S., Heredia-Guerroro, J. A., & Athanassiou, A. (2017). Advanced materials from fungal mycelium: Fabrication and tuning of physical properties. *Nature, 7*, 41292. https://doi.org/10.1038/srep41292

Hart, S. L. (2005). Innovation, creative destruction and sustainability. *Research-Technology Management, 48*(5), 21.

Hawksworth, D. L. (2001). The magnitude of fungal diversity: The 1.5 species estimate revisited. *Mycological Research, 105*(12), 1422–1432.

Hillier, D., Comfort, D., & Jones, P. (2017). The packaging industry and sustainability. *Athens Journal of Business and Economics, 3*(4), 405–426. ISSN 2241-794X.

Holt, G. A., Mcintyre, G., Flagg, D., Bayer, E., Wanjura, J. D., & Pelletier, M. G. (2012). Fungal mycelium and cotton plant materials in the manufacture of biodegradable molded packaging material: Evaluation study of select blends of cotton byproducts. *Journal of Biobased Materials and Bioenergy, 6*(4), 431–439.

Hopewell, J., Dvorak, R., & Kosior, E. (2009). Plastics recycling: Challenges and opportunities. *Philosophical Transactions of the Royal Society of London. Series B, Biological Sciences, 364*(1526), 2115–2126.

Howland, D. (2017). Amazon is scrambling to minimize packaging. *Retail Dive*. https://www.retaildive.com/news/amazon-is-scrambling-to-minimize-packaging/513634/

Iles, A., & Martin, A. N. (2013). Expanding bioplastics production: Sustainable business innovation in the chemical industry. *Journal of Cleaner Production, 45*, 38–49.

James, K., Fitzpatrick, L., Lewis, H., & Sonneveld, K. (2005). Sustainable packaging system development. In W. Leal Filho (Ed.), *Handbook of sustainability research*. Peter Lang Scientific Publishing.

Janicke, M. (2008). Ecological modernization: New perspectives. *Journal of Cleaner Production, 16*(5), 557–565.

Jay, J., & Gerand, M. (2015). *Accelerating the theory and practice of sustainability-oriented innovation*. MIT Sloan School working paper 5148-15.

Kates, R., Parris, T. M., & Leiserowitz, A. A. (2005). What is sustainable development?: Goals, indicators, values, and practice. *Environment, 47*(3), 9–21.

Kearney, A. T. (2012). *Plastics. The future for automakers and chemical companies*. https://www.atkearney.com/documents/10192/244963/Plastics-The_Future_for_Automakers_and_Chemical_Companies.pdf/28dcce52-affb-4c0b-9713-a2a57b9d753e

Kim, Y., & Darnall, N. (2016). Business as a collaborative partner: Understanding firms' socio-political support for policy formation. *Public Administration Review, 76*(2), 326–337.

Knapp, A. (2015, May 6). This entrepreneur is literally growing the future of manufacturing. *Forbes*. https://www.forbes.com/sites/alexknapp/2015/05/06/this-entrepreneur-is-literally-growing-the-future-of-manufacturing/#2abbc7f9210b

Lackey, R. (2007). Science, scientists, and policy advocacy. *Conservation Biology, 21*(1), 12–17.

Larson, A. L. (2000). Sustainable innovation through an entrepreneurship lens. *Business Strategy and the Environment, 9*, 304–317.

Levi, M., Cortesi, S., Vezzoli, C., & Salvia, G. (2011). A comparative life cycle assessment of disposable and reusable packaging for the distribution of Italian fruit and vegetables. *Packaging Technology and Science, 24*, 387–400.

Lewis, H., Fitzpatrick, L., Verghese, K., Sonneveld, K., & Jordon, R. (2007). *Sustainable packaging redefined*. http://nbis.org/nbisresources/packaging/sustainable_packaging_guidelines.pdf

Mason, M. (2011). The sustainability challenge. In J. Brady, A. Ebbage, & R. Lunn (Eds.), *Environmental management in organizations* (pp. 525–532). Earthscan.

Menesatti, P., Canali, E., Sperandio, G., Burchi, G., Devlin, G., & Costa, C. (2012). Cost and waste comparison of reusable and disposable shipping containers for cut flowers. *Packaging Technology and Science, 25*, 203–215.

Mironov, V., Trusk, T., Kasyanov, V., Little, S., Swaja, R., & Markwald, R. (2009). Biofabrication: A 21st century manufacturing paradigm. *Biofabrication, 1*(2), 1–16. https://doi.org/10.1088/1758-5082/1/2/022001

Mol, A. P. J. (1995). *The refinement of production: Ecological modernization theory and the chemical industry*. Van Arkel.

Niero, M., Hauschild, M. Z., Hoffmeyer, S. B., & Olsen, S. I. (2017). Combining eco-efficiency and eco-effectiveness for continuous loop beverage systems. *Journal of Industrial Ecology, 21*(3), 742–753.

Park, S., Lee, D. S., & Han, J. H. (2014). Eco-design for food packaging innovations. In J. H. Han (Ed.), *Innovation in food packaging* (pp. 537–547). Academic. https://doi.org/10.1016/C2011-0-06876-X

Pelletier, M. G., Holt, G. A., Wanjura, J. D., Bayer, E., & McIntyre, G. (2013). An evaluation study of mycelium based acoustic absorbers grown in agricultural by-product substrates. *Industrial Crops and Products, 51*, 480–485.

Peters, T. (2014). Sustaining the local: An alternative approach to sustainable design. *Architectural Design, 85*, 136–141.

Peters, A. (2018). Can mushrooms be the platform we build the future on? *FastCompany*. https://www.fastcompany.com/90246740/can-mushrooms-be-the-platform-we-build-the-future-on

Pielke, R. J. (2002). Better safe than sorry. *Nature, 419*(6906), 433–434.

Pierce, L. M. (2017). Amazon on creating commerce packaging that's great for all: Customers, companies and the environment. *Packaging Digest*. http://www.packagingdigest.com/optimization/amazon-on-creating-ecommerce-packaging-thats-great-for-customers-companies-and-environment-2017-04-14

Prahalad, C. K., & Hart, S. L. (2002). The fortune at the bottom of the pyramid. *Strategy+Business, 26*, 54–67.

PricewaterhouseCoopers. (2010). *Sustainable packaging: Threat or opportunity?* https://www.pwc.com/gx/en/forest-paper-packaging/pdf/sustainable-packaging-threat-opportunity.pdf

Raffensperger, C., & Tickner, J. (1999). *Protecting public health and the environment: Implementing the precautionary principle*. Island Press.

Richtel, M. (2016, February 16). E-Commerce: Convenience built on a mountain of cardboard. *The New York Times*. https://www.nytimes.com/2016/02/16/science/recycling-cardboard-online-shopping-environment.html

Rigamonti, L., Grosso, M., Moller, J., Sanchez, V. M., Magnani, S., & Christensen, T. H. (2014). Environmental evaluation of plastic waste management scenarios. *Resources, Conservation and Recycling, 85*, 42–53.

Schiffman, R. (2013). One minute with…Eben Bayer. *New Scientist, 218*(2921), 29.

Schumpeter, J. A. (1947). The creative response in economic history. *Journal of Economic History, 7*(2), 149–159.

Sealed Air. (2012). *Sealed air and Ecovative complete agreement to accelerate commercialization of new sustainable packaging material*. http://ir.sealedair.com/phoenix.zhtml?c=104693&p=irol-newsArticle_Print&ID=1706817

Singh, S. P., Chonhenchob, V., & Singh, J. (2006). Life cycle inventory and analysis of reusable plastic containers and display-ready corrugated containers used for packaging fresh fruits and vegetables. *Packaging Technology and Science, 19*, 279–293.

Starik, M., & Kanashiro, P. (2013). Toward a theory of sustainability management: Uncovering and integrating the nearly obvious. *Organization and Environment, 26*(1), 7–30.

Sustainable Packing Coalition. (2011). *Definition of sustainable packaging*. https://sustainablepackaging.org/wp-content/uploads/2017/09/Definition-of-Sustainable-Packaging.pdf

Tudryn, G. J., Smith, L., Freitag, J., Bucinell, R., & Schadler, L. (2018). Processing and morphology impacts on mechanical properties of fungal-based biopolymer composites. *Journal of Polymers and the Environment, 26*(4), 1473–1483.

United Nations. (1992). *Rio declaration on environment and development*. United Nations. http://www.unep.org/Documents.Multilingual/Default.asp?DocumentID=78&ArticleID=1163

United Parcel Service (UPS). (2017). *Packaging guidelines: Billable weight*. https://www.ups.com/us/en/help-center/packaging-and-supplies/determine-billable-weight.page

United States Department of Agriculture (USDA). (2013). Biodegradable packaging from cotton waste. *Agricultural Research, 61*(10), 16–18.

United States Environmental Protection Agency (USEPA). (2009). *Ecovative design: Growing America's green economy with research and innovation*. Office of Research and Development. https://archive.epa.gov/ncer/publications/web/pdf/ncse-greensulate.pdf

United States Environmental Protection Agency (USEPA). (2012). *Development and demonstration of a low embodied energy, construction material that replaces expanded polystyrene and other synthetic materials*. EPA contract number EPD10058.

United States Environmental Protection Agency (USEPA). (2017). *Ecovative design: Greensulate|growing America's green economy with research and innovation*. Office of Research and Development. https://archive.epa.gov/ncer/publications/web/pdf/ncse-greensulate.pdf

USDA. (2015a). *Driving the bioeconomy: Economic impact of the biobased products industry*. Presentation by Marie Wheat, Industry Economist, USDA BioPreferred Program, to the BIO World Congress on Industrial Biotechnology. 22 July 2015.

USDA. (2015b). *Development of scaled manufacturing for mycological soilless growth media*. National Institute of Food and Agriculture. Grant number 2015-33610-23814.

USDA. (2017, May 24). *U.S. Forest Service awards grants to expand and accelerate wood energy and wood products markets in 19 States*. U.S. Forest Service Press Release.

USEPA. (2014). *Advancing sustainable materials management: Facts and figures*. https://www.epa.gov/smm/advancing-sustainable-materials-management-facts-and-figures

USEPA. (2015). *Documentation for greenhouse gas emission and energy factors used in the waste reduction model (WARM)*. Office of Resource Conservation and Recovery. https://archive.epa.gov/epawaste/conserve/tools/warm/pdfs/WARM_Documentation.pdf

USEPA. (2016a). *Documentation for greenhouse gas emission and energy factors used in the waste reduction model (WARM)*. Office of Resource Conservation and Recovery. https://19january2017snapshot.epa.gov/sites/production/files/2016-03/documents/warm_v14_containers_packaging_non-durable_goods_materials.pdf

USEPA. (2016b). Advancing sustainable materials management: 2016 recycling economic information (REI) report. October 2016. EPA530-R-17-002.

Van Haaren, R., Themelis, N., & Goldstein, N. (2010). The state of garbage in America. *Biocycle, 51*(10), 16–21.

VentureWell. (2014). VentureWell idea to impact. *Ecovative*. https://www.youtube.com/watch?v=HqmCSml15jU

Verghese, K., & Lewis, H. (2005). *Sustainable packaging: How do we define and measure it*. Presented at the 22nd International Association of Packaging Research Institutes Symposium.

Verghese, K., & Lewis, H. (2007). Environmental innovation in industrial packaging: A supply chain approach. *International Journal of Production Research, 45*(18–19), 4381–4401.

Wikström, F., & Williams, H. (2010). Potential environmental gains from reducing food losses through development of new packaging – A life-cycle model. *Packaging Technology and Science, 23*, 403–411.

World Economic Forum, Ellen MacArthur Foundation and McKinsey & Company. (2016). *The new plastics economy: Rethinking the future of plastics*. http://www.ellenmacarthurfoundation.org/publications

York, R., & Rosa, E. A. (2003). Key challenges to ecological modernization theory. *Organization and Environment, 16*(3), 273–288.

Zeller, P., & Zocher, D. (2012). Ecovative's breakthrough materials. *Fungi, 5*(1), 51–56.

Younsung Kim is associate professor of the Department of Environmental Science and Policy at George Mason University. She is also the undergraduate program director for the department. Her research lies in collaborative governance and firms' voluntary policy tools designed to tackle today's most complex sustainability and climate challenges. Focusing on the private sector's role, she has investigated the efficacy of Environmental Management Systems and the motivations for firms' innovations and cross-sector partnerships for sustainability solutions. Her research was recognized by the Public and Nonprofit Management Division of the Academy of Management with the 2016 Best Article Award. Recently, she published "Integrated Market and Nonmarket Strategies: Empirical Evidence from the S&P 500 firms' Climate Strategies" (Business Politics, 2022). Prior to her career in academia, she worked for the Ministry of Environment in South Korea as Deputy Director and served on the World Bank's Carbon Finance Assist Team.

Daniel R. Ruedy is an environmental professional with 13 years of experience in environmental science, protection, and program development for leading organizations including the Department of Defense's (DoD) Strategic Environmental Research and Development Program (SERDP) and Environmental Security Technology Certification Program (ESTCP), U.S. Marine Corps (USMC), U.S. Environmental Protection Agency (EPA) and the Environmental Council of States' (ECOS) Interstate Technology and Regulatory Council (ITRC). Expansive knowledge of today's environmental challenges and programs including munitions/unexploded ordnance (UXO) detection and cleanup; pollution prevention; hazardous material and waste (HM/HW) management; research and development (R&D); technology transfer; remediation technologies; and law, regulation and policy.

Chapter 10
Leadership and Achieving Sustainable Solutions: Shifting Landscape of Green Business Practices

Sujo Thomas, Susmita Suggala, and Sudhir Pandey

Abstract Due to the emerging environmental issues and scarcity of resources, leaders in every business have started adopting green business practices worldwide. The preliminary concept of 'green business' was developed in the late twentieth century. The notion of green business has always been misunderstood as a challenge which cannot be overcome with basic practices undertaken by leaders. It becomes crucial for leaders in the organizations to explore the green business practices that cannot be universally applied and would be subjective on grounds such as culture, politics, and economic contrasts. Green transformation by businesses, to reach from current state to desired future state, is intended to achieve the long-term goal of incorporating sustainability. The leaders need to consider multiple forces (climate change, increased demand, etc.) that impact (raw material scarcity, public pressure, security concerns, etc.) businesses. The green business transformation process is comprehensive, which requires the active participation of stakeholders, planning and implementation. This chapter will focus on the drivers and barriers of business organizations while following green business practices. On the whole, this chapter will help in developing an understanding about the following: (a) understand leadership and sustainability, (b) explore the changing landscape of green business practices followed by leaders in various business organizations in the Indian context, (c) understand the organizational green transformation strategies adopted by leaders thereby leading to sustainable solutions and (d) explore the drivers and barriers towards implementation of green business practices, by the leaders in various industries.

Keywords Green business practices · Green environment · Sustainability · Green packaging · Green marketing

S. Thomas (✉) · S. Pandey
AMSOM, Ahmedabad University, Ahmedabad, Gujarat, India
e-mail: sujo.thomas@ahduni.edu.in; sudhir.pandey@ahduni.edu.in

S. Suggala
LJIMBA, Gujarat Technological University, Ahmedabad, Gujarat, India

© The Author(s), under exclusive license to Springer Nature Switzerland AG 2023

S. Dhiman (ed.), *Sustainable Development and Environmental Stewardship*,
https://doi.org/10.1007/978-3-031-28885-2_10

10.1 Introduction

Due to the emerging environmental issues and scarcity of resources, leaders in every business have started adopting green business practices worldwide. The preliminary concept of "Green Business" was developed in the late twentieth century. The notion of green business has always been misunderstood as a challenge which cannot be overcome with basic practices. The green initiative has now evolved and started becoming a good term in most industries. Many businesses have developed an environmental management system and waste management system, thereby integrating sustainable business operations practices. Sustainability itself has become a product which is exchanged, traded, bought, and sold. The leading global chains all around the world are putting 'sustainability'on top priority. Sustainability is becoming mainstream in business conduct, operation, and strategic concern for leveraging capital accumulation and ensuring competitive advantage. Initially, when the companies began applying sustainability as a value proposition, they did not serve it as a product but in public relations and corporate social responsibility or cause-related marketing (Kureshi & Thomas, 2020; Thomas & Kureshi, 2020). Sustainability is the need of the hour due to the continued climate change, loss of biodiversity, resource depletion and air pollution, impacting all of us as the earth's citizens.

Growth in a dynamic environment is a result of strong leadership which steers the organization to acclimatize with the current business trends, build and achieve a sustainable competitive edge. In this changing business landscape, the responsibility of achieving a constructing performance is shouldered by leaders who are focused to lead the organization responsibly. This constructive leadership is central to organizations which can weather the evolving trends and changing economic environment (Cabeza-Erikson, 2008). Constructive leadership analyses the past environmental trends, business culture, and other internal resources to shape and create new dynamic strategies to achieve long-term sustainable solutions for survival thereby fulfilling organizational goals. An able leader inspires change and influences the values, behaviour, performance, and attitude of employees or a group of members (Ganta, 2014). Leaders effectively harness the skills of their subordinates and inspire them to take challenges. They attain successful results by effectively comprehending the subordinate's psychology and support them to achieve their goals. Organizations, in the absence of effective leadership, are continually in peril as it lacks the foresight of future trends besides the absence of guidance and motivation to the employees (Atkinson, 1999).

Change is inevitable. Modification and variations in any form is a complication for an organization. It represents the beginning of transformation at both the individual and organizational levels. People find it difficult to accept the alterations or deviation from their routine activities which displaces them from their comfort zones exposing them to uncertainties which results in low acceptance by people (Lorenzoni et al., 2007). Companies with strong leadership steer their employees to accept changes as a need for sustainable competitive edge in the market (Du Plessis, 2007). Employee acceptance to change varies from person to person. The leader's approach and deliverance decide the rate of positive acceptance resulting into higher work expectancy, a consequence of perceptions and reactions of the employees. Employee satisfaction in work is directly related to the acceptance of change by people. Hence, the role of a leader is deemed essential. He plays a crucial role in managing a change in the working environment.

The consumption of natural resources is unsustainable and is creating environmental problems. With the rise in the resource deterioration amongst the nations concerning natural resources and the awareness about sustainable development, the term green has evolved in the business world. Green business is often used intangibly with sustainable business, whereas green business is a strategy to adopt a sustainable business long. For instance, from the customer perspective, there are various challenges to practice green business such as individualism and collectivism. Customers are one of the most critical green agents, and their behaviour drives the market. People who do not think about public benefit and take decisions considering personal matters lead to individualistic behaviour.

On the other hand, collective action encourages people to go green, adopt green lifestyle, and act in a coordinated manner. The additional barrier to the green practices is the people who behave individually under the demographics of age, education level, and income level. Government, acting as a green agent, also encourages business to adopt green practices by providing legislation and policies such as eco-labelling, eco-auditing, environment taxation, etc. Government and NGOs together work to uplift and support green practices in the corporate world. Businesses often choose to go green not just due to environment care, but due to the rise in revenues (product differentiation) and fall in cost (waste utilization). Likewise, many businesses prefer not to enter green business due to the practice legislation and high expenses that may diminish their revenues.

This chapter will focus on the drivers and barriers for business organizations while following the green business practices. This chapter will explore the landscape of green business practices followed by leaders in various business organizations towards achieving sustainable solutions and challenges associated with the same. This chapter will provide an opportunity to comprehend how green business practices followed by leaders would impact businesses. On the whole, this chapter will help in developing an understanding about the following: (a) Understand leadership and sustainability, (b) explore the changing landscape of green business practices followed by leaders in various business organizations in the Indian context, (c) understand the organizational green transformation strategies adopted by leaders thereby leading to sustainable solutions and (d) explore the drivers and barriers towards implementation of green business practices, by the leaders in various industries.

10.2 Leadership and Sustainability

Organizational culture is considered to be a significant contributor of effective leadership. Manufacturing companies encourage employee engagement and innovations in organizational processes and activities (Bass, 1994). Satisfaction is deemed a substantial factor for better performance. Manufacturing companies thereby pay more attention to the satisfaction of employees, rewarding them and boosting their morale to ensure that the productivity and performance indices are always kept high (Chandrasekara, 2019). Collective synergy, a consequence of positivity at workplaces, creates a strong bond between leaders and employees (Ionescu, 2014). This increases the loyalty and commitment of employees towards the group (Schein, 2010).

Freifeld (2013) believes in continuous learning. It helps employees to overcome their shortcomings after a thorough analysis. It also helps unearth better opportunities to overcome inflexibility enabling accomplishment of an action with greater ease. Training and development for official or religious groups is considered important to achieve individual and organizational goals by acquiring and widening their knowledge and by sharpening their skill base through continuous learning. Training enhances embracing changes in a constructive manner which leads to better employee productivity (Abou-Moghli, 2015). Training also increases the scope of innovations and creative solutions. Creativity, out-of-the-box thinking, and innovations are the basic pillars denoting sustainable environment. Leaders encourage group members to trust and support each other without negative interferences (Malloch, 2013). In technological companies, change is a common necessity to upgrade and adopt new trends for achieving an edge over its competitors. Regular practices supported by systematic, methodological indications point at sustainable practices. The social orders are feasible only if the facts are proven on a scientific background. Moving towards sustainable solutions requires scientific description analysis associating long-term vision, a leadership trait required by organizations to achieve their goals. Sustainable vision is a path which every organization takes and it requires an able leadership to build strategies to close the gap between the current trends and sustainable vision. Systematic leadership towards sustainability suggests use of systematic thinking adopting a step-wise approach for changing into sustainable society (Broman et al., 2013). Achieving sustainable solutions in the future suggests outlining local, national, and global changes in a controlled and systematic way to achieve sustainable future.

10.3 Role of Green Packaging and Green Marketing

Industries have started taking severe measures to attain sustainable development goals on an individual basis, and one such way where 'going green' is becoming a trend is the packaging. According to IMARC, on a global level, the green packaging industry's market size reached USD 168.2 billion in 2018 and is anticipated to be

USD 238.6 billion by 2024. According to IMARC, Europe has the most significant market share for green packaging, followed by Asia and the Middle East. The problem originated as early as 1950, where the companies supplied the plastics worldwide not knowing the consequences for the same. The company's main challenge is to get customers to buy and be aware of the packaging. From 1950, companies have a significant influence on consumers and encouraged people to use plastics. Due to the plastic-free environment research, the multi-national companies are now supplying and producing green products to protect the environment. Many renowned companies like Mars, Starbucks, and Nestle gained benefits from managing sustainability concerns. Some of the unique services are aspects like better control or product quality and logistics and stability of availability of resources.

Focusing on India specifically, the packaging industry has witnessed tremendous growth and change in preference for green packaging due to critical factors such as increasing income and changing lifestyle. According to Associated Chambers of Commerce and Industry of India, the green packaging sector grew at a fast pace of about 25% in 2011. Few such companies are Nestle, Pepsi, ITC, Hindustan Unilever, and Procter & Gamble. Tata Starbucks was one of the first companies to shift towards green packaging for their products by providing glassware cutlery and paper straw. Due to the rapid increase in the societal concern in the context of environment and sustainability, customers are engaging more and more in green products. Organizations have started practising green packaging along with their other business practices. This green packaging practice provides mutual benefits in the form of high growth to the producer and environment sustainability and improved society's quality of life. The customer base of green packaging is ever increasing because of constant support from government and organizations taking up the responsibility to use green packaging. Indian FMCG has a massive market for green packaging, but it is yet to adopt this practice due to the low awareness and cost sensitivity.

Many companies in India have also started using eco-labels. Eco-labels in the packaging of the company's products represent that they have made the packaging environmentally friendly. India had launched an eco-mark scheme in 1991 to influence the consumers to buy such kinds of products. The sustainability indicator concerning eco-labels usually has information concerned with aspects such as the industrial application or academic purpose, and hardly any information is provided for direct consumers. Thus, these labels are not that effective to encourage potential customers for green purchasing. The concept of 'green' is different to all economies to determine the economic dynamics and public consciousness. The green market challenge is that it is not much known, and due to this, its awareness by the companies is necessary.

Green marketing involves manufacturing the right quality products and how they are harmful, even in the long run. Green marketing consists of the marketing of ecologically reliable products to scale down its carbon footprint. Activities such as packaging innovation, product modification, revised advertisement, lowering production wastes, etc. can be classified as green practices. Green marketing by green businesses has resulted in a favourable outcome in India, some of the examples being: Honda launching its hybrid car, successful campaign of Idea cellular where they had the motive of spreading the use of mobile and save paper, IndusInd Bank promoting the usage of solar power, and ITC coming with eco-friendly business papers (Chauhan & Chaturvedi, 2015). Several electronic merchandise companies have been marketing their products with an 'Energy Star' illustrating that usage will reduce energy consumption, enabling consumers to save money. Godrej consumer products are selling their new product by adopting green practices and providing incentives to consumers to exchange their old and worn-out products with new, environmentally friendly products. Such measures attempted to convince consumers to adopt a green lifestyle and contribute their bit for our environment. NDTV (New Delhi Television Limited) endorses green practices through 'Greenathon' related to social and philanthropic activities to enhance environmental affairs awareness. Nerolac Paints has been practising green marketing as they have eliminated the use of perilous heavy metals from their paints. Businesses have started to focus on using bags made out of recyclable paper in place of plastic bags. Tanishq and Titan have efficiently adopted and implemented this corporate policy wherein the products would be delivered to consumers only in paper bags.

Some of India's big businesses, like Tata consultancy services, ONGC (Oil and Natural Gas Companies), ITC, Wipro and MRF, are continuously developing tactics of conservation of energy, water, and technology. Tata consultancy services have four core practices: green buildings, efficient working, green technology, and renewable resources instead of the non-renewable ones. ONGC uses paperless initiatives to save paper, and they eliminate carbon usage as much as possible. ITC also focuses on low carbon and chlorine usage to avoid harming the ozone layer. Wipro and MRF have continuously been increasing their sustainability practices over the years (Rai and Jambhulkar, 2018).

10.4 Sustainable Business Practices in Various Industry Sectors

This section provides insights into sustainable business practices in various industry sectors concerning green practices.

Automobile Industry The idea of sustainable development has been flourishing worldwide, and consumers have increasingly become aware of the surrounding they are living in. People have started accepting electric vehicles over conventional transportation methods and started saving fuel and energy in some parts of the world.

Tourism Industry According to Peattie (2001), the transition of service industries focusing on green practices started later in the 1990s, before that phase, only manufacturing industries were into green practices. According to Kasliwal and Agarwal (2019), the tourism sector is growing continuously, and the importance of environmentally friendly ways has become immensely important to save the resources sustainably. The tourism sector is dependent on extensive usage of environmental resources such as water, electricity, other non-durable and semi-durable goods. Still, their value addition is responsible for providing these goods, making the hotels use heavy usage of natural resources such as water and energy.

Hospitality Industry With the change in the conditions of the environment, the industries all over the world are adapting themselves towards green practices, and the hotel industry being just one of the many sectors adapting its way according to the increase in consumer preference in terms of being environment friendly (Bhattacharya, 2011). While the hotel industry has a good impact on GDP and economy, it harms the ground due to the heavy usage of resources like water and energy. There is a term known as 'green hotels' for the hotels that operate in an environmentally friendly and sustainable way, some of the factors being energy-efficient, water-efficient, and using eco-friendly technology.

According to the study conducted by Millar and Photakoun (2008), many of the people had positive views on the green hotel, and some of the opposing views had 'expensive' in their mind means to say that some of the consumers think green hotels are expensive. According to the study of Manaktola and Jauhari (2007), only 15% of the guests felt that there is a need to charge a high price for green hotel rooms while in America it was 30%. While many of them express their concern towards the environment, only a few do pay a premium price for such services. In a study conducted by Millar and Photakoun (2008), the consumer's preference was being tested upon various environmentally friendly ways some of them being light bulbs that save energy, low flow shower and toilets, recycling bins and critical cards, majority of the consumers had a positive response on these ways. However, some of them have an adverse reaction to some tactics like reusing shampoo and soap dispensers and using low flow showers. Another factor is that the consumers do not want any compromise in terms of the quality of services, meaning they do not want to replace their comfort with environment-friendly ways used in green hotels. According to the study conducted by Kasliwal and Agarwal (2019), awareness and education of customers in this area would make the customer differentiate the green hotels and would also be a helping hand for convincing the customers to pay a premium price for the environment-friendly technology of the hotels. Also, training to the employees must be provided to help them deliver green hotel services'value over the long run. Young people who are more aware of the degradation of environment and women generally being sensitive to these issues are more likely to acquire such services and are developing a sense where they are a bit convinced about paying a premium price of up to 3% of their total package for environmentally friendly practices. According to a study of Millar and Baloglu (2011), although the majority of the consumers in the current situation are unwilling to afford the higher cost for the environmentally friendly services provided by hotels, many websites are providing a difference between standard hotel rooms and green hotel rooms. This strategy offers them a competitive advantage and gives a different brand image to the hotels that operate in environmentally friendly ways.

According to Bansal and Roth (2000), the reasons businesses tend to operate in environmentally friendly ways are driving more towards a competitive environment, being acceptable, and holding responsibilities towards the society in which they operate. Also, in the long run, businesses find it profitable to save cost and provide value to the customers. It was also found that companies that go green on an average save 28% costs through saving energy, water, emission costs, operational and maintaining costs (Butler, 2008). According to the studies conducted by Rahman (2012) and Fernández-Robin et al. (2019), whether it employs environmentally friendly tactics or not and the extent to which they use these environment-friendly tactics vary.

The size of a hotel also matters according to research by Mensah (2006). If the size of a hotel is large, it would require more resource usage and more wastage. But this does not coincide with the research results by Fernández-Robin et al. (2019) where it was found that medium and large hotels were more into environmentally friendly operations. This was because small firms found it expensive to add such technology. After all, consumers do not value it sufficiently. Also, in the studies conducted by Rahman (2012), it was found that medium-size hotels save the highest and are best in terms of environmentally friendly operations.

Textile Industry The textile industry, in the context of the emerging economy, has been highly criticized for their activities which lead to wastage of non-renewable resources, heavy usage of chemicals and pesticides thus resulting in harming the environment (Alay et al., 2016). So, the textile industry has adopted the process of green supply chain management (GSCMP). However, the developing textile industry faces many barriers in the implementation stage. According to Tumpa et al. (2019), the most critical barrier is financial constraints. In emerging economies, the demand from customers is less due to lower awareness levels, and the economic gains of businesses are minimal, leading to less investment in GSCMP. The next critical obstacle includes lesser government support and lack of any regulations. The emerging economies fail to provide enough incentives to the textile businesses, resulting in lack of promotion and engagement in green business practices.

Moreover, the green textile materials in emerging economies are majorly bought from other countries due to local markets' unavailability. The cost incurred for the same is high, and therefore the textile industry fears making a further massive investment in greening supply chain management. Another critical barrier is the lack of top management commitment. The top management of most textile businesses does not support and commit to green business practices as they aim for short-term financial returns. The culture of the textile businesses, too, does not encourage change. Other barriers include technical hindrances and weak collaboration between supply chain partners due to complexities in the process.

Sharing Economy Platforms The sharing economy has been growing over the years with significant players like Airbnb, Uber, etc. While economy and society-related sustainable supply chain management (SSCM) activities, such as corporate social responsibility (CSR) and investment recovery (IR) in the context of sharing economy, have widely affected customer's intention to adopt sharing economy-based products/services positively, green SSCM procedures have not been affected (Hu et al., 2019). Their research found that the most critical barrier in promoting green practices of sharing economy-based products/services platforms is limited, which results in low customer awareness. Moreover, customers are willing to pay if they get incentives in return or else customers do not feel connected with environment protection. In such a situation, sharing economy platforms must provide monetary benefits and focus on advertisements which shed light on facts and personal values of green practices to attract customers.

Green Start-ups With the increase in environmental issues, green start-ups have started emerging in the growing green economy. Existing large firms, too, have begun shifting to sustainable practices. However, small green start-ups often face difficulties in cost efficiency, revenue growth, better organizational management, etc. (Demirel et al., 2019). According to Leoncini et al. (2019), environment-friendly technology plays a crucial role in determining the overall organizational growth. Most of the firms now at the maturity stage of the life cycle seem to pursue the green growth part. Young entrepreneurial green firms seem to divert from technology-based growth due to less finance. Hall et al. (2017) examine that while the new green ventures pro-actively shape the institutional environment as a core strategy, new trade challenges arise that work against the venture's interests. So, factors such as life cycle, institutions, availability of external finance, and knowledge are considered a significant determinant of green start-ups' performance and behaviour (Demirel et al., 2019).

Small and Medium Enterprises They contribute significantly to the nation's GDP and can be witnessed as the backbone of our nation's long-term success. Upon the astounding development of green businesses, SMEs have merged the green marketing approach with their core fundamentals. Green marketing practices can help enhance corporate reputation and bring about the consumers' attention. Green businesses undertake such methods which are crucial to efficient production and reduced material cost. Implementing green practices in SMEs can be considered one factor ensuring steady future progress with consumers becoming more ecologically conscious in recent years.

Food Industry Increase in population growth has been massive since the last decade. Such increase in population compels consumers to increase demand for food products. The FMCG companies concentrate on CSR activities, and CSR also has variants such as sustainable business practices, cause marketing, corporate philanthropy, and cause-related marketing (Kotler & Lee, 2008; Kureshi & Thomas, 2014; Thomas & Kureshi, 2016; Thomas et al., 2020). As compared to conventional food products, most consumers believe organic food is hygienic, safer, healthy, and tastier. Despite the favourable attitude of consumers being high for organic food, the actual purchases remain low. Studies have proven that even though consumers know about the additional benefits of organic food, their purchase intention remains low due to high prices, inadequate labelling, unavailability, and ineffective Point of Purchase displays. Developing nations are less inclined to take into account 'environmental quality' while purchasing food products.

On the contrary, consumers of developed nations who are high-income earners regard this factor while purchasing food products. Production and consumption of certain meat products contribute to air pollution, scarcity of water and land, and domestic waste into the environment. Students having a positive attitude towards organic food prefer consuming the same at home, at restaurants, and on campus. Green marketing practices by food outlets involve eco-friendly shopping atmosphere, eco-labelling, and organic food using in-store promotional instruments. Such exercises strengthen product representation and corporate social responsibility.

Hart (1997) sketched out three kinds of sustainable goals that most companies aspire to fulfil, prevent environmental pollution, encourage product modifications to make it reusable to ensure there is no ecological pollution throughout its lifecycle period, and use advanced technology that creates no harmful emissions. These are applicable at the societal and at the organizational level. A few barriers identified in the actual implementation of green practices are a lack of commitment from top management. Despite the rise in green practices, a lack of eco-literacy amongst its supply chain partners has been noted resulting into the unpreparedness of these suppliers to adopt these processes or engineer reverse logistics due to lack of adequate strategic planning. Organizations awarded with discounts or subsidies increase the cost of production compared to the effective implementation of green practices. On the other hand, there is no specific requirement from the customers, demanding products of a green supply chain management designed to protect the environment at every stage of its manufacturing process (Moktadir et al., 2018).

10.5 Conclusion

Corporates, big and small, have realized the significance of leadership and its impact in achieving sustainable solutions. The conservative insights of green business in large companies' eyes are considered the benefits emphasized in terms of innovation, revenue growth, and improved relationships with all stakeholders (Ambec & Lanoie, 2008; Porter & van der Linde, 1995). A leader by practicing green entrepreneurship is professed to provide a continuous flow of opportunities for the advanced and emerging economies. This has also opened developments for socio-economic environments across countries (Hall et al., 2010). It can hence be conveniently deduced that green businesses are making steady headway in all industries. The leadership plays a vital role in making strategic decisions to support the green business practices in their organizations, a step towards creating a sustainable global environment by reducing the negative impact on the society, environment, or economy. Green transformations are made possible by conditional effects on the green supply chain management, which is due to the moderating efforts on the environment performance factors at reasonable cost considerations which are collaborative activities for sustainable existence. A leader can influence by changing the standard practices and introducing new standards and certification systems which can ease the green supply chain management process. For instance, in the textile industries, though a gross earner, which has contributed heavily to the rising economies have earmarked the lack of sustainable supply chain management practices (Chang et al., 2018). The processes indicate the necessity of implementing green supply chain management processes and environmentally friendly production, consumption, and disposal units (Moktadir et al., 2018; Pandit et al., 2018).

The role of leadership in creating a sustainable green environment is not limited to the manufacturing industries but also imperative to services and other non-industrial units, implying that green practices are not limited to production. Leadership in HR practices can also propound greenways to be in the form of green recruitment, selection, performance management, exit interviews, and the pay and rewards. Green methods are implemented by aligning environmentally friendly practices. Such initiatives create high awareness and develop self-motivation amongst students and faculties who can encourage ecologically friendly alternatives at their workplaces and homes. Sometimes, for a leader practicing what you preach also promotes the culture within an organization and all work towards a common goal (Mukherjee & Singh, 2020). According to Oshodi & Aigbavboa (2017), the construction industry has not adopted green practices at the same rate as the other industries. However, the industry's corporate commitment is reflected by their ready acceptance of higher regulations (ISO 14001) for preserving the environment (Zhao et al., 2012) from high carbon emission, energy usage, and in the management and disposal of construction waste. This has prompted several transformational programmes setting targets for environmental sustainability (Yusof et al., 2019).

In case of lack of infrastructure to support sustainable product solutions; alternative techniques and solutions should be incorporated to implement and manage sustainability (Hunting & Tilbury, 2006). Recycling schemes, an alternative for online and in-store packaging should be encouraged in a slow and phased manner. Offers and discounts can also encourage people to carry their own bags or shopping containers. Green practices can also be incorporated in manufacturing processes, where cost reduction of materials can be levied if the materials checked do not have a long-lasting ability. Working for an environmental or a social cause reverberating the organizational goals is considered as the minimum effort an organization places in achieving green sustainable solutions. Early adopters of green marketing are projected to be winners in the long run. The late comers having lost the race are expected to spend more time, money, and efforts to come up with sustainable solutions to give it a competitive edge, as in the global scenario, there is an abundance of sustainable solutions spread across the market. Long-term sustainable strategies are demanded by manufacturers, retailers, and services organizations alike. Khare (2019) has rightly said that knowledge and human sensitivity as manufacturers and consumers can encourage sustainable consumption leading to identify and incorporate product life cycles to reduce carbon footprints in manufacturing, transportations, and stores.

Customer awareness is the need of the hour to save the environment, and its benefits on health can coerce companies' leaders to look up for sustainable manufacturing processes to bring about a revolution in production process, marketing strategies, and consumption patterns. All this clearly demonstrates the need for strong leadership that is capable of maintaining an equitable balance with the economy and environment (green practices) and marching the company towards sustainability.

Learnings from various industries provide diverse skills and capability of new companies adopting enhanced processes and taking small and big steps to convert green technologies into performance-oriented systems (Marin et al., 2015). Green entrepreneurship is an emerging field and efforts are invested in developing some low carbon environmental goods and services industry. According to the researchers Dean and McMullen (2007), low carbon emission in the environment has been viewed as a significant concern which is also seeing an upsurge in technological investments as there are multiple proposed solutions for a sustainable carbon-free environment (Esty & Winston, 2009). Technology plays a dual role where it is perceived to be a solution and burden for the environment. The world has always stood united on the need for a sustainable environment by bringing about a change in the current behaviour to reduce and replenish biodiversity and the finite resources to meet the present and the future generation requirements (Brundtland, 1987).

Lodging industry, high consumption of natural and non-renewable resources also contributes to environmental pollution. The size of the hotels is directly proportional to the scale of pollution. Independent hotels though cause less damage are also easily manoeuvred. Large hotel chains undergo multiple levels to have the changes implemented in their regular day-to-day chores. Kirk (1995) elaborates the challenges of not losing customers and integrating green practices in the services and working out solutions to help maintain customer perception, to keep the façade of the services, taken as an attempt to remain at the frontline in providing and supporting environmentally friendly services. Ecologically aware businesses are aware of the multiple processes where green practices can be implemented. Organizations have implemented the 4R rule, which stands for reduction (wastage or consumption), reusing, recycling, and recovery. The criteria for different cultures embracing green practices vary. Some consider costs over the benefits of implementing green operations, while others give more weightage to the generation of higher revenue to saving costs. A summary of the various processes that can implement green functions is labelling, packaging, processes, reduction of power wastage, less printing, eco cleaning, reduction of transportation fuel using public transport, saving electricity in running appliances and many more. Some companies embrace green operations, green products or bio-degradable resources for products based on the business environment's awareness levels (Čekanavičius et al., 2014).

Born green enterprises exploit the drawbacks of existing enterprises for sustainable environment-friendly processes resulting in higher awareness giving the green enterprises a competitive edge compared to corporate social responsibility initiatives. The ecological friendly approach provides recognition of its contribution to its impact on the social environment (Dean & McMullen, 2007; Pacheco et al., 2010; Criscuolo & Menon, 2015; Esty & Winston, 2009). Factors imploring the green start-ups are summarized as the ease and convenience of access to home-grown details and scientific facts and data in addition to the implementation of environmentally friendly policies implemented with the awareness of local governments and communities in addition to the environment-friendly going green ventures by established industries, governments, universities, and non-governmental organizations (Demirel et al., 2019). The drivers identified for green business practices are regulations, organization culture, social influencers, business ecological concerns, financial motivations, and competitive advantage (Aghelie, 2017). The top-level management is involved in the strategic implementation of green supply chain management (GSCM). THE GSCM in the following stages measures the performance of the suppliers and customers with complex and problematic eco designs. Organizational performance is related to environmental accounts. The eco designs are ineffective on the costs and functionality but have a substantial impact on eco designs, as Grote et al. (2007) elaborated. The organizational performance is found to have a strong association with operational performance, which results from quality, delivery, and the predetermined time and efficiency. The green information system is an effort by companies to fulfil their requirements without compromising the resources to enable them to carry out uninterrupted activities for future generations (Brundtland, 1987). Information technology has been the backbone of industrial growth; companies should ensure that the wastage of resources, energy inefficiency, maintaining quality, and reducing noise and heat emissions can build a sustainable green IT system. Sustainable development should not be seen as a cost of doing business. Incorporating thin computers at call centres have resulted in less energy consumption and less carbon waste dumped into the environment, similarly leaner packaging have encouraged companies to reduce shelf space and transportation costs making systems more efficient.

In an economy where competitive pressures are increasing day by day, businesses have to go through various tactics to save the environment and employ technology in their firm. The term 'green business' is carried out as a strategic management tool these days, which generates customers' value through environmentally friendly practices (Cohen, 2007). Green manufacturing has a progressive effect on the environment, society, and business itself (Sezen & Çankaya, 2013). Green manufacturing can also lead to cost-saving in a way it saves raw materials, provides more production out of the same input, reduces environmental legal expenses, and improves goodwill (Sezen & Çankaya, 2013). In India, policymakers have understood that environment and growth are interconnected. Therefore, the government has played a significant role in putting up legislation on businesses to protect the environment by reducing the usage of scarce natural resources and improving the efficiency by using technology (Ramirez et al., 2019).

Green businesses are developing and are being carried out in India broadly due to the government pressurizing the firms, green marketing, and social responsibility. Government is influencing to reduce the burden on the environment in different ways such as reducing the production of goods that harm the environment and modifying usage patterns of green products and non-green products thereby increasing the pattern of usage of green products and services (Aggarwal & Kadyan, 2014).

Adding light to the previous literature, which mainly focused on sustainable production and employee's behavioural changes, i.e. supply, a business also needs to analyse the demand side. Due to the businesses' inability to provide adequate information about the sustainability of products offered during the process causes a gap. The gap may be reduced when a consumer is informed and communicated about environmental concerns and social concerns. Businesses are designing their products with green elements expecting that consumers will purchase it based on environmental and individual benefits. Indian consumers are often hesitant in purchasing green products because of their low awareness about green marketing. Companies practicing green marketing are preferred over conventional companies, thus establishing a competitive advantage and consequently achieving their business objectives. Over the years, Indian consumers have been exhibiting considerable interest in green products and provided that they are aware of the conduct, elements, price and quality of various green products, the usage of such products would expand. Indian consumers believe in Ayurveda and herbal products for improved health and beauty. Consumers often experience a sense of satisfaction post purchasing green products as they feel that their duty as a citizen to conserve the environment has been fulfilled. Considering on a macro level, green marketing is a mere strategy that helps gain trust and goodwill of consumers and subsequently generate huge profits. Environmental issues mainly ozone depletion and global warming have made more and more companies aware of climate change. Many companies have also taken massive steps towards undertaking green marketing to combat environmental change. India has been following the same pathway as developed nations. Pressure from the government and ecological actions have induced businesses to go green. However, this progression continues to be government-proposed rather than business-initiated. Nevertheless, much advancement can hardly be anticipated in the coming years, unless consumers and industries turn green and initiate environmentally friendly behaviour.

References

Abou-Moghli, A. (2015). The role of organizational support in improving employees performance. *International Business Research, 8*(2), 198.

Aggarwal, P., & Kadyan, A. (2014). Greenwashing: The darker side of CSR. *Indian Journal of Applied Research, 4*(3), 61–66.

Aghelie, A. (2017). Exploring drivers and barriers to sustainability green business practices within small medium sized enterprises: Primary findings. *International Journal of Business & Economic Development, 5*, 41–48.

Alay, E., Duran, K., & Korlu, A. (2016). A sample work on green manufacturing in textile industry. *Sustainable Chemistry and Pharmacy, 3*, 39–46.

Ambec, S., & Lanoie, P. (2008). Does it pay to be green? A systematic overview. *The Academy of Management Perspectives*, 45–62.

Atkinson, P. (1999). Without leadership there is no change. *Management Services, 43*(8), 8–11.

Bansal, P., & Roth, K. (2000). Why companies go green: A model of ecological responsiveness. *Academy of Management Journal, 43*(4), 717–736.

Bass, B. M. (1994). *Improving organizational effectiveness through transformational leadership*. Sage.
Bhattacharya, S. (2011). Consumer attitude towards green marketing in India. *IUP Journal of Marketing Management, 10*(4).
Broman, G., Robèrt, K. H., Basile, G., Larsson, T., Baumgartner, R., Collins, T., & Huisingh, D. (2013). Systematic leadership towards sustainability. *Journal of Cleaner Production, 64*.
Brundtland, G. H. (1987). *Our common future: Report of the world commission on environment and development*. Oxford University Press.
Butler, J. (2008). The compelling "hard case" for "green" hotel development. *Cornell Hospitality Quarterly, 49*(3), 234–244.
Cabeza-Erikson, I. K. (2008). *Development of leadership capacities as a strategic factor for sustainability*. Digitala Vetenskapliga Arkivet.
Čekanavičius, L., Bazytė, R., & Dičmonaitė, A. (2014). Green business: Challenges and practices. *Ekonomika, 93*, 74–88.
Chandrasekara, W. S. (2019). The effect of transformational leadership style on employees job satisfaction and job performance: A case of apparel manufacturing industry in Sri Lanka. *International Journal of Economics, Commerce and Management, 7*(7), 385–391.
Chang, R.-D., Zuo, J., Zhao, Z.-Y., Soebarto, V., Lu, Y., Zillante, G., & Gan, X.-L. (2018). Sustainability attitude and performance of construction enterprises: A China study. *Journal of Cleaner Production, 172*, 1440–1451.
Chauhan, H. P. S., & Chaturvedi, R. (2015). Green marketing initiatives in India: An overview. *International Journal of Management Research & Review, 10*(5), 834–844.
Cohen, A. (2007). Commitment before and after: An evaluation and reconceptualization of organizational commitment. *Human Resource Management Review, 17*(3), 336–354.
Criscuolo, C., & Menon, C. (2015). Environmental policies and risk finance in the green sector: Cross-country evidence. *Energy Policy, 83*, 38–56.
Dean, T. J., & McMullen, J. S. (2007). Toward a theory of sustainable entrepreneurship: Reducing environmental degradation through entrepreneurial action. *Journal of Business Venturing, 22*(1), 50–76.
Demirel, P., Li, Q. C., Rentocchini, F., & Tamvada, J. P. (2019). Born to be green: new insights into the economics and management of green entrepreneurship. *Small Business Economics, 52*, 759–771.
Du Plessis, M. (2007). The role of knowledge management in innovation. *Journal of Knowledge Management, 11*, 20.
Esty, D. C., & Winston, A. (2009). *Green to gold: How smart companies use environmental strategy to innovate, create value, and build competitive advantage*. John Wiley & Sons.
Fernández-Robin, C., Celemín-Pedroche, M. S., Santander-Astorga, P., & Alonso-Almeida, M. D. M. (2019). Green practices in hospitality: A contingency approach. *Sustainability, 11*(13), 3737.
Freifeld, L. O. (2013). Emerging training leaders. *Training, 50*(3), 20–31.
Ganta, V. C. (2014). Leadership during change and uncertainty in organizations. *International Journal of Organizational Behaviour & Management Perspectives, 3*(3), 1183.
Grote, C. A., Jones, R. M., Blount, G. N., Goodyer, J., & Shayler, M. (2007). An approach to the EuP directive and the application of the economic eco-design for complex products. *International Journal of Production Research, 45*(18–19), 4099–4117.
Hall, J. K., Daneke, G. A., & Lenox, M. J. (2010). Sustainable development and entrepreneurship: Past contributions and future directions. *Journal of Business Venturing, 25*(5), 439–448.
Hall, R., Edelman, M., Borras, S. M., Scoones, I., White, B., & Wolford, W. (2017). Resistance, acquiescence or incorporation? An introduction to land grabbing and political reactions 'from below'. *In Global Land Grabbing and Political Reactions 'from Below'* (pp. 1–22). Routledge.
Hart, S. L. (1997). Beyond greening: Strategies for a sustainable world. *Harvard Business Review, 75*(1), 66–76.

Hu, J., Liu, Y. L., Yuen, T. W. W., Lim, M. K., & Hu, J. (2019). Do green practices really attract customers? The sharing economy from the sustainable supply chain management perspective. *Resources, Conservation and Recycling, 149*, 177–187.

Hunting, S. A., & Tilbury, D. (2006). Shifting towards sustainability. In *Th e Australian research Institute in Education for sustainability (ARIES) for the Australian Government Department of the Environment and heritage*. ARIES.

Ionescu, V.-C. (2014). Leadership, culture and organizational change. *Manager, 20*, 65–71.

Kasliwal, N., & Agarwal, S. (2019). Green marketing initiatives and sustainable issues in hotel industry. In *Green business: Concepts, methodologies, tools, and applications* (pp. 512–529). IGI Global.

Khare, A. (2019). Green apparel buying: Role of past behavior, knowledge and peer influence in the assessment of green apparel perceived benefits. *Journal of International Consumer Marketing*, 1–17.

Kirk, D. (1995). Environmental management in hotels. *International Journal of Contemporary Hospitality Management, 7*(6), 3–8.

Kotler, P., & Lee, N. (2008). *Corporate social responsibility: Doing the most good for your company and your cause*. John Wiley & Sons.

Kureshi, S., & Thomas, S. (2014) *Cause related marketing – An Indian overview*, W.P. no. 2014-04-03. Indian Institute of Management Ahmedabad.

Kureshi, S., & Thomas, S. (2020). Testing the influence of message framing, donation magnitude and product category in a cause related marketing context. *Journal of Marketing Communication, 26*(3), 268–289.

Leoncini, R., Marzucchi, A., Montresor, S., Rentocchini, F., & Rizzo, U. (2019). 'Better late than never': The interplay between green technology and age for firm growth. *Small Business Economics, 52*(4), 891–904.

Lorenzoni, I., Nicholson-Cole, S., & Whitmarsh, L. (2007). Barriers perceived to engaging with climate change among the UK public and their policy implications. *Global Environmental Change, 17*(3–4), 445–459.

Malloch, K. a. (2013). Developing high-level change and innovation agents: Competencies and challenges for executive leadership. *Nursing Administration Quarterly, 37*(1), 60–66.

Manaktola, K., & Jauhari, V. (2007). Exploring consumer attitude and behaviour towards green practices in the lodging industry in India. *International Journal of Contemporary Hospitality Management, 19*(5), 364–377.

Marin, G., Marzucchi, A., & Zoboli, R. (2015). SMEs and barriers to eco-innovation in the EU: Exploring different firm profiles. *Journal of Evolutionary Economics, 25*(3), 671–705.

Mensah, I. (2006). Environmental management practices among hotels in the greater Accra region. *International Journal of Hospitality Management, 25*, 414–431.

Millar, M., & Baloglu, S. (2011). Hotel guests' preferences for green guest room attributes. *Cornell Hospitality Quarterly, 52*(3), 302–311.

Millar, J., & Photakoun, V. (2008). Livestock development and poverty alleviation: revolution or evolution for upland livelihoods in Lao PDR. *International Journal of Agricultural Sustainability, 6*(1), 89–102.

Moktadir, M. A., Rahman, T., Rahman, M. H., Ali, S. M., & Paul, S. K. (2018). Drivers to sustainable manufacturing practices and circular economy: A perspective of leather industries in Bangladesh. *Journal of Cleaner Production, 174*, 1366–1380.

Mukherjee, I., & Singh, U. K. (2020). Delineation of groundwater potential zones in a drought-prone semi-arid region of east India using GIS and analytical hierarchical process techniques. *Catena, 194*, 104681.

Oshodi, O., & Aigbavboa, C. (2017). Theoretical analysis of the metrics for measuring the sustainability of infrastructure projects. In *Carbon footprint and the industrial life cycle* (pp. 325–337).

Pacheco, D. F., Dean, T. J., & Payne, D. S. (2010). Escaping the green prison: Entrepreneurship and the creation of opportunities for sustainable development. *Journal of Business Venturing, 25*(5), 464–480.

Pandit, D., Joshi, M. P., Sahay, A., & Gupta, R. K. (2018). Disruptive innovation and dynamic capabilities in emerging economies: Evidence from the Indian automotive sector. *Technological Forecasting and Social Change, 129*, 323–329.
Peattie, K. (2001). Towards sustainability: The third age of green marketing. *The Marketing Review, 2*(2), 129–146.
Porter, M. E., & Van der Linde, C. (1995). Toward a new conception of the environment-competitiveness relationship. *Journal of Economic Perspectives, 9*(4), 97–118.
Rahman, I., Reynolds, D., & Svaren, S. (2012). How "green" are north American hotels? An exploration of low-cost adoption practices. *International Journal of Hospitality Management, 31*(3), 720–727.
Rai, S., & Jambhulkar, S. (2018). An analytical study of Green business practices in India with specific reference to selected Indian companies. *International Journal in Management & Social Science, 6*(8), 123–135.
Ramirez, C., Navas, L., Delgado, Á., González, M., Caicedo, L., & Mejía, M. (2019). Promoting entrepreneurship through a community learning model – Case study: Green businesses. *Systemic Practice and Action Research, 32*, 629.
Schein, E. H. (2010). *Organizational culture and leadership*. John Wiley & Sons.
Sezen, B., & Çankaya, S. (2013). Effects of green manufacturing and eco-innovation on sustainability performance. *Procedia - Social and Behavioral Sciences, 99*, 54–163.
Thomas, S., & Kureshi, S. (2016). A study of cause related marketing initiatives and consumer insights – Evidences from Indian context. *South Asian Journal of Management, 23*(2), 109–130.
Thomas, S., & Kureshi, S. (2020). Consumer skepticism towards cause related marketing: Exploring the consumer tendency to question from emerging market perspective. *International Review on Public and Nonprofit Marketing, 17*, 225–236.
Thomas, S., Kureshi, S., & Vatavwala, S. (2020). Cause-related marketing research (1988 – 2016): An academic review and classification. *Journal of Nonprofit & Public Sector Marketing, 32*(5), 488–516.
Tumpa, T. J., Ali, S. M., Rahman, M. H., Paul, S. K., Chowdhury, P., & Khan, S. A. R. (2019). Barriers to green supply chain management: An emerging economy context. *Journal of Cleaner Production, 236*, 117617.
Yusof, N. A., Tabassi, A. A., & Esa, M. (2019). Going beyond environmental regulations—The influence of firm size on the effect of green practices on corporate financial performance. *Corporate Social Responsibility and Environmental Management, 27*(1), 32–42.
Zhao, Z.-Y., Zhao, X.-J., Davidson, K., & Zuo, J. (2012). A corporate social responsibility indicator system for construction enterprises. *Journal of Cleaner Production, 29-30*, 277–289.

Sujo Thomas is a marketing faculty at Amrut Mody School of Management, Ahmedabad University. He has over 19 years of experience in academics and industry. His areas of research interests include cause-related marketing, non-profit marketing and online retailing.

Susmita Suggala holds a Ph.D. in Marketing and has been closely associated for the last 18 years with the marketing area. She has been associated with L.J. Integrated MBA program at School of Management Studies, LJ University. Her current research interests include healthcare marketing, sustainable practices, digital networks and relationship marketing.

Sudhir Pandey is a communication faculty at Amrut Mody School of Management, Ahmedabad University. He is passionate about developing material that equip business school graduates with contemporary employability skills such as problem-solving, collaboration, socializing, written communication and managing self.

Chapter 11
The Recovery and Takeover of the Water Pedagogical Forest as an Urban Green Innovation Public Space for Knowledge and Learning

José G. Vargas-Hernández and M. C. Omar C. Vargas-González

Abstract This study aims to analyze a successful soft approach to new forms of protest as applied to the recovery and takeover of the Water Pedagogical Forest as an urban green innovation public space for knowledge and learning.. The method used is the empirical analytical based on the facts of the specific case under study framed by the theoretical and empirical analysis of the literature reviews and the reflection applied to the situation. The analysis of this case concludes that the new forms of movements and protest actions are shifting toward using more soft power, non-violent, long-term processes and regulated through the institutional and legal framework being capable of processing the conflicts and managing economic efficiency, social justice and inclusion, political participation, and sustainable environmental interests of all the stakeholders involved.

Keywords New forms of protest · Water Pedagogical Forest · Recovery · Takeover

---

Some small portions of the chapter were previously published by the author in *Project GIAGEM-The Role of Green Innovation Areas in Revitalizing German and Mexican Cites*, as part of the scientific report.

J. G. Vargas-Hernández (✉)
Postgraduate and Research Division, Instituto Tecnológico José Mario Molina Pasquel y Henríquez, Unidad Zapopan, Zapopan, Jalisco, México
e-mail: jose.vargas@zapopan.mariomolina.tecnm.mx; jose.vargas@zapopan.tecmm.edu.mx

M. C. O. C. Vargas-González
Instituto Tecnológico de México, Campus Cd. Guzmán, Cd. Guzmán, Jalisco, México
e-mail: omar.vg@cdguzman.tecnm.mx

© The Author(s), under exclusive license to Springer Nature Switzerland AG 2023

S. Dhiman (ed.), *Sustainable Development and Environmental Stewardship*,
https://doi.org/10.1007/978-3-031-28885-2_11

11.1 Introduction

In times of sociopolitical conflict, protest is an extra-institutional form of political behavior. Protest actions and other forms of engagement have direct impact on the policy decisions, legislation, political parties, and the politics in general. Variation of the terms protest, rally, demonstration, march, manifestation, petition, strike, etc., indicates some specific form of protest actions (Fisher et al., 2019, p. 5).

Soft forms of protest and movement actions lead to hybrid forms of authority linked to shifting forms of governance, which nonetheless involves persistent attempts to reinforce state power. Although governance has shifted toward the ideal soft use of power, management and civic self-responsibilities in movements and protest actions, it continues to pursue aims of strengthening the state use of power as a facilitator of wellbeing, development sustainability, and resilience through intensified commodification of nature.

Social movements and civil organizations, with more charismatic authority, stimulate a more diverse range of protest responses, based on repudiating the claims of rulership, while others may manage into the prevailing power discourse with the aim of subsequently softening or redirecting the movement and protest actions.

The first instance conceptualizes the social movements and protest, to continue with the analysis of some theoretical approaches followed by the forms of protest and finally the outcomes. After this framework of reference, the analysis turns to the case: the Pedagogical Forest of Water, which is divided in the following sections: the geographical, spatial, and temporal dimensions, the actors in conflict, protest, and movement actions, and finally the successful outcomes.

11.2 Conceptualizing Social Movements and Protest

Social movements and protest are the result of discontent with existing economic, political, ideological, and other conditions, relying on societies as a human product that can be transformed (Buechler, 2011).

Social movements are organized collectivities continuously acting in collective action outside of the institutional and organizational channels to challenging and defending of extant authority, institutionally or culturally based in the world order, society, culture, community, organization, or group of which they are a part (Snow et al., 2019, p. 10). The essential elements of the concept of social movements are to be a form of collective action to challenge or defending the existing authority in organized manner with temporal continuity. A contemporary evolving definition of protest when drawing differences of historical comparisons has implications of accuracy and consistency.

Social movement and protest are distinct concepts (Diani, 1992), although they can be used as interchangeably referring to extra institutional and common political action. Social movements are defined as the collective challenges with common purposes, social solidarities, and sustained interaction with elites, opponents, and authorities (Tarrow, 2011, p. 9; Bishara, 2021). A new type of social movement is transnational on the move (Rizzo Lara, 2021)

The online media and social networking adoption as resources of protest and use by social movements is a relevant trend (Borbáth & Gessler, 2020; Bremer et al., 2020; Milkman, 2017; Tufekci, 2017). The conceptualization of new social mobilizations emphasizes the use of social networks used in virtuality with the support of added information and communication technologies (Shirky, 2011). Political protest on media is facilitated using framing and sharing of information and concepts during the space and timespan of the political protest. The concept of framing allows people to perceiving and making sense of information aligned with their beliefs and values (Goffman & Berger, 1986).

Framing is used to articulate concepts and define specific concerns of the protest which explain the scope and limitations of the mobilizations (Snow et al., 1986). Framing has the categories of thematic defined as covering background information and episodic focusing on an event and individual created for coding purposes (Entman, 1993).

Interventions of actors' opportunities in social movements to engage in politics may occur as part of the politics of scaling among, across, and within scales (Bebbington & Bury, 2013; Perreault, 2014; Sayre, 2015), evolving into relationally dispersed hierarchies, horizontal networks, and other interactive webs (Brenner, 2009; Swyngedouw, 2004) with fluidity of articulations and avoid differentiated power relationships (Jessop et al., 2008). The scalar politics theory supports networked scales in governance analysis separating the entities in nested spheres from the political construction (Bulkeley, 2005; Sayre, 2009; Smith, 1992; Swyngedouw, 2004).

Social movement research stresses the centrality of space and time of the study of contentious politics of the protest dynamics (Della Porta, 2016; Koopmans, 2004; McAdam & Sewell, 2001), waves of contention and the concept of protest cycles (Koopmans, 2004; Tarrow, 2011). A contrasting perspective focuses on long-term process of a unique and singular event (McAdam & Sewell, 2001, p. 101; Moore, 2011).

Widespread discontent and the desire to challenge the power of authorities can persist before any resistance and protest activity which can erupt in space settings as the result of protest activism of collective opinion that is not disaffected but demand change.

A critical concept for the contentious politics is upscaling of the protest. The contentious politics approach to protest process acknowledges that engagement of progressive or reactionary mobilization and political activism are supported by concepts and tools while government authority weakens.

The concept of mobilization of protest availability is based on a model on the supply and demand (Klandermans, 2004) determined by the efforts of people to protest, ideological and psychological dispositions, access to mobilization networks and organizations, and the available biographical structure in place and time (Rainsford & Wambach, 2021). The ideological position of social movements to attract people is defined by Klandermans (2004) as the ideological aspect of supply. The analytical framework developed by Rainsford and Wambach (2021) merges the theory of supply and demand for protest with structural and agentic factors to be able to predict protest participation.

The concept of biographical availability structure (McAdam, 1986) refers to the absence of individual constraints on willingness for protest participation (Beyerlein & Hipp, 2006). The concept of mobilization availability (Klandermans, 2004) captures the differences between old and new political protests and the characteristics of individual protesters.

11.3 Theoretical Approaches

A theoretical framework used to analyze the relationship between political behavior to protest outcomes is relevant to more democratic contexts. The theory of the supply and demand of protest used by Klandermans (2004) argues the substantial change based on the supply of protest. According to this theory, supply of protest participation is referred to the opportunities to protest provided by organizers, grass roots activists, and social movement organizations. In the building of an analytical framework, Rainsford and Wambach (2021) contextualize the theoretical insights in relation to Global Climate Strikes (GCSs) by notions of supply and demand of protest with the dominant forms of protest.

The cleavage theory is defined as a division of society confronted in two dichotomous positions on two sides determined by the social structure (Lipset & Rokkan, 2001), which shows the differences and a deep division that configures alignments with political parties (Castromil, 2020). The social identity theory (Tajfel & Turner, 1979) argues that protestors can be identified by alternative group of available categorizations by deploying tactical persuasive repertoires (Wouters, 2019; Mazumder, 2018). Collective identity of a protest movement refers to the building and development of deeper bonds of trust relationship and feelings of solidarity among protestors.

The sociology of mobilizations studies the forms of protest, patterns, and regularities based on the collective action and the sociohistory of protests with a limited argumentation used by actors in different contests by holding a specific contentions repertoire (Rennes, 2011). The theoretical framework on contentious politics developed by Tilly (2000, 2001) supports the connection between social movements and democratization of nation states.

The core tenets of social movement theory became the sociological approach on social movements and protest concerning mobilization processes more than the concerns of the political context and outcomes (Walder, 2009). The paradigm of social movement theory has been referenced in mobilization processes and movement outcomes by the most cited work of McAdam (1982), McAdam and Sewell (2001), McCarthy and Zald (1977), Tarrow (1998), Tilly (1978).

Social movement and geography theory have studied place and space arguing that space structures the networks and relationships (Lefebvre, 1991; Martin & Miller, 2003, p. 143). An anarchist theory of change and propaganda of the deed is being advanced by environmental activists as a differentiated form of protest which makes a proper account of leading to diversity of aims which may have positive or negative impact on political agendas and public opinion.

Collective action and social movements theories intend explaining the determinants of mobilization (Olson, 1965; Opp, 2009; Tilly, 1978) stressing the opportunities to mobilize structures, framing, and resources (McAdam & Snow, 1997; McAdam & Sewell, 2001; Opp, 2009; Staggenborg, 2016; Tarrow, 1998; Tilly & Tarrow, 2015). The framing theory explains the perception of information by the interactions between people and their surrounding environment. Social movement theory has as it tenet the political opportunity and resource mobilization (Eisinger, 1973; Lipsky, 1968; Meyer & Lupo, 2010; Vassallo, 2018), the political opportunity structures (Tarrow, 1998), and resource mobilization (McCarthy & Zald, 1977).

The resource mobilization theory of social movement activism involving more emotional and identitarian is supported by the collective action theory (Zald et al., 1987) despite the utility economic assumptions limitations of the rational choice model. The collective action theory is based on economic micro-foundations of cost benefit models and connecting political behavior to social science theory to provide an interactive account of protest movements (Hirschman, 1970, 1974).

Theorizing the dynamics involved in the different forms of protest needs to be broadening and deepening. The effects of different modalities of protest events on movement dynamic and outcomes have been theorized to determine their interactions dynamics between special and temporal dimension (Bishara, 2021; Della Porta, 2008; p.27). The spatial and temporal dimensions of different forms of protest generative power theories have effects on contentious politics, movement broadening, expansion, and deepening. The broken window theory (Kelling & Wilson, 1982) refers to the graffiti's disruptiveness and breeding more grafitti (Myre, 2003; Milon, 1999), which reinforces the potential of graffiti's resistance. The political opportunity structure theory takes the form of protest focusing on cultural and expressive dimension as the substance of action for post citizenship, as in the case of protest graffiti.

Some methods have been outlined by Chun et al. (2019) using grounded theory approaches based on sampling and thematic exploration to be codified and categorized (Charmaz, 2006). The theoretical use of an explanatory design method is based on deploying a survey and delving deeper into specific topics using focus groups (Ivankova et al., 2006). The systematic field observations on the events of protest actions and other forms of collective action are a component of research building.

Building event databases to document the protests, providing standardized coding of their occurrence, characteristics, and other features is a theoretical and methodological innovation to research on contentious politics. Research relying on media reports to develop protocols and document theoretically critical features of protest events. All these methods can be coupled with comparative analysis to answer broader theoretical questions and the effects on political outcomes.

11.4 Forms of Protest

The timeframes underlying political and social conflicts among the actors involved lead to varied forms of protest action. Influences on protest participation vary across different forms of ritual demonstrations and reactive protests as the dominant forms of protest. Different forms of protest actions affect the internal dynamics of social movements and have different effects on broadening and deepening outcomes depending on the geographical space, temporal, and duration configurations (Bishara, 2021). The temporality of protest calls for various times and different forms of contention.

One social movement can engage in various forms of protest actions, subject to the potential achievements, the perceived advances of success, resources of members, and the political constraints. Political experience and interests are influential in the decision to take part in various forms of initiative-taking or reactive demonstrations and protest actions. Social movements have different forms of organizations between tight and loose.

Forms of protest actions require organizational capacity supported for an existing or created organizational structure and the development and expansion of organizational infrastructure. Single and multiple-site and extended forms of protest actions require coordination and deep ties among participants (Fu, 2017) established through collective experiences. Regarding the spatial dimensions of the forms of protest actions, it can placed in a single site involving the occupation of one space and enabling physical outreach to accommodate broader meetings over a period or extending over multiple sites at multiple periods of time. Multiple-site protest actions lead to broadening the social base of the movement. However, extended single-site protest actions offer the opportunity to foster collective identities and organizational development, which lead to deepen the movements (Bishara, 2021).

Authoritarian governments may become wary of extended single-site protest actions as a form of protest which can limit the repertoire of contention. An authoritarian regime may tolerate extended or shorter protest actions and repress them on others, depending on considerations such as the nature of demands, the identity of protesters, peaceful or violent tactics, organized groups and movements, places, and size of protesters, etc. (Cai, 2010; Carey, 2009; Davenport et al., 2011; Goldstone & Tilly, 2001; Li, 2019).

The various forms of contention can be transferred by a social movement from one site to another (Tarrow, 2005). Authoritarian and totalitarian states always generate various forms of protest and resistance actions including internal exiles depending on the intensity of repression. Repressed individualized and passive forms of protest tends to provide the possibilities of eventual emergence of more assertive, experienced, and coordinated forms of public claim-making and more potential forms of protest (Tilly, 2010).

The different forms of protest constitute the movement's repertoire. The repertoires of protest are the non-conventional forms of action interrupting the routines to attract the attention of the public and influencing power elites (Della Porta, 2020). Typical forms of protest actions are violent acts, boycotts, strikes, obstructions, etc. Street demonstrations and protest actions are some of the visible forms of opposition to government policies.

A potent form of protest pattern is those that appeal for the restoration of entitlements. Other protest actions may be progressively planned while the most irrational forms of protest will be resisted. More democratic authorities allow and stimulate a diverse protest response based on the repudiation or in the prevalence of a discourse to soft and redirect the protest at restoring the status quo. Protest graffiti is a visual artifact of a depersonalized approach to indirect forms of personalization and challenging messages to give political meaning, significance, and identity. Another sociological perspective to study collective mobilization in the interplay between political and cultural, symbolic, and instrumental are the infrapolitical forms of protest and mobilization, such as the use of biographies in social movement sociology.

Climate change is an issue for constant and diverse forms of protest actions and for making comparisons. The organization Global Climate Strikes (GCS) supports accessible diverse forms of protest actions and encouraging and emotionally engaging young people to get involved in mobilization and environmental activism (Rainsford & Wambach, 2021). Rainsford and Wambach (2021) have concluded that young people have more structural and agentic availability than the adults on older forms of climate protest due to the difficulties in connecting with more adult-centric forms of climate protest. The ideological factors are relevant in motivating young people to participate in various forms of climate protest, but more adult-oriented. Young people are motivated by several factors to participate in conventional and less conventional forms of climate change protest actions.

11.5 Protest Outcomes

Research on protest actions and social movements explores the power of contentious practices and experiences in shaping internal and external movement outcomes. However, research has focused more on external outcomes related to the effectiveness of tactical repertoires of protest actions on achieving social movement goals (Taylor & Van Dyke, 2004, p. 284). Protest experiences and actions can affect the internal social movement outcomes such as generate organizational infrastructures (Frenzel et al., 2014) building collective identities (Beissinger, 2002, p. 22; Gamson, 1992; Hunt & Benford, 2004, p. 433; Melucci, 1989; Taylor & Whittier, 1992), create networks (Tarrow, 1998), and enhance commitment to some common causes (Hunt & Benford, 2004; Tarrow, 1998). The different forms of protest events itself shape the internal movement dynamics.

Research on protest outcomes includes changes in public opinion, policy, voting, behavior of political elites, and political transition to democratic contexts (Chenoweth & Stephan, 2011; Haggard & Kaufman, 2016; Kadivar, 2018). Public opinion is the mediating factor determining outcomes (Giugni, 1998). Protest actions have political outcomes mediated by public opinion and media attention.

An analysis of the effects and outcomes of modalities and forms of protest actions should minimize complexity to isolate the sustaining movement outcomes. The outcome mobilizes people against the ruling institutions, governments, and parties while being repressed (Sosa, 2018; Sosa & Irías, 2018). Protest duration and geographic space may affect, but do not determine repression. Although the forms of protest actions and governance shift over time, they require statecraft to consolidate authority and controlling environments to prevent social mobilization.

There is a recent trend of convergence between sociology and political science on protest and movement outcomes. Protest outcomes were not relevant for explaining and analyzing in earlier political science. Protest actions have witnessed a resurgence of interest in political science shared with sociology and focusing on the outcomes of protest. Sociology tends to prioritize the explanation and analysis of protest actions and mobilization processes over political outcomes.

Sociology has favored the mobilization processes over outcomes. Political sociology channels social forces into political outcomes (Sartori, 1969) of protest actions and collective actions, such as the attitudinal and electoral outcomes. Political attitudes of protestors are more politically involved and they are experienced than the nonparticipants due to their personal and organizational networks.

A political sociology of protest action may combine mass opinion, legislative behavior, and the relationships between social movements and the media in such a way that protest mobilization combines frames and tactics, media, legislative speech, and electoral outcomes (Wasow, 2020). Sociology is reluctant to talk about public opinion of protest and movement outcomes by diffusing the consequences beyond the terms of success or failures (Oliver et al., 2003: 220). Political scientists and sociologists have a shared interest in the attitudinal dimensions of protest outcomes.

The diverse protest action aims require recognition of diverse protest outcomes, one of which, the desire for publicity, is constant correlating with media attention and other political outcomes (Andrews & Caren, 2010; Seguin, 2016).

Outcomes are contingent to the processes of protest and movement actions and reactions characterized by contentious claims making (Giugni, 2007; Oliver & Myers, 2003; Tilly, 1998). The outcomes of protest actions are the cost-benefit analysis of behavior constraints of protestors drawing on the influence of discursive and attitudinal source and their bad outcomes. On the analysis of protest actions and political outcomes, the inflicting costs are relevant at the time of raising awareness and become sensitive to the aims of protestors. Cost considerations should be incorporated into an institutional and political sociology analysis of protest outcomes. Institutions are relevant to analyze political outcomes.

Focusing on the collective action and social mobilization, sometimes may be runs counter to any account of outcomes. From a longer term perspective, the outcomes triggered by the nonviolent and mass protests across the Arab world are regressive. The mobilization processes must be analyzed over the political correlates of movement emergence and outcomes. Protest outcomes may lead to legislative outcomes, considering that rational political leaders take advantage of the preferences of voters as the majoritarian incentives to respond.

Protest actions exert a direct effect on political outcomes through the common mechanism of sourcing information to legislators or informative cue (Gillion, 2013; Wouters & Walgrave, 2017) about the issues and concerns of citizens in a limited information environment (Andrews & Seguin, 2015; Gillion, 2013). Considering that each area has specific types of outcomes, the political sociology of protest outcomes are based on the characteristics of movements, legislative speeches, and the outcomes of political campaigns.

Democratic transition processes can be outcomes precipitated and accompanied by protests. The types of outcomes related to protest actions are relevant to more democratic contexts. Eventually protests actions of social movements become progressive and achieve outcomes such as the case of the Fourth Transformation of México that has raised the hopes of large population segments that resented corruption as the result of institutional capture. However, social movements promoting nonviolent and spontaneous mass protest actions may become regressive in the long-term perspective, such as the case of the Arab Spring movement.

The form of protest action differentially affects the outcomes. Short and multiple-site actions can expand the base of movement, extend, and broaden while single-site protest actions may deepen movements building organizational capabilities and fostering collective identities (Bishara 2021). The power of different forms of protest actions experiences themselves to deepen and broaden movements. There are concerns with explaining the political contexts of mobilization processes and political outcomes of social movements.

11.6 The Pedagogical Forest of Water: Contextual, Geographical, and Temporal Dimensions

The Water Pedagogical Forest is a municipal project in Zapopan, Jalisco, México, that seeks the recovery of the public owned land known as Arroyo La Campana to serve the community and the environment through the restoration of the natural water and socio-ecology system, the consolidation of infrastructure for visitors and the promotion of an environmental culture, through the sustainable management of the water resource, the conservation of the socio-ecosystem and environmental services.

The sector in which this project is being developed by all the stakeholders involved is part of Colomos III, a Municipal area of Hydrological Protection approved by the Congress of Jalisco, which has an impressive hydrological value, with a beautiful spring with a couple of water streams as consolidated sources of healthy water supply and a source of human life. The Pedagogical Forest of Water is an area covered by 45 hectares of land which is being developed to become the real element of practice and promotion of an environmental culture favoring its protection, restoration, conservation, and responsible preservation was done to make the public space and urban area of innovation with a rational use to make the Water pedagogical Forest a public place for citizens an venue for pedagogical and cultural events and other environmental services (Prieto, 2013).

This park is an important opportunity provided by the ecosystem, being a living laboratory and formidable scenario for meaningful learning. The pedagogical sense is based on a fundamental principle, which should regulate the cognitive and the awareness work of learners and the general population: "Educating with water and for it" (Ruiz Barajas, 2016).

This is a strategic environmental project that benefits the sustainable management of water resources, provides improvements to the use of water supply and quality, promotes the conservation of the ecosystem and its biodiversity, proposes comprehensive solutions for flood management, and promotes environmental culture. The surface of hydrological conservation benefits 64,000 inhabitants in direct way and more than 20 colonies of Zapopan.

11.7 The Actors in Conflict

The municipal authority of Zapopan is currently prevented from conducting any improvement work for the Pedagogical Forest of Water, because some individuals have undertaken trials to fight for the ownership of that property and have announced measures to defend them from its field of action. The Territorial Strategy for Urban Prosperity Zapopan 2030 is an interdisciplinary planning exercise that identifies Zapopan as a municipality of great value and serious threat.

It is essential to understand how Zapopan wants to be a prosperous territory and it proposes three strategies for taking action: The first one is having a sustainable management of natural capital and its ecosystems, the second one is having a system of connectivity and integration of territorial networks which is based on the direct experience of the ecological environment that allows us to immediately assimilate the importance of the rivers in which surface waters flow and the springs that are outcrops of groundwater that come from hydrological and geomorphological flows of the aquifer, as well as natural sources of water for human consumption and supply for the inhabitants and especially children and young people of a growing metropolitan area. The third one is fight for be a "Compact City" model with the tendency to reconcentrate population (Ruiz Barajas, 2016).

The forest forms part of an estate in the "Arroyo La Campana Colomos III" has an area of 36.4 hectares that was declared a natural protected area by the local Congress in July 2014 to conserve the hydrological zone threatened by real estate developments. This was allowed by The Technical Committee of Colomos III, headed also by a counselor, and composed of councilors, municipal offices (Ecology and Parks and Gardens), and representatives of civil society, among them the academic Jaime Eloy Barajas, from the University of Guadalajara.

This agreement represents the participation of various organizations, groups, and citizens committed and compromised to the forest that for almost 20 years have worked in the forest restoration and have demanded the recovery of those areas that have been illegally invaded and that are part of the original polygon of the lands that were acquired, as public patrimony, by the government of the state, between the end of the nineteenth century and principles of the twentieth century, in order to protect them and to preserve them by the presence of springs for water supply of the city.

The (Collective Pro-Water Pedagogical Forest) CCPROBPA is an independent civil organization of governments and political parties that, among its objectives, seeks to promote the protection, preservation, restoration, and conservation of the Water Pedagogical Forest and the hydrological system to which it belongs, as well as to promote the integration and citizen participation in favor of this forest and of other socioenvironmental causes (CCPROBPA, 2017).

Citizen members of civil society who have participated in the reforestation of the soils of this area have been involved in this work, adding to the work of weed removal, cleaning, improvement and conservation of soil, planting, and tree improvement as ahuehuetes, magnolias, sapote, and willows.

The Collective Pro-Water Pedagogical Forest and the Government of Zapopan agreed on the protection of the forest and announced a proposal to create a Citizen Council between organized civil society and the authorities. This collective is a non-governmental organization that was composed by people committed to the improvement of environmental conditions: neighbors, universities, and students that form a multidisciplinary and interdisciplinary group with the purpose of strengthening the necessary actions to consolidate the biophysical and ecological characteristics of the ecosystem. In this way, organized citizens participate with the municipal authority as an important challenge of environmental and patrimonial rescue (Ruiz Barajas, 2016).

A Decentralized Public Organism has the responsibility to administrate and guarantee the correct use of the forest with the contributions of the City Council. Each with 2 million pesos and with donations from citizens who are concerned about the protection of space. The Government Board that heads this Decentralized Public Organism is composed exclusively of citizens and members who are selected based on a public call. They evaluate the proposals for forest management and they also have an exemplary management for the authorities in the transparency of the use of resources.

The Collective Citizen Pro-Pedagogical Woodlands of Water provides the formative and organizational experience based on fundaments from pedagogy for integrality, integral sustainability, community, and complex thinking and chaordic dynamics, concretized by the model of urban native forest and the methodology of participative restoration of the forest through the construction of the socioenvironmental fabric (Gutiérrez Rosete Hernández, 2019)

11.8 Protest and Movement Actions

The project Pedagogical Forest of Water integrates the citizen demand that began a couple of decades ago with protest actions for patrimonial rescue and environmental restoration. The space owned by the state government since 1898, which was granted as a loan to the municipality of Zapopan in 1993, was in a state of deterioration and with systematic invasions that violated public property and its environmental value.

This place was in a state of several deterioration and systematic invasions that violated public property and therefore its environmental value. Since the second half of the last century, the area where the Agua Pedagogical Forest is located, which is originally part of an area of more than 200 hectares, has been in the greed of speculation and corruption with the support of complicities and oversights that facilitate dispossession for the benefit of real estate interests. These practices have turned the Bosque del Agua property into a no-man's land subject to invasions and illegal dispossession, predatory deforestation practices, sand extraction, rubble deposits, and garbage.

The Pedagogical Forest of Water has been under siege from the predominance of real estate capital in collusion with sectors of the government and political power with highly speculative interests to build projects with exclusive reserves or closed subdivisions and build residential, business, and service towers priced at soaring prices. The lands of the pedagogical forest of the water are threatened and continuously affected by recurrent fires caused by strong real estate interests that see it as a loot for the development of subdivision projects and residential towers, as well as for large commercial and business complexes with high value of speculation.

Invasions for the purpose of dispossession, gentrification and the loss of public space, green areas, forests, water, etc. have been encouraged on the grounds of the Pedagogical Forest of Water. These predatory actions have led to the change of land use of substantial portions of the property to become a public space with forestry potential and conservation and water supply into exclusive residential subdivisions, shopping centers, apartment towers. This situation ends with converting the land use of green areas and infrastructure of public spaces into exclusive urban areas for housing, business, services, and private entertainment (Gutiérrez Rosete Hernández, 2019). The entire property of the Pedagogical Forest of Water is in litigation processes promoted by actors who want to appropriate it through corrupt practices.

The common denominator is that in the face of the practices of damage and dispossession of public property motivated by real estate developers with predatory speculative economic interests in real estate in collusion with colluded politicians, it has led to the creation of defense and resistance movements of residents, organized groups and collectives that defend, protect, rescue, and conserve the public spaces and common goods of the Water Pedagogical Forest that has a vital socio-environmental value for the inhabitants of Zapopan. Social pressure sustained demands and efforts from various social organizations, the area where the forest is located has been declared as a Municipal Hydrological Protection Area in 2014 and as a State Hydrological Protection Area in 2018, which gives it greater legal certainty of coverage as a protected area, but without neglecting or lowering the guard against the attempts of dispossession (Gutiérrez Rosete Hernández, 2019).

The Colectivo Ciudadano Pro Bosque Pedagógico del Agua (Ccprobpa) is a civil organization, non-profit, autonomous, and independent of the state, political parties and electoral political organizations, business organizations, and religious institutions. The Collective emerged in 2013 with antecedents dating back to 1998 when citizens, civil organizations, ecologists, and defenders of the environment began to worry about the recovery of the property as a public space. Its objectives are the protection, preservation, restoration, and conservation of the Water Pedagogical Forest and the hydrological system to which it belongs, while promoting processes of participatory, comprehensive, collective, intersectoral, collaborative, and solidarity organization that contribute to other related socio-environmental causes and the construction of a sustainable society.

The Citizen collectivity Pro Pedagogical Forest of Water (Colectivo Ciudadano Pro Bosque Pedagógico del Agua) with the collaboration and support of various educational institutions, civil organizations, various social and private organizations of corporate volunteers and neighborhood groups, environmental groups, and various civilians have been organized to resist the attacks of developers, protect, and restore by applying a participatory forestry methodology from an integral sustainability perspective centered on the native urban forest model. As a citizen project, it is a space for training and field practices for reforestation, care, and forest maintenance. In addition to promoting voluntary participation, it seeks to make the link between the forest and water explicit, with a pedagogical and participatory sense (Ruiz, 2019).

Students, professionals, retirees among other volunteers, meet every Saturday morning to conduct different conservation activities in the forest, such as irrigation, planting, land treatment, among others, inviting society to be part of, and help, the cause. The municipal government of Zapopan called on citizens to seize the Water Pedagogical Forest, for the coexistence of families, since it assures that is the best weapon of citizens to guarantee their permanence (Gobierno de Zapopan, 16 enero, 2016).

11.9 Successful Outcomes

The outcome of this defense movement that conducts actions of defense, resistance, recovery, restoration, maintenance, conservation, and care of the public space Bosque Pedogógico del Agua has yielded satisfactory results so far, even though the conflicts derived from the disputes over property ownership subsist. The resistance actions have been long-term, in a peaceful way and without the use of violence, using soft power, through institutional, legal, and political mechanisms, social involvement, and in the implementation of pedagogical practices for integrality.

From being a vacant no man's land, subjected to the deposit of garbage and rubble and to the extraction of sand and materials for construction, initiates a process of rescue, defense and forest restoration that becomes a wooded environment, while it becomes a space for training and environmental education, as well as organized social participation, and in an example of recovery and appropriation of a common good of nature, of resistance and alternative.

The Ccprobpa collective promotes the process of appropriation as a common good and of restoration of the Pedagogical Forest of Water, from a model of native urban forest, as well as the training of forest promoters in two closely related axes: learning to restore forest and learning to organize collectively. The collective's fundamental principles are integral sustainability, collective organization, participation, collaboration, autonomy, intersectionality, continuity, and multiplicativity, which imply the confluence of various civil, public, private, neighborhood and citizen sectors who collaborate, voluntarily and as a social service, in the forest restoration process.

The participation of professionals and students in the remarkably diverse academic areas is relevant in their voluntary involvement with neighbors and citizens of social and civil organizations in joint tasks of the organization, restoration, and conservation processes of the place, under a management plan. Other activities conducted are those of dissemination, promotion, linkage, technical training, legal and operational advice, solidarity support with other groups, organizations, and civil networks with related projects for the defense, restoration, maintenance, and conservation of forests and other green areas.

Under the paradigm of integral sustainability, conditions of autonomy, self-management, and self-sufficiency of means and restoration processes are supported, based on the principles of pedagogy for integrality that promote sustainability from the satisfaction of social needs in conditions of sufficiency, dignity., equality, well-being, and good living and through the responsible and sustainable use of the potential of nature's means for life. The Collective recovers and applies the foundations of pedagogy for integrality, integral sustainability, communality, complex thinking and chaordic dynamics that the participatory forest restoration methodology through the construction of a socio-environmental fabric lays the foundations that guide the experience of the Ccprobpa as a comprehensive educational and training experience.

The case of the Colectivo Ciudadano Pro Bosque Pedagógico del Agua implies an innovative formative and organizational experience, well thought out, felt, experienced, and lived in the gestation of a collective process that arises from the vocation and the union of wills of people, groups, and organizations that, around the common objective of restoring the Water Pedagogical Forest, converge to promote, precisely, the training of forest promoters in forest restoration and in collective organization processes, as well as to build an experience of construction of a socio-environmental fabric. The restoration process continues despite the continuous attempts of dispossession, fires, invasions, and sabotage of the progress achieved, by those who obey political and economic interests linked to real estate projects that pursue profit-making purposes.

11.10 Concluding Remark

The analysis of this chapter concludes that the new forms of movements and protest actions are shifting towards using more soft power, non-violent, long-term processes, multifaceted, regulated through the institutional and legal framework being capable to process the conflicts and manage economic efficiency, social justice and inclusion, political participation, and sustainable environmental interests of all the stakeholders involved.

This case based on the Pedagogical Forest of Water has learning implications to think, feel, and live in collective processes of horizontal organization, teamwork and in assemblies with decision-making by consensus, solidarity, collaboration, and cordiality in carrying out joint tasks such as, in this specific case, the defense, resistance, restoration, protection, maintenance, and conservation of environmental natural resources, which as a common good must be accessible and beneficial to the entire population. This set of soft actions of the movement and the protest imply mutual encounters, interconnections, and appropriations of the fabric of the socio-ecosystem based on the territory, environment, space, and time with the social and community fabric.

11.11 Chapter End Reflections/Take-Aways

1. A successful soft approach to new forms of protest is applied with some implications to the recovery and takeover of public spaces such as the Water Pedagogical Forest.
2. The new forms of movements and protest actions are shifting toward using more soft power, non-violent, long-term processes, multifaceted, regulated through the institutional and legal framework being capable of processing the conflicts and managing economic efficiency, social justice and inclusion, political participation, and sustainable environmental interests of all the stakeholders involved.
3. Social movements and civil organizations, with more charismatic authority, stimulate a more diverse range of protest responses, based on repudiating the claims of rulership, while others may manage into the prevailing power discourse with the aim of subsequently softening or redirecting the movement and protest actions.
4. The set of soft actions of the movement and the protest imply mutual encounters, interconnections, and appropriations of the fabric of the socio-ecosystem based on the territory, environment, space, and time with the social and community fabric.
5. The case based on the Water Pedagogical Forest, has learning implications to think, feel, and live in collective processes of horizontal organization, teamwork and in assemblies with decision-making by consensus, solidarity, collaboration, and cordiality in carrying out joint tasks such as, in this specific case, the defense,

resistance, restoration, protection, maintenance, and conservation of environmental natural resources, which as a common good must be accessible and beneficial to the entire population.

References

Andrews, K. T., & Caren, N. (2010). Making the news: Movement organizations, media attention, and the public agenda. *American Sociological Review, 75*(6), 841–866.

Andrews, K. T., & Seguin, C. (2015). Group threat and policy change: The spatial dynamics of prohibition politics, 1890–1919. *American Journal of Sociology, 121*(2), 475–510.

Bebbington, A., & Bury, J. (2013). Political ecologies of the subsoil. In A. Bebbington & J. Bury (Eds.), *Subterranean struggles: New dynamics of mining, oil, and gas in Latin America* (pp. 1–25). University of Texas Press.

Beissinger, M. R. (2002). *Nationalist mobilization and the collapse of the soviet state*. Cambridge University Press.

Beyerlein, K., & Hipp, J. (2006). A two-stage model for a two-stage process: How biographical availability matters for social movement mobilization. *Mobilization: An International Quarterly, 11*, 299–320. https://doi.org/10.17813/maiq.11.3.8p1758741377684u

Bishara, D. (2021). The generative power of protest: Time and space in contentious politics. *Comparative Political Studies, 54*(10), 1722–1756. https://doi.org/10.1177/0010414020970227

Borbáth, E., & Gessler, T. (2020). Different worlds of contention? Protest in Northwestern, Southern and Eastern Europe. *European Journal of Political Research, 59*(4), 910–935.

Bremer, B., Hutter, S., & Kriesi, H. (2020). Dynamics of protest and electoral politics in the Great Recession. *European Journal of Political Research, 59*(4), 842–866.

Brenner, N. (2009). Open questions in state rescaling *Cambridge Journal of Regions, Economy and Society, 2*(1), 123–139. https://doi.org/10.1093/cjres/rsp002

Buechler, S. M. (2011). *Understanding social movements: Theories from the classical era to the present*. Paradigm Publishers.

Bulkeley, H. (2005). Reconfiguring environmental governance: Towards a politics of scales and networks. *Political Geography, 24*(8), 875–902. https://doi.org/10.1016/j.polgeo.2005.07.002

Cai, Y. (2010). *Collective resistance in China: Why popular protests succeed or fail*. Stanford University Press.

Carey, S. C. (2009). *Protest, repression, and political regimes: An empirical analysis of Latin America and sub-Saharan Africa*. Routledge.

Castromil, A. R. (2020). *La teoría del cleavage (Lipset y Rokkan)*. https://bit.ly/2QL9KTr

CCPROBPA. (2017). *Comunicado de prensa del Colectivo Ciudadano Pro Bosque Pedagógico del Agua*. Colectivo Ciudadano Pro-Bosque Padagógico del Agua.

Charmaz, K. (2006). *Constructing grounded theory: A practical guide through qualitative analysis*. Sage.

Chenoweth, E., & Stephan, M. J. (2011). *Why civil resistance works: The strategic logic of nonviolent conflict*. Columbia University Press.

Chun, T., Ylona, M. B., & Francis, K. (2019). Grounded theory research: A design framework for novice researchers. *SAGE Open Medical, 7*, 205031211882292. https://doi.org/10.1177/2050312118822927

Davenport, C., Soule, S. A., & Armstrong, D. A., II. (2011). Protesting while black? The differential policing of American activism, 1960 to 1990. *American Sociological Review, 76*(1), 152–178.

Della Porta, D. (2008). Eventful protest, global conflicts. *Distinktion: Journal of Social Theory, 9*(2), 27–56.

Della Porta, D. (2016). *Where did the revolution go?* Cambridge University Press.

Della Porta, D. (2020). Social movements. In D. Berg-Schlosser, B. Badie, & L. Morlino (Eds.), *The Sage handbook of political science* (Vol. 2). Sage.

Diani, M. (1992). The concept of social movement. *The Sociological Review, 40*(1), 1–25.

Eisinger, P. K. (1973). The conditions of protest behavior in American cities. *American Political Science Review, 67*(1), 11–28.

Entman, R. M. (1993). Framing: Toward clarification of a fractured paradigm. *Journal of Communication, 43*, 51–58.

Fisher, D. R., Andrews, K. T., Erica, N. C., Heaney, C. M. T., Nathan, T. L. L., & Pressman, P. J. (2019). The science of contemporary street protest: New efforts in the United States. *Science Advances* 2019-10-04 5(10): eaaw5461 AID - 10.1126/sciadv.aaw5461 [doi] https://www.science.org/doi/full/10.1126/sciadv.aaw5461

Frenzel, F., Feigenbaum, A., & McCurdy, P. (2014). Protest camps: An emerging field of social movement research. *The Sociological Review, 62*(3), 457–474.

Fu, D. (2017). Disguised collective action in China. *Comparative Political Studies, 50*(4), 499–527.

Gamson, W. A. (1992). The social psychology of collective action. In A. D. Morris & C. M. Mueller (Eds.), *Frontiers in social movement theory* (pp. 53–76). Yale University Press.

Gillion, D. Q. (2013). *The political power of protest: Minority activism and shifts in public policy*. Cambridge University Press.

Giugni, M. G. (1998). Was it worth the effort? The outcomes and consequences of social movements. *Annual Review of Sociology, 24*, 371–393.

Giugni, M. (2007). Useless protest? A time-series analysis of the policy outcomes of ecology, antinuclear, and peace movements in the United States, 1977–1995. *Mobilization, 12*(1), 53–77.

Gobierno de Zapopan. (16 enero, 2016). *Defenderá Zapopan por todas las vías al Bosque Pedagógico del Agua*. Gobierno de Zapopan. https://www.zapopan.gob.mx/defendera-zapopan-por-todas-las-vias-al-bosque-pedagogico-del-agua/

Goffman, E., & Berger, B. (1986). *Frame analysis: An essay on the Organization of Experience*. Northeastern University Press.

Goldstone, J. A., & Tilly, C. (2001). Threat (and opportunity): Popular action and state response in the dynamics of contentious action. In R. R. Aminzade, J. A. Goldstone, D. McAdam, E. J. Perry, W. H. Sewell Jr., S. Tarrow, & C. Tilly (Eds.), *Silence and voice in the study of contentious politics* (pp. 179–194). Cambridge University Press.

Gutiérrez Rosete Hernández, J. G. (2019). Aprender a restaurar, aprender an organizarnos. El caso del Colectivo Ciudadano Pro Bosque Pedagógico del Agua. *Rev. Eletrônica Mestr. Educ. Ambient. Rio Grande, Edição especial Educação Ambiental e Movimentos Sociais Populares na América Latina.* 2 Congreso Nacional de Educación Ambiental para la Sostenibilidad, pp. 209–227.

Haggard, S., & Kaufman, R. R. (2016). *Dictators and democrats: Masses, elites and regime change*. Princeton University Press.

Hirschman, A. O. (1970). *Exit, voice and loyalty: Responses to decline in firms, organizations and states*. Harvard University Press.

Hirschman, A. O. (1974). Exit, voice, and loyalty: Further reflections and a survey of recent contributions. *Social Science Information, 13*(1), 7–26.

Hunt, S. A., & Benford, R. D. (2004). Collective identity, solidarity, and commitment. In D. A. Snow, S. A. Soule, & H. Kriesi (Eds.), *The Blackwell companion to social movements* (pp. 433–459). Blackwell Publishing Ltd.

Ivankova, N. V., Creswell, J. W., & Stick, S. L. (2006). Using mixed-methods sequential explanatory design: From theory to practice. *Field Methods, 18*(1), 3–20. https://doi.org/10.1177/1525822X05282260

Jessop, B., Brenner, N., & Jones, M. (2008). Theorizing sociospatial relations *Environment and Planning D, 26*(3), 389–401, https://doi.org/10.1068/d9107

Kadivar, M. A. (2018). Mass mobilization and the durability of new democracies. *American Sociological Review, 83*(2), 390–417.

Kelling, G. L., & Wilson, J. Q. (1982, March). Broken windows. The police and neighborhood safety. *The Atlantic Monthly*.

Klandermans, B. (2004). The demand and supply of participation: Socialpsychological correlates of participation in social movements. *The Blackwell Companion to Social Movements*, 360–379.

Koopmans, R. (2004). Protest in time and space: The evolution of waves of contention. In D. A. Snow, S. A. Soule, & H. Kriesi (Eds.), *The Blackwell companion to social movements* (pp. 413–432). Blackwell Publishing Ltd.

Lefebvre, H. (1991). *The production of space*. Blackwell.

Li, Y. (2019). A zero-sum game? Repression and protest in China. *Government and Opposition, 54*(2), 309–335.

Lipset, S. M., & Rokkan, S. (2001). Estructuras de división, sistemas de partidos y alineamientos electorales. In A. Batller (Ed.), *Diez textos básicos de ciencia política* (pp. 231–274). Ariel.

Lipsky, M. (1968). Protest as a political resource. *American Political Science Review, 62*(4), 1144–1158.

Martin, D., & Miller, B. (2003). Space and contentious politics. *Mobilization: An International Quarterly, 8*(2), 143–156.

Mazumder, S. (2018). The persistent effect of U.S. civil rights protests on political attitudes. *American Journal of Political Science, 62*(4), 922–935.

McAdam, D. (1982). *Political process and the development of black insurgency, 1930–1970*. University of Chicago Press.

McAdam, D. (1986). Recruitment to high-risk activism: The case of freedom summer. *The American Journal of Sociology, 92*, 64–90. https://doi.org/10.1086/228463

McAdam, D., & Sewell, W. H. (2001). Temporality in the study of social movements and revolutions. In R. Aminzade, J. A. Goldstone, D. McAdam, E. J. Perry, W. H. Sewell, S. Tarrow, & C. Tilly (Eds.), *Silence and voice in the study of contentious politics* (pp. 89–125). Cambridge University Press.

McAdam, D., & Snow, D. A. (Eds.). (1997). *Social movements: Readings on their emergence, mobilization, and dynamics*. Roxbury Publishing Company.

McCarthy, J. D., & Zald, M. N. (1977). Resource mobilization and social movements: A partial theory. *American Journal of Sociology, 82*(6), 1212–1241.

Melucci, A. (1989). In J. Keane & P. Mier (Eds.), *Nomads of the present: Social movements and individual needs in contemporary society*. Temple University Press.

Meyer, D. S., & Lupo, L. (2010). Assessing the politics of protest: Political science and the study of social movements. In B. Klandermans & C. Roggeband (Eds.), *Handbook of social movements across disciplines* (pp. 111–156). Springer.

Milkman, R. (2017). A new political generation: Millennials and the post-2008 wave of protest. *American Sociological Review, 82*(1), 1–31.

Milon, D. (1999). *L'étranger dans la ville: du rap au graff mural* (p. 1999). PUF.

Moore, A. (2011). The eventfulness of social reproduction. *Sociological Theory, 29*(4), 294–314. 341.

Myre, C. (2003). *True legend. Artistic expressions, development of an urban counterculture and getting fame: 20 years of Graffiti in the Urban Ghetto*. Righter.com.

Oliver, P., & Myers, D. (2003). The coevolution of social movements. *Mobilization: An International Quarterly, 8*(1), 1–24.

Oliver, P. E., Cadena-Roa, J., & Strawn, K. D. (2003). Emerging trends in the study of protest and social movements. In B. A. Dobratz, L. K. Waldner, & T. Buzzell (Eds.), *Political sociology for the 21st century: Research in political sociology* (Vol. 12, pp. 213–244). JAI Press.

Olson, M. (1965). *The logic of collective action: Public goods and the theory of groups*. Harvard University Press.

Opp, K.-D. (2009). *Theories of political protest and social movements: A multidisciplinary introduction, critique, and synthesis*. Routledge.

Perreault, T. (2014). What kind of governance for what kind of equity? Towards a theorization of justice in water governance. *Water International, 39*(2), 233–245.

Prieto, S. (2013). *Restauran Bosque Pedagógico del Agua*. Retrieved from http://www.informador.com.mx/jalisco/2013/481045/6/restauran-bosque-pedagogico-del-agua.htm

Rainsford, E., & Wambach, A. (2021). Cash and class: Intergenerational transmission of values and capital and the consequences for social mobility in the UK. In *Intergenerational transmission and economic selfsufficiency* (pp. 183–208). Cham. https://doi.org/10.1007/978-3-030-17498-9_8

Rennes, J. (2011). Forms of Protest: The Sociology of Mobilizations and Theories of Argumentation. *A contrario, 16*, 151–173 https://www.cairn-int.info/journal.

Rizzo Lara, R. L. (2021). La Caminata del Migrante: a social movement. *Journal of Ethnic and Migration Studies, 47*(17), 3891–3910. https://doi.org/10.1080/1369183X.2021.1940111

Ruiz, J. E. (2019). Los Colomos. La reconstrucción socioambiental de un hábitat. Revista Impulso. Temas de la Tierra. Núm. 1, p. 51–62, ene-may. Fundación Impulso.

Ruiz Barajas, J. (2016). *Bosque Pedágogico del Agua*. Clavijero, comunidades de saberes. ITE.

Sartori, G. (1969). From the sociology of politics to political sociology. *Government and Opposition, 4*(2), 195–214.

Sayre, N. S. (2009). In Scale. N. Castree, D. Demeritt, D. Liverman, B. Rhoads (Eds.), *A companion to environmental geography* (pp. 95–108), Wiley-Blackwell.

Sayre, N. S. (2015). Scales and politics. In T. Perrault, G. Bridge, & J. McCarthy (Eds.), *The Routledge handbook of political ecology* (pp. 504–515). Routledge.

Seguin, C. (2016). Cascades of coverage: Dynamics of media attention to social movement organizations. *Social Forces, 94*(3), 997–1020.

Shirky, C. (2011). The political power of social media: Technology, the public sphere, and political change. *Foreign Affairs, 90*(1), 28–41.

Smith, N. (1992). Geography, difference and the politics of scale. In J. Doherty, E. Graham, & M. Malek (Eds.), *Postmodernism and the social sciences* (pp. 57–79). St. Martin's Press.

Snow, D. A., Burke Rochford, E., Jr., Wonden, S. K., & Benford, R. D. (1986). Frame alignment processes, micromobilization, and movement participation. *The Sociological Review, 51*(4), 464–481.

Snow, D. A., Soule, S. A., Kriesi, H., & McCammon, H. J. (Eds.). (2019). *The Wiley Blackwell companion to social movements* (2nd ed.). Wiley Blackwell. https://www.wiley.com/en-us/The+Wiley+Blackwell+Companion+to+Social+Movements%2C+2nd+Edition-p-9781119168553

Sosa, Y. (2018). La protesta estudiantil en la Universidad del Valle entre 1968-1971 y 2007-2011: una lectura comparada. *Fórum, 13*, 33–74. https://doi.org/f97d

Sosa, E., & Irías, G. (2018). Después del fraude electoral: Las posibles salidas a la actual crisis política en Honduras. *CESPAD, 35*.

Staggenborg, S. (2016). *Social movements* (3rd ed.). Oxford University Press.

Swyngedouw, E. (2004). Scaled geographies: Nature, place, and the politics of scale. In E. Sheppard & R.B. McMaster (Eds.), *Scale and geographic inquiry: Nature, society, and method* (pp. 129–153). Blackwell.

Tajfel, H., & Turner, J. C. (1979). An integrative theory of intergroup conflict. In W. G. Austin & S. Worchel (Eds.), *The social psychology of intergroup relations* (pp. 33–47). Brooks/Cole.

Tarrow, S. (1998). *Power in movement: Social movements and contentious politics*. Cambridge University Press.

Tarrow, S. (2005). *The new transnational activism*. Cambridge University Press.

Tarrow, S. G. (2011). *Power in movement: Social movements and contentious politics* (Rev. & updated 3rd ed.). Cambridge University Press.

Taylor, V., & Van Dyke, N. (2004). Get up, stand up: Tactical repertoires of social movements. In D. A. Snow, S. A. Soule, & H. Kriesi (Eds.), *The Blackwell companion to social movements* (pp. 262–293). Blackwell Publishing Ltd.

Taylor, V., & Whittier, N. E. (1992). Collective identity in social movement communities: Lesbian feminist mobilization. In A. D. Morris & C. M. Mueller (Eds.), *Frontiers in social movement theory* (pp. 104–129). Yale University Press.

Tilly, C. (1978). *From mobilization to revolution*. Addison-Wesley Pub. Co.

Tilly, C. (1998). From interactions to outcomes in social movements. In M. Giugni, D. McAdam, & C. Tilly (Eds.), *How movements matter*. University of Minnesota Press.

Tilly, C. (2000). Spaces of contention. *Mobilization: An International Quarterly, 5*(2), 135–159.

Tilly, C. (2001). Mechanisms in political processes. *Annual Review of Political Science, 4*(1), 21–41.
Tilly, C. (2010). *Regimes and repertoires*. Chicago University Press.
Tilly, C., & Tarrow, S. (2015). *Contentious politics* (second edition, fully revised and updated). Oxford University Press.
Tufekci, Z. (2017). *Twitter and tear gas: The power and fragility of networked protest*. Yale University Press.
Vassallo, F. (2018). The evolution of protest research: Measures and approaches. *Political Science & Politics, 51*(1), 67–72.
Walder, A. G. (2009). Political sociology and social movements. *Annual Review of Sociology, 35*, 393–412.
Wasow, O. (2020). Agenda seeding: How 1960s black protests moved elites, public opinion, and voting. *American Political Science Review, 114*(3), 638–659.
Wouters, R. (2019). The persuasive power of protest: How protest wins public support. *Social Forces, 98*(1), 403–426.
Wouters, R., & Walgrave, S. (2017). Demonstrating power: How protest persuades political representatives. *American Sociological Review, 82*(2), 361–383.
Zald, M. N., McCarthy, J. D., & Gamson, W. A. (1987). *Social movements in an organizational society*. Routledge.

José G. Vargas-Hernández is research professor at Instituto Tecnológico Mario Molina Unidad Zapopan, and earlier at University Center for Economic and Managerial Sciences, University of Guadalajara. Member of the National System of Researchers of Mexico. Professor Vargas-Hernández has a Ph.D. in Public Administration and a Ph.D. in Organizational Economics. He has been visiting scholar at Carleton University Canada, University of California Berkeley and Laurentian University, Canada. He holds a Ph.D. in Economics, Keele University; Ph.D. in Public Administration, Columbia University; studies in Organizational Behavior at Lancaster University and has a Master of Business Administration. He has Published 9 books and more than 300 papers in international journals and reviews (some translated to English, French, German, Portuguese, Farsi, Chinese, etc.) and more than 300 essays in national journals and reviews. He has obtained several international awards and recognitions. He has also experience in consultancy. His main research is in organizational economics and strategic management. He teaches for several doctoral programs.

M. C. Omar C. Vargas-González serves as professor and head of the Department of Systems and Computing at the Technological National of Mexico Campus Ciudad Guzmán and professor of Telematics Engineering at the University Center of the South of the University of Guadalajara, with a master's degree in Computer Systems. He has received training in Entrepreneurship and Multidisciplinary Innovation at Arizona State University (2018) as well as a diploma in Ecosystems of Innovation and Entrepreneurship from the Harvard University School of Business. He currently conducts research in diverse areas such as Entrepreneurship, Economics, Statistics, Mathematics and Information and Computer Sciences, has collaborated in the publication of more than 15 articles in magazines and directs innovation and technological development projects.

Index

© The Editor(s) (if applicable) and The Author(s), under exclusive license to Springer Nature Switzerland AG 2023

S. Dhiman (ed.), *Sustainable Development and Environmental Stewardship*, https://doi.org/10.1007/978-3-031-28885-2

MIX
Papier aus verantwortungsvollen Quellen
Paper from responsible sources
FSC® C105338

If you have any concerns about our products,
you can contact us on
ProductSafety@springernature.com

In case Publisher is established outside the EU,
the EU authorized representative is:
Springer Nature Customer Service Center GmbH
Europaplatz 3, 69115 Heidelberg, Germany

Printed by Libri Plureos GmbH
in Hamburg, Germany